THE ASTROBIOLOGICAL LANDSCAPE

Astrobiology is an expanding, interdisciplinary field investigating the origin, evolution and future of life in the universe. Tackling many of the foundational debates of the subject, from discussions of cosmological evolution to detailed reviews of common concepts such as the 'Rare Earth' hypothesis, this volume is the first systematic survey of the philosophical aspects and conundrums in the study of cosmic life. The author explores the increasing number of crossover problems that highlight the relationship between astrobiology and cosmology, and presents some of the challenges of multidisciplinary study, not least prevalent misunderstandings of the fundamental ontological and epistemological assumptions about the relationship between biological complexity and its cosmological background. Modern physical theories dealing with the multiverse add a further dimension to the debate. With a selection of beautifully presented illustrations and a strong emphasis on constructing a unified methodology across disciplines, this book will appeal to graduate students and specialists who seek to rectify the fragmented nature of current astrobiological endeavour, as well as to curious astrophysicists, biologists and SETI researchers.

MILAN M. ĆIRKOVIĆ is a Research Professor at the Astronomical Observatory of Belgrade, and a Research Associate of the Future of Humanity Institute, Oxford University. His primary research interests are in the fields of astrobiology (habitability, SETI studies, catastrophic episodes in the history of life), astrophysical cosmology (baryonic dark matter, star formation, future of the universe), as well as philosophy of science (risk analysis, observation selection effects, epistemology). He has co-authored more than a hundred research papers.

Cambridge Astrobiology

Series Editors

Bruce Jakosky, Alan Boss, Frances Westall, Daniel Prieur and Charles Cockell

Books in the series

1. Planet Formation: Theory, Observations and Experiments
   Edited by Hubert Klahr and Wolfgang Brandner
   ISBN 978-0-521-18074-0

2. Fitness of the Cosmos for Life: Biochemistry and Fine-Tuning
   Edited by John D. Barrow, Simon Conway Morris, Stephen J. Freeland and Charles L. Harper, Jr.
   ISBN 978-0-521-87102-0

3. Planetary Systems and the Origin of Life
   Edited by Ralph Pudritz, Paul Higgs and Jonathan Stone
   ISBN 978-0-521-87548-6

4. Exploring the Origin, Extent, and Future of Life: Philosophical, Ethical and Theological Perspectives
   Edited by Constance M. Bertka
   ISBN 978-0-521-86363-6

5. Life in Antarctic Deserts and other Cold Dry Environments
   Edited by Peter T. Doran, W. Berry Lyons and Diane M. McKnight
   ISBN 978-0-521-88919-3

6. Origins and Evolution of Life: An Astrobiological Perspective
   Edited by Muriel Gargaud, Purificación Lopez-Garcia and Hervé Martin
   ISBN 978-0-521-76131-4

7. The Astrobiogical Landscape: Philosophical Foundations of the Study of Cosmic Life
   Milan M. Ćirković
   ISBN 978-0-521-19775-5

THE ASTROBIOLOGICAL LANDSCAPE

Philosophical Foundations of the Study of Cosmic Life

MILAN M. ĆIRKOVIĆ

Astronomical Observatory of Belgrade

CAMBRIDGE
UNIVERSITY PRESS

CAMBRIDGE
UNIVERSITY PRESS

University Printing House, Cambridge CB2 8BS, United Kingdom

One Liberty Plaza, 20th Floor, New York, NY 10006, USA

477 Williamstown Road, Port Melbourne, VIC 3207, Australia

314-321, 3rd Floor, Plot 3, Splendor Forum, Jasola District Centre, New Delhi - 110025, India

79 Anson Road, #06-04/06, Singapore 079906

Cambridge University Press is part of the University of Cambridge.

It furthers the University's mission by disseminating knowledge in the pursuit of education, learning and research at the highest international levels of excellence.

www.cambridge.org
Information on this title: www.cambridge.org/9780521197755

First published 2012

A catalogue record for this publication is available from the British Library

Library of Congress Cataloging in Publication data
Cirkovic, Milan M.
The astrobiological landscape : philosophical foundations of the study of cosmic life / Milan M. Cirkovic.
p. cm. – (Cambridge astrobiology)
Includes bibliographical references and index.
ISBN 978-0-521-19775-5 (hardback)
1. Exobiology. I. Title.
QH326.C57 2012
520 – dc23 2012008168

ISBN 978-0-521-19775-5 Hardback

Contents

Acknowledgements

I am indebted to many people, some of whom were blissfully unaware of their help in writing this book. They include Zoran Knežević, Zoran Živković, Steven J. Dick, Damian Veal, Anders Sandberg, Nick Bostrom, Jelena Andrejić, Carlos Cotta, Max Tegmark, Predrag Ivanović, Dejan Rajković, Irena Diklić, Christopher Chyba, Kim Sterelny, Srdjan Samurović, Goran Đorđević, George Dvorsky, Robin Mackay, Svetlana Lukić, Seth Baum, Karl Schroeder, Aleksandar Obradović, Vesna Milošević-Zdjelar, Sunčica Zdravković, Cosma Shalizi, Bojana Pavlović, Petar Grujić, Fred Adams, Dušica Božović, Momčilo Jovanović, Goran Milovanović, Branislav Šimpraga, Predrag Nikolić and Robin Hanson. Invaluable comments on various parts of the manuscript by Srdja Janković, Ivan Almár, Paul Birch, Nikola Tucić, Martin Beech, Massimo Pigliucci, Eva Kamerer and Paul Gilster are also hereby acknowledged. Stephen Webb was not only the author of one of the most insightful popular science books on astrobiological subjects, but has also been a source of advice and support during the work on this project. The help, in its early stages, of Clive Horwood is warmly appreciated. Sadly, a dear friend and collaborator, Robert J. Bradbury, suddenly passed away in February 2011, as the book was nearing completion. Thus, I was deprived of his incisive criticism, and I can only hope that he would have recognized the echoes and influences of some of his thoughts on the subject matter. Among several co-authors of related papers, a special place is reserved for Branislav Vukotić, an excellent graduate student and colleague. Key technical help was provided by the great artist and friend, Slobodan Popović Bagi, whose wonderful drawings served as both illustrations and inspiration for further thinking and refinement of the discussion of often abstract and complex issues; they are a constant reminder that we still need new manoeuvres in the everlasting battle for clarity. It is also a great pleasure to express sincere gratitude to Dušan Inđić, Joanna Henderson, Duška Kuhlmann, Ljugomir Aćimović, Edi Bon and Branislav K. Nikolić for their kind technical help. Of course, my outstanding thanks go to Zona Kostić, whose radiance illuminated the critical period in the writing of this book; in more than one way all grains of wisdom to be found here belong to her.

I would like to thank Vince Higgs, Natasha Pulley, Claire Poole, Peter Sinclair and their colleagues at Cambridge University Press for their kind help, encouragement and many useful suggestions. As usual, the Astronomical Observatory of Belgrade provided key academic resources and a friendly working environment during work on this project. I am also grateful to the Future of Humanity Institute of Oxford University, which offered me hospitality for some of the time devoted to this book, and provided one of the most pleasant and stimulating environments for creative scientific work on the planet. This is an opportunity to thank the *KoBSON* Consortium of Serbian libraries, which enabled at least the partial overcoming of the gap in obtaining scientific literature during the tragic 1990s.

It is hardly possible to determine precisely the intellectual genealogy of a single thought/sentence, not to mention a book, but an attempt should nevertheless be made. Three major literary influences have been those of Stanislaw Lem, Olaf Stapledon and Howard P. Lovecraft. Among thinkers and artists not explicitly quoted who have exerted the strongest influence on me during the long gestation of this manuscript were William Butler Yeats, Leonard Cohen, Diana Krall, Philip K. Dick, the brothers Strugatsky, Nick Cave, Borislav Pekić, William S. Burroughs, Tom Waits, Richard Buckminster Fuller, Alastair Reynolds, Ernesto Sabato and Warren Spector. I do hope that this modest output of mine will be at least a faint shadow of their invaluable input. *Ars longa, vita brevis.*

Many inadequacies and weaknesses certainly remain; for all of those I take full responsibility.

Introduction

At the time of writing, NASA has announced that the *Kepler* observatory – a small, man-made satellite trailing the Earth on its orbit around the Sun – has made the first discovery of five terrestrial-size planets around other stars in the Galaxy, as well as a planetary system (provisionally named Kepler-11) containing no less than six planets.[1] While this newsflash may not sound revolutionary in isolation – have we not grown accustomed to many spectacular 'firsts' coming from new astronomical instruments? – it may mark, in retrospect, an important milestone in the scientific and intellectual movement, bringing the question of life and intelligence in their general cosmic context to the forefront of scientific research. This development, dealing with sometimes surprisingly old questions, is embodied in a new discipline – astrobiology. After several hundreds of planetary systems have been discovered in the last two decades, which include planets around pulsars, planets in the halo of the Milky Way, and possibly even planets in *another galaxy*,[2] we are now witnessing a clear convergence towards a Galactic set of habitable, Earth-like planets.

This book investigates several philosophical and methodological issues related to the ongoing 'astrobiological revolution' (*c.*1995–today), and the surge in both professional and public interest in the search for life and intelligence beyond Earth. We are lucky enough to live in an epoch of great progress in this nascent discipline, which deals with three canonical questions: How does life begin and develop? Does life exist elsewhere in the universe? What is the future of life on Earth and in space?

A host of fascinating discoveries has been made during the last two decades or so, some of the most important being: a discovery of a large number of extrasolar planets; the existence of many extremophile organisms, possibly comprising the 'deep hot biosphere' of Thomas Gold; the discovery of subsurface water on Mars and the huge ocean on Europa, and possibly also on Ganymede, Callisto and Enceladus; controversial evidence concerning Martian methane and microfossils in the Martian meteorite ALH84001; the unequivocal discovery of a large variety of amino acids and other complex organic compounds in meteorites; modelling organic chemistry in Titan's atmosphere; the quantitative treatment of the Galactic Habitable Zone; numerical elucidation of astrobiologically relevant ecological models, such as James Lovelock's 'Daisyworld'; the development of a new generation of panspermia theories, spurred on by experimental verification that even terrestrial microorganisms easily survive conditions of an asteroidal or cometary impact; elucidation of mass extinction episodes in Earth's history; rapid progress in understanding biogenesis, etc. In addition, there is a great deal of activity on the organizational, managerial and public outreach level, reflected in the setting up of new specialized institutes and university programmes at both undergraduate and graduate level, launching of several new research journals (*Astrobiology*, *International Journal of Astrobiology*, *Planetary Science*, etc.) and monograph series, coupled with the reorientation of some of the older publishing outlets, as well as a host of popular journals, web portals and blogs, maintaining vibrant interest in astrobiological topics both inside and outside of academia.

The epistemological and methodological basis of astrobiological studies presents us, however, with a hornet's nest of issues that have not been, with few exceptions, tackled in the literature so far. It is not surprising, therefore, that seemingly paradoxical situations and controversial conclusions arise from time to time, as is usual in young scientific fields, coupled with confusion which does not always stay limited to the lay public. Thus, the aim of this book is to frame the relevant questions about philosophical and methodological aspects of the astrobiological enterprise, rather than to provide answers. Perhaps it is too early even for speculative and tentative answers; but as in the prototypical case of the dog which did not bark, even the absence of answers tells us something very important about the puzzle itself.

The composition of the book reflects a symbolic shift both (1) from the 'origins' (distant cosmological past, large spatio-temporal scales) towards the present-day and near-future practice of astrobiology, and (2) from the abstract and theoretical towards issues that are more empirical. I will try to demonstrate that in a young field, as astrobiology certainly is, foundational philosophical and methodological questions can play a very stimulating and inspirational role. This parallels the development of physical cosmology since Einstein, especially in the crucial and formative 1929–1965 period; in my view, astrobiology is today in a position similar to the one cosmology was in at the time of Friedmann, Eddington, Hubble, Lemaître or Hoyle. The overarching analogy – recognizable, for instance, in the title of Steven J. Dick's seminal study *The Biological Universe*[3] – is useful both in heuristic and metaphoric terms; it may further the already huge popular appeal of astrobiology as well. Therefore, after introducing the astrobiological revolution since 1995 and some of its foundational concepts (Chapter 1), I consider the relationship between cosmological and astrobiological enterprise through various puzzles that bridge the gap between these realms. Some of them have been resolved (or only noticed *post festum*), like Dirac's large-number coincidences or the Davies–Tipler argument, others are still very much with us (Fermi's paradox), while fresh ones have been unearthed only through recent and excessively complex research (like the problem of Boltzmann's brains). Such amalgamative problems present not only research opportunities, but also point toward wider scientific synthesis, or consilience. One of the central themes of this first part of the book (Chapters 2–4) is the concept of observation selection effects, which helps us not only to resolve some of the encountered conundrums, but also dispels further confusions and controversies, like the ones surrounding anthropic reasoning. In particular, the notion of the astrobiological landscape, introduced in Chapter 3, offers a convenient platform for unifying the (Earth-) specific and the (universe-) general. Conversely, the leitmotif of the second part of the book (Chapters 5–8) is the *continuity thesis*: roughly speaking, the idea that physical, chemical, biological and perhaps even cultural, evolutions are parts of the same evolutionary continuum. Here, a tradition of thought starting in its modern version with J. B. S. Haldane has already brought about important insights; for example, as Iris Fry has persuasively shown in several articles and a brilliant monograph, *The Emergence of Life on Earth*,[4] some variant of the continuity thesis is necessary for the scientific consideration of biogenesis. In a generalization of this thesis, a proper response to many popular arguments of the opponents of astrobiological and the Search for ExtraTerrestrial Intelligence (SETI) studies can be found, as will be repeatedly discussed in the book. Notably, the generalization will include the undermining of gradualism, the old-fashioned view that the 'present is the key to the past' in terms of the tempo and mode of evolutionary processes. A new perspective on the classical anti-SETI arguments follows, like the argument from biological contingency, Fermi's paradox, and Carter's 'anthropic' argument – and this new perspective is more optimistic as far as the practical aspect of SETI searches is concerned. In addition, astrobiology provides us with potentially powerful insights into the nature of *terrestrial* biological evolution itself, as well as the antidote to the Popperian scepticism (too often misused by creationists and other pseudo-scientists) towards the contingent or 'lawless' nature of evolutionary biology. The concluding chapter will offer a glimpse of what we could expect in terms of synthesis if the expanding trend of multidisciplinary effort centred on astrobiology continues.

But danger also lurks in bringing such philosophical perspectives to the fascinating issues of contemporary astrobiology. One should be wary of an almost reflexive tendency in works of philosophy to present them as though the authors believe them to be the final word on their subject. This comforting illusion would be self-indulgent, even in much better developed fields than astrobiology is at present; sadly, this has not prevented some authors from writing in this way, and we shall encounter such negative examples later on. In my view, the very *openness* of the subject cannot be overemphasized. Thus, the present book should be understood, in the literal sense, as a *philosophical exploration* of perplexing issues arising from contemporary research on the origin, existence and future of life in its widest cosmological context. Conceptual completeness is overrated, anyway, even in well-established realms of knowledge. Half-baked ideas that cohere in tone and attitude have more often been fruitful seeds of novelty and sources of inspiration than heavy volumes of well-developed 'grand systems' – and no apology should be sought for that. Those who insist on completeness in the tangled reality of the history of ideas look, more often than not, akin to Shigalyev, a tragicomic character in Dostoevsky's *Demons*: a disturbingly persuasive fool, who argues that if people do not devote exactly ten weeks to listening to his universal theory of society and liberty, they can go home and forget about political activism, since there can be no viable alternative to his programme.

Consequently, some of the views I present are still crude sketches, far from being polished and refined, or incorporated into a complete and elegant whole. As the analogy with cosmology will show, this remains in the future, even for more advanced fields of study. Nevertheless, if there is no place for the last word in this field, there is still ample place for the first words to be said on many issues, including not only the main line of argument, but many leads and side issues as well. Philosophical quest starts with an idea about the destination, but it necessarily changes all the time; instead of misplaced zeal for perfect, streamlined consistency, I believe the fluid nature of the quest simply enriches its flavour.

The same strategy applies to the level of technical complexity of the book. As emphasized by Sir Arthur Eddington long ago – and yet sorely misunderstood by scientists on the one hand, and science writers and journalists on the other – science writing evolves (just like everything else):[5]

> Science has its showrooms and its workshops. The public to-day, I think rightly, is not content to wander round the showrooms where the tested products are exhibited; the demand is to see what is going on in the workshops.

Nothing is new under the Sun. It cannot be overemphasized how much the old myth of clear, sharp, antiseptic division between scientific research writing and popular science writing has long ago been debunked by intellectual giants of Eddington's calibre. While the scientific research discourse may – and should – be entirely logical, fair-minded and based on careful empirical and/or theoretical analysis, it is still fashioned as rhetorical, persuasive discourse. And yet, this myth remains perniciously influential today, exercising its impact in a particularly repulsive form in science education and fundraising. It is incumbent on us scientists to fight what Stephen Jay Gould, another fine stylist in contemporary science, labelled scientific selection for poor writing.[6] While it is certainly more difficult to correct for than other selection effects discussed in this book, it is certainly rewarding to try. While some parts of the book (in particular Chapters 3, 4 and 7) may look superficially more technical, a reader can skip the mathematical or tougher philosophical parts and, if interested, return to them at a later time. Some suggestions for further reading and points of entry into the existing literature are clearly indicated in the notes. Throughout the book, I use illustrative and graphically marked scenarios, either as thought experiments or examples from the literature (both fictional and discursive).

A final note on the use of history. On the pain of anachronism, I give some of the concepts and phenomena their modern labels for the sake of compactness and better understanding. This is a book about astrobiology and not a history of science, although, of course, historical scholarship has a large role to play in this youthful realm. Therefore, I label Alfred Russel Wallace's 1903 study 'astrobiological', although astrobiology clearly did not exist as either a word or a concept in the first years of the twentieth century. Let the purists of 'anti-Whiggish' historiography or postmodern relativists be offended, but it is exactly an illustration of the fact of scientific progress that is at (epistemic) stake here and offers the best prospect for understanding the astrobiological synthesis that is crystallizing as we speak. No apology for progress is ever necessary. As a great poet of idealistic optimism wrote almost two centuries ago:[7]

> The lightning is his slave; heaven's utmost deep
> Gives up her stars, and like a flock of sheep
> They pass before his eye, are numbered, and roll on!
> The tempest is his steed, he strides the air;
> And the abyss shouts from her depth laid bare,
> Heaven, hast thou secrets? Man unveils me; I have none.

1

Astrobiology

'The Colour Out of Space?'

And what shall I love if not the enigma?

Giorgio de Chirico

In April 1897, *Pearson's Magazine*, a rather influential London literary publication, although launched only about a year earlier, published one of the eeriest prologues ever to appear in the world of *belles lettres*. The author was a 31-year-old former cloth retailer and biology student by the name of Herbert George Wells, who two years before had created a mini-sensation with his first novel, *The Time Machine*, controversial for both its outrageously speculative scientific premise and for its radical social criticism. Now, he did it again, having started the new novel, *The War of the Worlds* (to be published in book form the following year), with this dramatic warning:[1]

> No one would have believed in the last years of the nineteenth century that this world was being watched keenly and closely by intelligences greater than man's and yet as mortal as his own; that as men busied themselves about their various concerns they were scrutinised and studied, perhaps almost as narrowly as a man with a microscope might scrutinise the transient creatures that swarm and multiply in a drop of water. With infinite complacency men went to and fro over this globe about their little affairs, serene in their assurance of their empire over matter. It is possible that the infusoria under the microscope do the same. No one gave a thought to the older worlds of space as sources of human danger, or thought of them only to dismiss the idea of life upon them as impossible or improbable. It is curious to recall some of the mental habits of those departed days. At most terrestrial men fancied there might be other men upon Mars, perhaps inferior to themselves and ready to welcome a missionary enterprise. Yet across the gulf of space, minds that are to our minds as ours are to those of the beasts that perish, intellects vast and cool and unsympathetic, regarded this earth with envious eyes, and slowly and surely drew their plans against us.

The famous progressive rock version by Jeff Wayne produced in 1978, gives an even more fascinating introduction by condensing Wells' second to sixth sentences into 'Few men even considered the possibility of life on other planets' – rather frightening in the superb narration of Richard Burton.[2] It is even more pertinent from the point of view of the present book. In *The War of the Worlds*, Wells, once a pupil of Thomas Henry Huxley, the legendary 'Darwin's bulldog', struck a perfect balance between dramatic and philosophical discourse. The then reigning Kant–Laplace theory about the formation of the Solar System predicted that the planets' ages correlate with their distance from the Sun, so Mars was considered older than the Earth, which would, in turn, be older than Venus, and so on. The Copernican principle – and naturalism regarding biogenesis! – suggested that, if Mars is habitable at all (and many influential astronomers thought so), it is likely to be the home of a biosphere older in comparison to the terrestrial one. The same Copernican principle, coupled with naturalism with regard to the origin of intelligence (or *noogenesis*), led Wells to assume the existence of Martians as an intelligent species older then humans. The hallmark Victorian belief in progress in both biological and cultural domains led Wells, and many other thinkers of his day, to translate this greater age into greater intelligence and into greater capacity for manipulating nature, i.e., more advanced technology. However, more advanced technology needs not, and here Wells parted company with many of his optimistic contemporaries, pacify essentially biological – or *socio*biological – aggressive instincts of a dominant species. Coupled with the climatic and ecological degradation of their home world (also stemming from the Kant–Laplace theory conjoined with the dominant paradigm of Lyellian gradualism), these instincts led the Martians to undertake the interplanetary expansion and colonization of the nearest habitable ecosystem – our Earth. As noticed by Wells' protagonist, who is perpetually torn between paralyzing fear and an irrepressible curiosity, while Martian invaders brought horrible destruction and death to humans, they did not seem to act any more irrationally than humans do when clearing a forest in order to cultivate land or irrigating a swamp to build housing. Such actions are not regarded as obviously morally repugnant even today, in this epoch of heightened ecological awareness. In the end, the invasion from Mars fails, but not due to any action of humans – supposed pinnacles of the terrestrial evolution. Instead, the Martians, who are of course well adapted to their own biotic and abiotic environment, are defeated by the simplest terrestrial life forms – bacteria to which they had evolved no resistance, bacteria that have lived on our planets for billions of years, thus prompting again the question whether it is sensible to talk about progress in the context of biological evolution.[3] Consider how deep is the gold mine of philosophical issues (and I mention just the most obvious ones) contained in what is still occasionally – and ignorantly – dismissed as 'just' a science-fiction thriller![4] And it is a contingent fact of history that as a consequence of Wells' writings more than a few men have hitherto 'considered the possibility of life on other planets'.

In contrast, consider the plots of recent movies – also *fin de siècle*, as was Wells' novel – like *Smilla's Sense of Snow* (1997) or *X-Files: Fight the Future* (1998): a prominent role in both is played by an ancient meteorite that fell to Earth in times past and brought microscopic alien life forms to our planet (both influenced by Robert Wise's 1971 classic *Andromeda Strain*, based on the 1969 novel by Michael Crichton). This has been for quite a long time, since Lord Kelvin and Svante Arrhenius, known as the panspermia hypothesis, one of the hotly debated topics in contemporary astrobiology. Now microorganisms, bacteria and viruses, are the invaders from space, if anything more threatening than before. The details of science are, of course, wrong (an interesting question for science, technology and society studies: why is it *so* difficult to get the science right in any major film?), but the general idea is the same as the one underlying the current efforts of researchers, technologists, and even politicians, to institute efficient planetary protection protocols. The famous Article IX of the Outer Space Treaty, adopted by the United Nations in 1967, explicitly puts the same fear and caution in legalistic terms, by proposing that parties to the treaty[5]

> shall pursue studies of outer space including the Moon and other celestial bodies, and conduct exploration of them so as to avoid their harmful contamination and also adverse changes in the environment of the Earth resulting from the introduction of extraterrestrial matter, and where necessary, shall adopt appropriate measures for this purpose.

This admirably non-anthropocentric statute (it lists adverse consequences for other celestial bodies first and those for the Earth after; with good reason we shall return to it in Chapter 6) is just as useful a gauge of our thinking as are the motion pictures mentioned above. Like the discussion of extraterrestrial life at the end of the nineteenth century, in the cultural context it was unavoidably framed by the Schiapparelli–Lowell 'discovery' of Martian canals, as well as debates on Darwinism vs. other theories of evolution and, last but not least, the late-Victorian anxiety about the conflict of civilizations, so analogous discussions at the end of the twentieth century are coloured by our fear of deadly pandemics, as well as the post-Cold War anxiety about the conflict of civilizations. The difference – and a very real one – consists of the ongoing *astrobiological revolution*, which has opened wide prospects for an objective assessment of the perennial questions about life and intelligence in their cosmic context.[6] Scientists are understandably reluctant to talk about revolutions in what is usually perceived as day-to-day research work. But an avalanche of both observational and theoretical results from various fields, starting about 1995, being incorporated into a wider synergistic whole, together with large-scale organizational changes and restructuring, give any observer at least some indications that we are living through a real revolutionary epoch. That the revolution could become even more radical, as more and more fields and themes are involved and interconnected, is one of the central topics of this book.

The Canonical Three

But what is astrobiology in the first place? One should not seek a formal definition for many reasons, some of which touch upon philosophical issues, and others are similar to the famous US Supreme Justice Potter Stewart's statement on obscenity: 'I know it when I see it.'[7] Is astrobiology a research activity recognizable on sight? Some of the standard textbooks avoid the question of definition entirely, and pass on to the exposition of a circle of topics that certainly belong to the field.[8] The rationale here is quite clear: after all, the formalization of knowledge – which includes giving precise definitions – usually comes at the *end* of the original research in a given field, not at the very beginning. The history of science is full of examples: consider why we feel Euclid's definitions ('a point is that which has no part', 'a line is a breadth-less length') amusing, even laughable, today. The reason for such a reaction of ours – and, indeed, even of Euclid himself, who did not use the definitions at all in the further discourse on geometry! – is that a definition is useless if it does not reduce a more complex concept to a simpler one. Since, for example, the concept of a 'part' is arguably not simpler than the concept of a 'point', Euclid's definition does not help our understanding at all. Because it is clear that simplification cannot proceed indefinitely, it turns out – and the history of philosophy and mathematics confirmed this long ago – some concepts need to be left undefined, as 'primitives' of any formal system. Similarly, the proper definition of many other important concepts – even if they can be properly reduced to simpler entities – has had to wait for a long time before the adequate *theory* of simpler entities was developed. A particularly illuminating example is the concept of *number*, which was properly defined in the modern sense only after the development of axiomatic set theory in the first decades of the twentieth century – which obviously does not imply that Archimedes or Fermat or Gauss or any other mathematician of old did not know what they were working with. Contrary to the sad prejudice which is forcefully instilled in primary and high-school pupils, formal strictness is much less important in 'real' science than in its cardboard (or too often, textbook) version.

In the realm of astrobiology, the strength of the dilemma can be appreciated when we realize that there are literally dozens of definitions of *life* – which, after all, has been the subject matter of biological sciences for centuries, if not millennia.[9] Like the concept of number, life seems so familiar to us that an intuitive view of it is satisfactory for the vast majority of practical problems. One of the most brilliant minds of modern science, the Austrian physicist Erwin Schrödinger, put it in the title of his epochal 1944 booklet: *What Is Life?*[10] In contrast to mathematical entities, in the case of life it is the complexity of associated phenomena that causes difficulties for the definitional enterprise. The road Schrödinger and most subsequent researchers took is, therefore, to state the list of properties a system needs to possess in order to be called alive: biochemistry based on polymers such as proteins, metabolism, imperfect reproduction, etc. However, these 'list definitions' are, as in other fields, vulnerable to counterexamples, so that a significant amount of the ongoing discussion has been caused by questions such as, 'Are viruses alive?' 'Are prions?' 'Are mineral assemblages?' In order to surmount this difficulty, the so-called 'NASA definition' adopted at one of the first astrobiological scientific meetings in 1994, states simply that *Life is a self-sustaining chemical system capable of Darwinian evolution.*[11] This has been criticized on the basis that it presumes a theory of life (for instance, excluding life based on the strong force, which was speculated about in the fictional context), and presupposes a complete understanding of processes comprising 'Darwinian evolution'. Both are serious criticisms, closely connected to the issues I shall repeatedly address in the present book, notably the need to fight anthropocentrism. While Wells' invading Martians are legitimately alive according to the NASA definition, at the very end of this chapter we shall encounter a fictional example of a life form that eminently defies this definition. The normative justification offered by practising astrobiologists is that the exclusion of non-chemical or non-Darwinian entities aspiring to the status of being alive is justified by a constructive belief that such life forms are not possible. This, in turn, motivates some of the critics of the entire astrobiological endeavour, such as biologist Jack Cohen and mathematician Ian Stewart, to charge astrobiology with being narrow-minded and conservative.[12] However, it is generally accepted that the NASA definition and any particular refinement are *necessarily provisional*, and will evolve as the underlying theory evolves.

The general lesson is that only when it comes to life in a sufficiently novel and strange context – such as when we are discussing biogenesis (the origin of life), or artificial life, or life on other worlds – that the definitional questions come to the fore. Similar reasoning (but understandably more loaded with wider practical, societal and political baggage) applies to the philosophical enterprise of defining *intelligence*: until the advent of fields touching upon foundational issues, such as artificial intelligence and SETI, few people even paused to ask what, exactly, if anything, is that thing we call intelligence (or consciousness or self-awareness or any number of similar high-level mental phenomena). Thus, we are likely to run into trouble if we try to define astrobiology through a second-order definition, since the concept of life itself is problematic in this respect.

Happily enough, this has been widely recognized in research circles (although not as often or as easily amongst science writers and journalists), and the mainstream approach is nowadays to try to build the understanding of the nature of astrobiological endeavour around wide questions that endeavour is supposed to answer. That is the strategy adopted by NASA in producing its famous 'Astrobiology Roadmap', the first version of which was drafted in 1998, and repeated in the 2003 and 2008 editions.[13] The 'Canonical Three' are usually listed as:

1. How does life begin and develop?
2. Does life exist elsewhere in the universe?
3. What is the future of life and intelligence on Earth and in space?

A recent anthology of studies about some of the basic issues in the philosophy of astrobiology has clearly identified the tripartite nature of the definition in the very title of the Roadmap: *Exploring the Origin, Extent and Future of Life*. Its introductory chapter, by Constance Bertka of the American Association for the Advancement of Science (AAAS), gives a simplified, vernacular version of the canonical questions as: Where do we come from? Are we alone? Where are we going?[14] Provided that basic care is taken with the interpretation of the delicate 'we' (life forms and intelligent observers, which for the purposes of the second question need to be further specified as having evolved on Earth, though *not necessarily originating* on it), the simplified version is equivalent to the canonical version above.

And yet, the questions seem to pose interesting dilemmas in their own right. For instance, research into the origin of life falls squarely into the astrobiological domain. This is one of the fields where tremendous progress has been made in the last three decades, leading to new and fruitful concepts such as the RNA-world, which is thought to precede the emergence of metabolism in the first self-replicators. However, the Canonical Three tend to hide that (i) we still do not know to what extent our biogenesis has been influenced by biotic or prebiotic transport from elsewhere in the universe (mixing the first two of the questions); and (ii) we do not know how prevalent 'our' type of biogenesis is in the wider spatial, temporal and ensemble-wise context. Suppose, for instance, that near-future space missions discover microfossils of simple life forms on Mars, very similar to terrestrial bacterial fossils, or to those alleged in the controversial August 1996 study by David McKay and collaborators.[15] Since inferring a biochemical detail from microfossils is at best an extremely difficult task, in this scenario (let me call it Areoparadise Lost; I shall return to it in a later chapter), we may still be unable to decide which of the three hypotheses is likely to give the best account of the data: (A) that life originated on Earth and was transported to Mars via meteorites; (B) that life originated on Mars and was transported to Earth via meteorites; or (C) that life emerged independently on Earth and Mars. (I neglect here the possible complication from a 'higher order' panspermia hypothesis that life emerged neither on Earth nor on Mars, but perhaps in an interstellar cloud and was transported to both planets via interstellar panspermia.) The last hypothesis implies that life similar to the terrestrial one is likely to emerge quickly wherever the conditions are hospitable for it; this would be an argument for a form of 'strong convergence' of viable life forms to something resembling the known terrestrial type. So (C) would, if correct, suggest that the answer to the Canonical #2 would be that we are very likely not alone; but the hypotheses (A) and (B) would preclude any conclusion regarding this question, just as a person seeing her own reflection in a mirror could not infer anything about being alone in the apartment or not. This circumstance does not entail that the probabilities of (A) and (B) being true are equal; it is a difficult empirical question related to the reconstruction of early conditions on Earth and Mars, a correct theory of biogenesis, etc. to decide if it is more likely that life first started on early Earth or early Mars. In contrast, an interesting – and obviously partly philosophical – question is to what degree would each of these alternative hypotheses impact on our wider conclusions; for example, whether we should be optimists or pessimists regarding the probability of success of SETI projects?

This is an example of the 'foundational' ambiguities that are part and parcel of astrobiological research, obviously possessing a philosophical dimension. Other such instances become obvious when we look into the intersection of the Canonical Three with classical philosophical puzzles; questions such as *What is intelligence? Is there an objective difference between future and past?* spring to mind. There is no doubt, however, that in practice the Canonical Three nicely circumscribe the activities intuitively thought of as belonging to the astrobiological realm. The same applies to more formal metrics, such as indexical listings, PACS codes, the scope of relevant research journals and calls for funding, etc. One easily perceives that a surprisingly wide range of disciplinary themes are admitted as belonging to the nascent astrobiological endeavour even by formal (i.e., mostly conservative) criteria of scientific practices and organizations.

Prides and prejudices

The discussion can be illustrated here by a specific 'live' research problem. It has been known for quite some time that important biochemical molecules comprising all living matter on Earth can be found in the form of asymmetric stereoisomers, thus exhibiting *chirality*. Amino acids, the basic building blocks of proteins, are almost exclusively found in form of L-enantiomeres (levo-), building left-handed proteins, while carbohydrates used by living beings, notably sugars, are D-enantiomeres (dextro-). In the non-living world, reflection symmetry is always preserved: on the molecular level, non-living nature does not differentiate between L- and D-forms, and all known prebiotic synthesis pathways produce chiral molecules in 50:50 mixtures. Yet living cells display the most exquisite selectivity. Why would this left–right symmetry be broken in the case of the living world, since it is one of the most general and ubiquitous symmetries in nature? Why is biology asymmetric, while underlying physics and chemistry are symmetric?[16] The origin of this *biological homochirality* is one of the outstanding puzzles in studies of the origin of life and astrobiology in general; homochirality was even used by Pasteur to *define* life.[17] One of the active lines of research suggests that amino acids contained in carbonaceous meteorites, like the Murchison meteorite, already possessed an L-enantiomere bias, so that prebiotic Earth could have been seeded by biased molecules, which early life forms inherited. But this approach just shifts the problem to the origin of chiral asymmetry in meteorites. According to an intriguing recent hypothesis, left-handed molecules could have been concentrated if circularly polarized synchrotron light from a rapidly rotating neutron star selectively photolysed right-handed amino acids in a protoplanetary nebula. Thus, a preponderance of their left-handed twins would be created, and subsequently transported to the early Earth. Of course, this and other hypotheses postulating an astrophysical origin for the observed asymmetry have their alternatives in those hypotheses postulating a chance local physical environment on Earth, like asymmetric crystalline surfaces of some clays and minerals. The exciting debate continues at full speed.[18]

Without entering into technical details, it is easy to perceive that symmetry ↔ asymmetry, living ↔ non-living, local ↔ global, *in situ* ↔ transported, etc. are important axes along which we 'parse' natural phenomena, each having significance far outranging the specific problem. This is a hallmark of topics interesting to philosophers of science as well. In addition, the processes of spontaneous symmetry-breaking are currently thought to have generated the predominant part of the complexity of the entire observable universe, including both its physical and biological features. To what extent such symmetry-breakings constitute an argument for the underlying continuity of 'cosmic evolution'[19] is a provocative issue I will be returning to in subsequent chapters.

As in the case of all scientific fields, astrobiology presents a constant and complex interplay between theory and data, hypotheses- and model-building on the one side, and experimental and observational work on the other. The essence of the process – at least in those aspects relevant for the present book – can be captured by the ancient metaphor of Ouroboros, the snake eating its tail (Figure 1.1). A favourite alchemical illustration of old, present in *The Dialogue of Cleopatra and the Philosophers* (a second-century manuscript), and a symbol of self-reflection, is particularly apt here, since even more than in mature fields, the shape and scope of theoretical work is determined by actual reflection upon the often-unexpected observational results, and vice versa. The two semicircular arrows and the 'natural' direction we intuitively ascribe to a bilaterally symmetric animal, like a snake, in the figure are intentionally drawn in opposite directions, in order to bring intuitively closer the Heraclitean dictum that 'the beginning and the end of a circle are one and the same'.

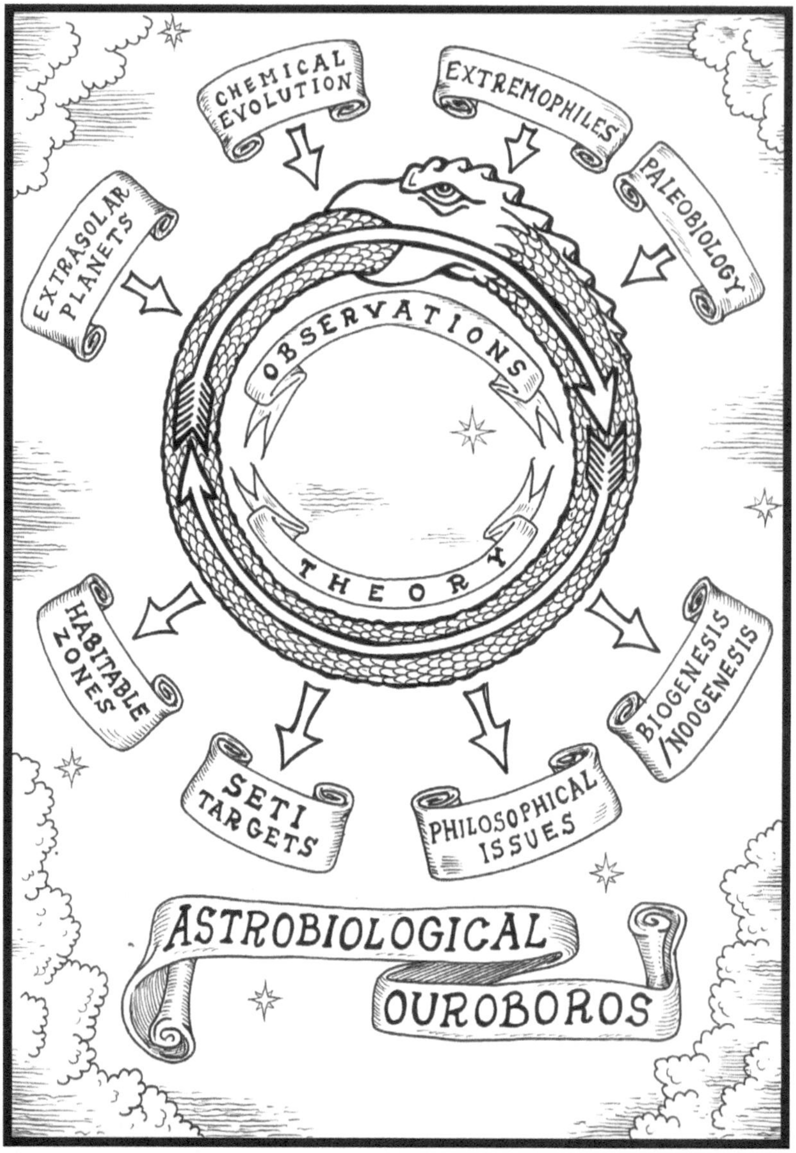

Figure 1.1. A schematic representation of some of the intertwined aspects of astrobiology in practice. The specific multidisciplinary nature of the field, coupled with its immaturity, has led to an imbalance toward observations so far, but it is possible to argue that it is a transient stage of development. (Courtesy of S. Popović.)

That is, real scientific activity is an entangled mess of both theoretical and observational threads, often connected by nothing stronger than our intuition about where a piece of the puzzle should be in the future grand picture. Thus, in using the metaphor of Ouroboros, we should not be too seduced by it into believing that we could 'truly' determine its beginning and end, head and tail. Instead, resembling the mystery of chirality itself, it is a complex network of causes and effects, to which we are too close in both historical and epistemological terms to recognize even a rough global pattern of flow of ideas and knowledge. A purely descriptive *historical* line of development of astrobiological ideas is also difficult to sketch (Figure 1.2), since the pattern is excessively complex and nonlinear.

The timeline of astrobiological development – such as a subjective one given in Figure 1.2 – teaches us a lesson of the moment: the pace of discovery is accelerating. This dynamic, similar to other accelerating trends in science and technology, like Moore's law in computer science, poses some interesting questions. In particular, what is the internal dynamic supporting such a tempo of accumulation of knowledge? In computer science, we see that, in retrospect, key discoveries such as integrated circuits, dynamical random access memory, CMOS transistors, have been basic technological ('internal') factors in maintaining Moore's law, together with external factors such as expanding markets for information processing, globalization of economy, etc. To determine these factors in the case of astrobiological development remains a challenge; in contrast to earlier periods, studied by distinguished historians of science such as Michael Crow and Steven Dick, there has so far been very little work done on the ongoing astrobiological revolution.[20]

This is a far cry from saying that astrobiology has not been subjected to considerable criticism, extending from research level through philosophical to popular-science and public outreach levels. Consider the following theses, all of which appear in the contemporary scientific or philosophical discourse, either in published sources or in informal discussions at meetings and conference dinners, and are uttered by informed and educated non-specialist members of the general public. But are they true or false?

- The entire astrobiological enterprise is, so far, an inquiry without a subject, considering that we have no evidence whatsoever of any kind of extraterrestrial life, intelligent or not.

False! *The subject of astrobiology is cosmic life, not just extraterrestrial life (disregarding that the notion of 'extraterrestrial' is today hard to define clearly, since the Earth is not a closed-box system). Thus, it is eminently clear to modern-day researchers as well as philosophers of science that studies on, for example, biogenesis on early Earth, or mass extinctions in the history of life and their occasional*

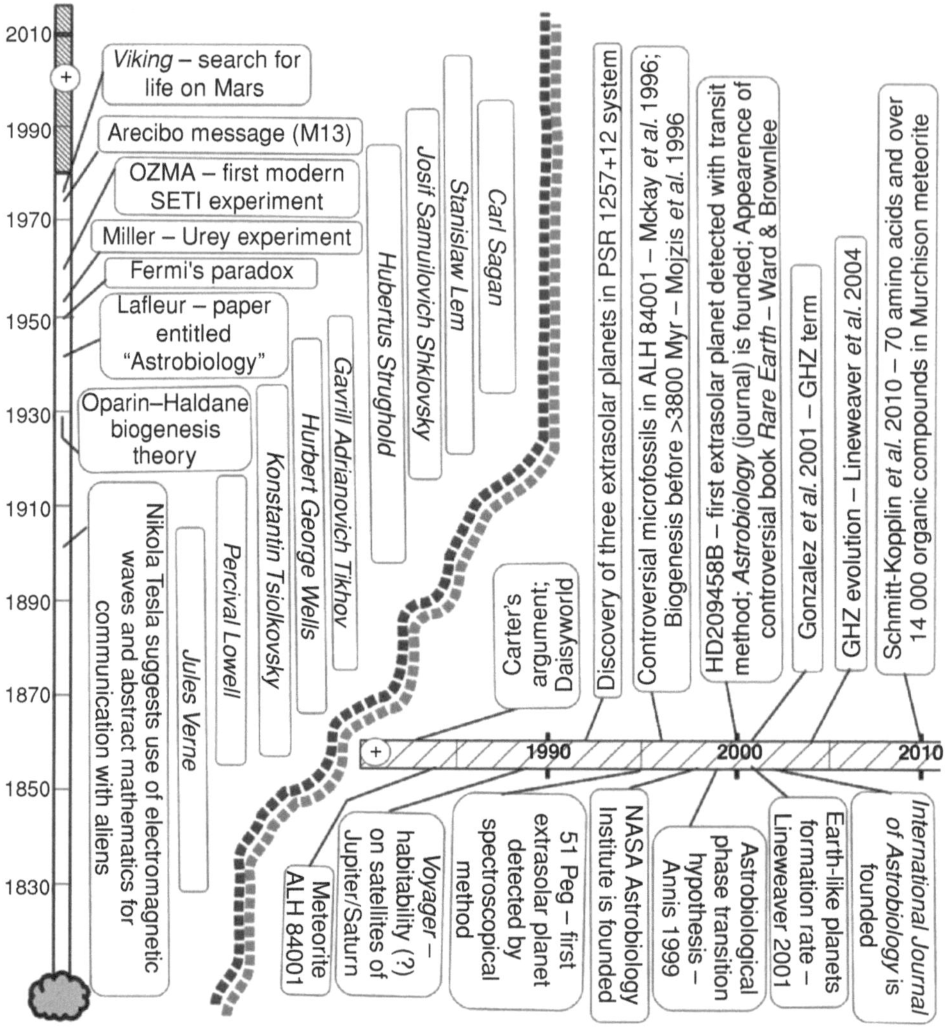

Figure 1.2. A timeline of some crucial events – subjectively chosen, of course – in the prehistory and history of astrobiology, together with lifetimes of some of the significant players in the early (roughly 1903–1995) period. The accelerating pace of the development in the last *c.*15 years is clearly visible. (Courtesy of B. Vukotić.)

extraterrestrial causes, are also squarely located in the astrobiological domain. To the extent that such research deals with testable hypotheses and its comparison to existing evidence, there is no anomaly in astrobiology compared to other scientific fields. It is visible, inter alia, *from the timely appearance of books such as* Astrobiology of Earth,[21] *or many chapters, reviews, and research articles on the apparently 'terrestrial' astrobiological topics. The difficult problem of visibility and accessibility of the primary literature in this epoch of 'information explosion' should not impact on the status or the integrity of any particular field; astrobiology cannot be an exception in this regard.*

- Astrobiological hypotheses are untestable.

False! *Many astrobiological hypotheses are not only testable in principle, but have, in fact, been falsified. For instance, the hypothesis of Alfred Russel Wallace about the location of the Earth and the Solar System in the context of the Milky Way galaxy, suggested in 1903, based on what essentially is an astrobiological argument, has been falsified by the great advances in observational and theoretical astronomy in the 1920s.*[22] *On the other hand, many hypotheses regarding organic matter in meteorites, for example, have been verified by novel and more sensitive methods of biochemical analysis, which have discovered more and more complex organic compounds.*[23] *There are many similar examples from recent research – and it is, in contrast, highly mysterious how something that obvious could ever be neglected. The solution probably lies either in artificial and unjustified restriction of the scope of astrobiological research (thus, consequently, missing the central point of the forthcoming synthesis), or in relict positivist – and duly anthropocentric – views on what constitutes an adequate test. Just as the revolution in massive discoveries of extrasolar planets, following the path-breaking events of 1995, was largely unannounced and unexpected in the mainstream astronomy of the 1970s and 1980s,*[24] *it would be premature to discard the possibility that dramatic development of observational techniques would not enable us to detect habitable or even inhabited planets around other stars. The accelerating trend of discovery also leads to a trend of accelerating returns, at least in some sectors of the astrobiological whole, so we should expect the invention of new and radical methods, enabling new arrays of empirical tests.*

- Astrobiology is dull and unoriginal, since it considers only the terrestrial kind of life and neglects a vast realm of quite different possibilities.

False! *In fact, astrobiology has from its very inception (and in the works of several illustrious early precursors) encompassed thinking about extremely diverse forms of life. Authors as different as J. B. S. Haldane, Konstantin E. Tsiolkovsky, Carl Sagan, Gerald Feinberg, Stanislaw Lem, Sir Fred Hoyle, Edwin E. Salpeter and Steven Benner, have discursively discussed ideas about life very different from the terrestrial one, in the sense of being based on different biochemistry and incompatible with any terrestrial ecological system. This is too often conveniently ignored. A very disturbing sign of the times and the reigning intellectual standards are that outdated charges are repeated over and over again without deeper analysis. To give just one example, back in 1964, the great palaeontologist George Gaylord*

Simpson argued that SETI projects are misguided, since the probability of detecting humans or humanoids on other planets are negligible.[25] *The argument is based on the arbitrary, and in fact hitherto undermined, assumption that the particular part of biological morphospace corresponding to intelligent or communication-capable beings is congruent with the part of the morphospace corresponding to humanoids. Since the latter is admittedly very small, it follows that the former is also small, and the chances of SETI's success are, henceforth, minuscule. But there are in fact no empirical or theoretical reasons why the congruence should hold!*[26] *And yet the criticism is repeated almost verbatim decades later by Frank Tipler, and yet again more recently by Elling Ulvestad – thus contributing, like a self-fulfilling prophecy, to the impression that there is nothing new in SETI studies.*[27] *Such epicycles need to be dismantled, and it is a philosophical inquiry into the various aspects of astrobiology that offers the best vehicle for the demolition work. Attempts to circumscribe the realm of astrobiology and limit it to the search for a terrestrial kind of life, or for life on terrestrial planets, or for carbon-based life, or to life based on chemical reactions, or even to life based exclusively on our particular low-energy physics are unproductive and will ultimately fail – but the bulk of the work remains to be done.*

Notice, however, how antithetical this is to the criticism about 'inquiry without a subject' cited above; as many wise people have observed, when X is attacked from diametrically opposed sides, there must be something of worth in X!

- Astrobiology forces a rich naturalist view of life into a straitjacket of mathematical sciences like physics or astronomy.

False! *In fact, classical naturalist disciplines, like evolutionary theory and ecology, have been more relevant to astrobiology than 'reductionist' molecular biology insofar as the distinction is worth making at all; this will likely change in the future, as we obtain a much better insight into the biochemistry of the universe at large, but it is still witness to a unifying nature of the new field. Search for life in the universe is necessarily reductionist in the ontological sense, but in no way does it imply any other form of reductionism. We shall see some instances that could be construed as* counter*examples for methodological reductionism in this book. In addition, astrobiology in fact opens a staggeringly huge realm of possible diversity for life – it is exactly in this diversity that the narrative of classical natural history may thrive and flourish, perhaps forever beyond a* complete *mathematical theory. Some of this spirit has again been captured in science fiction; the appropriately titled novel* Natural History *by Justina Robson is a nice example.*[28]

- Astrobiology rests on shaky philosophical and methodological foundations.

True! *This is something that could and should be rectified through further work. The pace of developments has been so rapid in the last two decades that it is almost unavoidable to talk about an 'astrobiological revolution' – and in times of revolution, urban planning, deep philosophical grounding, and gardening suffer, for good or ill. To this temporal constraint, one should add another problem located along the orthogonal dimension of disciplinary span: just as in similar fields with large (multi)disciplinary spans, like climate science or future studies, it is disturbingly difficult to find the courage to discuss questions that may possibly, even if improbably, lead to the rethinking of each individual discipline's sacred methodological tenets. Such reluctance shows something of a 'siege mentality' of multidisciplinary endeavours, ruled by the common-sense reasoning that, since we have expended much effort in bringing us together, we should not jeopardize this fragile unity by asking possibly awkward questions. A subsequent chapter will discuss why it is – with many other common-sense approaches in science – wrong, since the astrobiological synthesis is,* pace *relativists of various colours, firmly grounded in nature itself. But, as usually happens, reluctance to pose awkward questions leaves us facing awkward facts, and one such fact is that a gold mine of philosophical and methodological issues in astrobiology is hardly touched by a pick or a drill. The book you are holding is, hopefully, a small step in exactly that direction.*

So, why does astrobiology need philosophy? Steven Benner wrote in conclusion of his insightful paper on the definition of life:[29]

> Many (and perhaps most) of those scientists who are aware of philosophers on their periphery find their approach not particularly useful. In part, this is undoubtedly because philosophers too often deliver complex, abstruse, and perhaps nihilistic answers to questions that scientists view as concrete . . . We do what we generally do when a reality is too complex to meet our constructive needs: we ignore it and continue with a simpler, if arguably false, view.
>
> For astrobiologists, a need remains for some pragmatic philosophies of science, if only in the training of our youth. This may best come from those who are practicing astrobiologists . . . or philosophers closely connected with them . . . I suspect that an understanding dynamic between theory, observation, and definition will be important to these.

The relationship of the universe and life has for too long been a province of religion and mysticism, so it is not surprising that a sort of cultural reflex has arisen in many scientific circles that regards research projects in these kinds of foundational and 'deep' questions as ill-founded, superficial, and generally suspicious. While careful scrutiny is certainly necessary in dealing with such questions, it is in part *philosophical* scrutiny that is required. While Benner's pessimistic view may yet turn out to be correct, it is neither necessary, nor should we refrain from philosophical criticism based on a rather poor reputation of the philosophy of science in some circles. Some of the examples presented in this book aim at casting a more positive light on the whole enterprise.

And there is more to it in actual practice. Peter Ward, a distinguished astrobiologist and one of the authors of the 'Rare Earth' hypothesis (see Chapter 6), recalls in a somewhat amusing manner his conflicts with the SETI community and his own puzzlement as to how he could be regarded a sceptic when it comes to the search for extraterrestrial life when he is PI on a project which is officially sanctioned as a part of NASA's astrobiology initiative.[30] The differences he mentions, for instance, between him and Jill Tarter are quite legitimate differences, but his puzzlement can – and *should* – be explained as a consequence of the deep philosophical divide existing within the astrobiological community as a whole – and, to a degree, even within science in general. Philosophical discourse can help mediate and facilitate critical and fruitful dialogue among different currents of thought within the huge edifice of astrobiological endeavour.

I would go even further and claim that paeans often written – with so many perfectly good reasons – to Charles Darwin and the whole magnificent building of evolutionary thought in biology, are at best incomplete without astrobiological input. This can be seen when, for example, philosopher Timothy Shanahan writes, in the introductory section of his book on Darwinism:[31]

> No other scientific theory has had such a tremendous impact on our understanding of the world and of ourselves as has the theory Charles Darwin presented in that book [*On the Origin of Species*]. This claim will undoubtedly sound absurd to some familiar with the history of science. Surely the achievements of Copernicus, Galileo, Newton, Einstein, Bohr, and other scientists who developed revolutionary views of the world are of at least equal, if not greater, significance. Aren't they? Not really. Although it is true that such scientific luminaries made fundamentally important contributions to our understanding of the physical structure of the world, in the final analysis their theories are about *that* world, whether or not it includes life, sentience, and consciousness. Darwin's theory, by contrast, although it encompasses the entire world of living things, the vast majority of which are not human, has always been understood to have deep implications for our understanding of *ourselves*.
>
> *[emphases in the original]*

His emphasis on the 'world' should give us a reason to pause. What, exactly, is the 'world' whose understanding is impacted by the Darwinian theory, and how extensive is the 'world of living things' to which the theory is applicable? The 'world talk' is clearly at the crossroads of philosophy, cosmology and evolutionary biology, the crossroads that astrobiologists naturally tend to regard as home.

Our very disciplinarian culture is a consequence of the 'unfinished business' of the Copernican revolution. Chemistry is determined, in the limit of ontological reductionism, by a particular form of low-energy physics. It is necessary to qualify this particular set of laws as being low-energy, since modern physics is largely devoted to transcending these laws and finding the 'real' laws, describing in particular the early epochs of the universe, where complete unification of fundamental physical forces could be manifested. We speak of biology as a science of life and psychology as a science of mind, but rarely – if ever – in practice do we qualify or even tacitly understand it as anything but *terrestrial life* and *human mind.* But the Copernican revolution should have taught us – if anything – that Earth (and, consequently, its biosphere) is an infinitesimal speck, and that humans should not elevate themselves on a pedestal of being unique, special, or particularly important. Copernicanism in the narrow sense[32] tells us that there is nothing special about the Earth or the Solar System or our Galaxy within large sets of similar objects throughout the universe. In a somewhat broader sense, it indicates that there is nothing particularly special about us as observers: our temporal or spatial location parameters, or our location in other physical, chemical and biological abstract spaces, etc., are typical or close to typical.[33] But how do we proceed to verify that? Although Copernicanism was a guiding light of the great scientific revolutions – and some whose scientific credentials are often doubted, like Freudian psychoanalysis – its status is still somewhat ambiguous. As we shall see in Chapters 3 to 6, some of the best available elucidations linking Copernicanism to the wider physical picture can be correctly ascribed to astrobiology. Moreover, this is a historical consequence of the evolutionary pathway (and our language does point in the right direction here) of human thought about the universe.

Astrobiology already has, for some time, been playing the role of the standard-bearer of Copernicanism, without any specific intention in this regard. As I shall try to show, this could and should be a more intentional and forceful role, one which could not only contribute to the better elucidation of philosophical questions in the philosophy of physics and biology, but also achieve a wider interdisciplinary dialogue and multidisciplinary synthesis. The consequences of this *extended mandate* are far from being fully understood. In addition, they stand in stark contrast to the misunderstanding, confusion and scepticism that still greet the mention of astrobiology in many circles.

Some confusions arise at the very basic level of broad outlines. Other issues demanding clarification – or whose clarification is too often lost in the noise and excitement – is related to concepts and terms, including their limits. This category is bound to be tightly related to the underlying theories of physics, chemistry, etc. A prototype of this misunderstanding is the question I have often been asked in popular lectures on astrobiology: *If you say that we are searching for life in circumstellar habitable zones, what is all that fuss about Europa, Titan or Enceladus, since those bodies definitely do not belong to the Solar System's habitable zone?* Although the question is, of course, easy to answer, and there are textbooks and popular books giving a detailed discussion of the answer and its ramifications,[34] it is still illustrative of the confusion stemming from insufficient attention devoted to insight into the key concepts. Concepts are always rooted in theory, and the double confusion stems from both (i) the sad fact that this old epistemological adage is still insufficiently clear in scientific and lay circles, and (ii) the current situation in which the view of *specifically astrobiological* theories is foggy, even among the researchers. Similar confusions surround – to a higher degree – concepts such as those of the Galactic Habitable Zone, panspermia hypotheses, mass extinction episodes, as well as many key concepts in SETI studies, such as 'water hole' or 'communication window', and many others.

The third, and perhaps the ultimate source of misunderstanding as to the nature of astrobiological endeavour, stems from those deep 'implicit' assumptions about what is the proper subject matter of scientific research and related activities. I have mentioned the 'conservative' criticism of astrobiology that needs to be answered as strongly as possible, since it positively impedes progress in many areas, especially space missions, whose astrobiological components require substantial economic and societal efforts. Here we face another barrier, which is often considered as an impolite topic in scientific circles, namely the role of imagination and a presumed lack of it. In many ways, the ongoing story of astrobiological revolution is the story of imagination (re)gaining its rightful place in scientific endeavour, in an age and atmosphere in which the status of science in wider culture and society leaves much to be desired.

The steam engine was invented in order to pump water out of several coal mines in northern England. It has performed its task in this respect very well, and some of the first models were called the 'Miner's Friend', but hardly anybody apart from historians of technology is aware of that fact today. What we talk about when we mention the steam engine today is how it transformed the entire world by serving as a vehicle – in more than one sense – of the industrial revolution. Similarly, many features of the living world are invented by evolution for one purpose and then adopted for some other function. In a memorable neologism coined by Stephen Jay Gould, we talk about *exaptations* in such cases, as contrasted with usual adaptations (characters selected for their current utility). In the history of science and technology we are often dealing with exaptations, such as the steam engine. The same phenomenon could apply to whole branches of science. It is, I contend, the evolutionary trajectory of astrobiology itself, while emerging to unify advancing studies of several disciplines related to life in its most general cosmic context, to perform a different, and perhaps vastly more general role of being a herald of wider unity – or *consilience* – in science, as well as elucidating a vision of the possible cosmic future of humanity (or posthumanity).

Copernicanism and the promise of synthesis

Thus the stage is set. In the remaining chapters, I shall try to muster support for the following tightly interrelated theses:

1. The relationship between cosmology and astrobiology is much deeper than it is usually assumed – besides a similarity in the historical model of development of these two disciplines, there is an increasing number of crossover problems and thematic areas that stem from considerations of Copernicanism and observation selection effects.
2. Such a crossover area is both visualized and heuristically strengthened by the introduction of the *astrobiological landscape*, describing complexity of life in the most general context. Modern physical theories dealing with the multiverse add an additional level of detail to what is orthodoxly perceived as astrobiological enterprise, encapsulated in the *Archipelago of Habitability*.
3. Even in its orthodox version, within the well-defined confines of the Milky Way, modern astrobiology offers the prospect of both foundational support and a vast extension of the domain of applicability of the Darwinian biological evolution.
4. There is continuity between cosmology, the study of life in its cosmic context, and the study of intelligence, as encompassed by SETI studies – and some of the philosophical arguments act here to undermine traditional scepticism.
5. All these issues suggest that we need to look at astrobiological research as embedded in a bigger picture, something which I call the *extended mandate* of astrobiology, through which a deeper understanding and even consilience can be achieved across a wide spectrum of both scientific and extra-scientific fields (such as the arts).

Although the sketchiness here follows the lack of certainty and sparse nature of the results available thus far, and although the seemingly simple formulation may hide enormous tasks ahead, none should be discouraged by this; after all, pioneers of the Copernican revolution forged ahead with much less. Even meagre successes in this field justify a huge boost in optimism about pushing outward the boundaries of human cognition and understanding. And, of course, I shall approach *all* issues from the point of view of naturalism, which is at least in part a consequence of the development of a modern philosophy of biology. As David Hull notes: 'Creationists however, have made one important contribution to the philosophy of science. They have driven home how fundamental naturalism is to science (all science) and not just evolutionary biology.'[35] Nowhere more than in astrobiology is this requirement more important.

I have begun this chapter with an alien invasion described in fiction, and let me conclude it with another – but hardly more dissimilar. The title of this chapter is, of course, an *homage* to the celebrated story of Howard Phillips Lovecraft,[36] best known as the horror writer, but also a popular-science writer, a superb amateur astronomer, and a deep thinker whose works contain several astrobiology-related themes.[37] In the eponymous story, written in March 1927, a surveyor for the new reservoir to be built in New England encounters a bleak terrain where nothing will grow; the locals call it the 'blasted heath'. The surveyor, seeking an explanation for the term and for the cause of the devastation, finally finds an old man, Ammi Pierce, living near the area, who tells him an incredible tale of the events that occurred in 1882. A meteorite had landed on the property of one Nahum Gardner and his family. Scientists who examine the objects find that its properties are bizarre: the substance refuses to cool, displays spectroscopic bands never seen before, and fails to react to conventional solvents. Within the meteorite, a 'large coloured globule' is found by drilling: 'The colour . . . was almost impossible to describe; and it was only by analogy that they called it colour at all.' When tapped with a hammer, it bursts. The meteorite itself, continuing anomalously to shrink, finally disappears.

Henceforth, increasingly odd things occur. Nahum's harvest yields apples and pears huge in size, but they prove unfit to eat; plants and animals undergo peculiar mutations; Nahum's cows start to give bad milk. (Recall those 'adverse changes in the environment of the Earth resulting from introduction of extraterrestrial matter' which Article IX of the Outer Space Treaty warns about.) Then Nahum's wife goes mad, 'screaming about things in the air which she could not describe'; she is locked in an upstairs room. Soon all the vegetation starts to crumble to a greyish powder. Nahum's son Thaddeus goes mad after a visit to the well, and his other sons also break down. Then there is a period of days when Nahum is not seen or heard from. Ammi summons courage to bring policemen, a coroner, and other officials to the place, and after a series of bizarre events they see a column of the unknown colour shoot into the sky from the well; but Ammi sees a small fragment of it returning to Earth. The grey expanse of the 'blasted heath' grows by about an inch per year, and no one can say when it will end.

Lovecraft himself considered *The Colour Out of Space* his best writing, but more as an 'atmospheric study' than a classical short story. The alien presence is unknowable, not just to the characters, but to the reader as well. If this life form has any motives, goals, background, etc., they are never revealed. In fact, it is never definitively established that the alien entity is alive or whether it possesses intelligence, intentionality, or any other mental capacity. As Lovecraft scholars S. T. Joshi and David Schultz cogently note: 'It is precisely because we cannot define the nature – either physical or psychological – of the entities in *The Colour Out of Space* (or even know whether they are entities or living creatures as we understand them) that produces a sense of horror.'[38] The ambiguity persists even when we close the book;

the feeling that our very language is insufficient to deal with the full spectrum of Otherness is hard to shake off.

The Colour forcefully addresses the question many students and educated people ask when first confronted with the astrobiological revolution: why are we dealing with the search for a terrestrial kind of life? The same question is often posed from a sceptical or contrarian perspective, as a challenge: *Why don't you search for different kinds of life, say the one based on XYZ instead of on carbon and water? Is that not tantamount to giving a special position to our form of life, contrary to the Copernican spirit you boast of following?* This is a key question that cuts to the core of the effort to make astrobiology independent of our particular anthropocentric and geocentric biases. Lovecraft masterfully posits the situation in which our conventional wisdom and age-old recipes do not work any longer – and suggests that this is indeed what an open-minded researcher is likely to find; a situation to which astrobiology has to be prepared. When Lovecraft's protagonist concludes, 'Do not ask me for my opinion. I do not know – that is all', he does not relinquish any attempt at speculation. Indeed, in the very next paragraph, he dares to suggest that

> [i]n terms of matter I suppose the thing Ammi described would be called a gas, but this gas obeyed the laws that are not of our cosmos. This was no fruit of such worlds and suns as shine on the telescopes and photographic plates of our observatories. This was no breath from the skies whose motions and dimensions our astronomers measure or deem too vast to measure. It was just a colour out of space – a frightful messenger from unformed realms of infinity beyond all Nature as we know it; from realms whose mere existence stuns the brain and numbs us with the black cosmic gulfs it throws open before our frenzied eyes.

How really different – barring the lack of equations – is that from the discussions of habitability of universes without weak force, or with different numbers of macroscopic dimensions of space or born in a cold Big Bang?[39] A conscious effort at overcoming anthropocentrism is crucial for both cases.

This is the bold spirit of astrobiology. Establishing conditions for habitability vis-à-vis terrestrial life – even the extreme ends of its ecological spectrum – is an empirical matter; discussing and modelling of conditions for sufficiently different life forms, or even life forms conceivable under different laws of (low-energy) physics, is speculation. Both are – history and philosophy of science teach us – clearly and unavoidably necessary for any successful scientific enterprise. By taking only a slice of the whole mix, as is usually done in popular and journalistic science, or in opinionated presentations of the detractors, we are not only losing the flavour of the enormous complexity of the whole, but we are also missing and downplaying new and unexpected directions for original research.

The author who wrote that '[t]he aeons and the worlds are my sport, and I watch with calm and amused aloofness the anticks of planets and the mutations of universes',[40] and who was called 'a literary Copernicus' in a fine essay by Fritz Leiber,[41] is a good companion to have on board on this journey. I shall occasionally return to some other examples of astrobiological topics in Lovecraft's opus, as well as by other fiction authors, notably Olaf Stapledon and Stanislaw Lem, not only to illustrate the relevant points better, but since here, as elsewhere, artistic imagination has often stumbled upon philosophically and scientifically fruitful ideas.

2

Cosmology, life and duration of the past

If grey eyed Athena loved you
the way she did Odysseus in the old days
in Troy country, where we all went through so much . . .

Homer

Living beings wondering about the place of life in the universe they inhabit – this is a picture that cannot escape self-reference, the Hofstadterian 'eternal golden braid.'[1] The origin and evolution of matter on the largest scales, including the totality of physical existence has, at least since 1917, been the province of physical cosmology – as contrasted with mythological or religious or philosophical cosmologies of past epochs. At least until very recently, it did not include any account of observers, regarding the totality of the universe from some idealized 'nowhere' or the 'Archimedean point'; of course, there is no such detached point in reality. But, as I shall show in this chapter, there has been an undercurrent – even an ancient one – of connecting the properties of the physical world and the domain of life.

In his seminal study, *The Biological Universe*, and in several papers, the historian of science Steven J. Dick develops a thesis that the concept of extraterrestrial life, especially in its 'pluralist' form, is sufficiently comprehensible to qualify as a cosmological worldview – the one emerging in the seventeenth century with the success of the Copernican revolution, and the one which has at the end of the twentieth century become scientifically testable. I shall attempt to show how important and fruitful this thesis of Dick's is, especially when (i) developed further along the modern lines set in place by the astrobiological revolution, and (ii) some of its distant, but logical, consequences are brought into focus. Recent revolutionary developments have created conditions for leaving outdated metaphysical baggage behind and joining efforts in both the physical and biological domains to create 'biophysical cosmology' as a successful multidisciplinary enterprise.[2]

With further advances of contemporary astrobiology, and results such as those on the temporal and spatial distribution of habitable planets, Dick's thesis can become even more interesting and inspiring.[3] The purpose of its generalization will become clear in the further course of this book, especially when observation-selection effects and contemporary views on the nature of physical reality are taken into account.

Wallace's valiant attempt

In 1903, more than four decades after publishing a treatise proclaiming biological evolution by natural selection, the man who has largely remained to this day *[i]n Darwin's Shadow* (the title of the recent fine biography), Alfred Russel Wallace, wrote a fascinating booklet entitled *Man's Place in the Universe*, indicatively subtitled *A Study of the Results of Scientific Research in Relation to the Unity or Plurality of Worlds.*[4] In it, the co-discoverer of evolution sketched a truly vast, cosmic vision unifying terrestrial and cosmological views and offering a provocative counterpoint to the then prevailing naively optimistic views on all-pervading life. Wallace's bold foray highlights the relationship of cosmology and astrobiology at their very beginning as scientific magisteria and it is worth looking at in some detail, not only for the sake of history. It has been located in the realm of predominant – but *nonscientific* – 'pluralism' or optimism with regard to life and intelligence on other worlds. In contrast to his predecessors, Wallace had in his hands a powerful weapon – the understanding of biological evolution as a *universal* phenomenon. It was effectively the first attempt at wide-ranging astrobiological synthesis. It contained, however, one notably pre-modern element – teleology.

As was told repeatedly between 1859 and 1869, Wallace changed his mind regarding the power of natural selection to create the seemingly redundant and non-adaptive complexity of the human mind.[5] While not commited to any particular form of supernaturalism, in Darwin's words, he still 'called in an additional and proximate cause in regard to Man'. In spite of the criticisms from Darwin – who published *The Descent of Man* in 1871 – and other distinguished naturalists, Wallace maintained this supernaturalism until the end of his long and productive career. In the first decade of the twentieth century, he became intensely interested in what will many decades later be called astrobiology; the outcome of that interest were two books, the already mentioned *Man's Place in the Universe* and, four years later, *Is Mars Habitable?*[6] As commented by modern historians, Wallace's astrobiological excursion was neither accidental, nor disconnected with his wider interests, including non-scientific ones, especially spiritualism (although he was rightly praised for keeping metaphysics and pseudoscience *almost* entirely out of his reasonable and sober discourse). In this respect, he was firmly grounded in his own time; in other respects, however, he was surprisingly modern, prefiguring subsequent research for decades. Perhaps his best claim for the latter consisted of perceiving the key importance of a cosmological background for his conclusions about life. Ironically, he demonstrated it in making his own grave mistakes, which followed from relying on a factually incorrect cosmological view.

The most authoritative model of the universe of the pre-Hubble era was what was called the Kapteyn model (or the Kapteyn universe), named after the great Dutch observational astronomer, Jacobus Cornelius Kapteyn.[7] He effectively brought to a logical conclusion the 150-year long 'star gauges' programme of William Herschel, perhaps the single most inspiring line of research in the history of observational astronomy. The main idea was seductively simple: to determine the shape of our stellar system – and, in the view of practical astronomers, the entire material world – by counting stars in small, precisely defined patches around different lines of sight on the celestial sphere. Herschel and his successors in the premier league of astronomy in the 1750–1900 period (like Sir John Herschel, William Herschel's son and Darwin's infamous critic, William Huggins and several members of the Struve astronomical dynasty) invested a huge amount of telescope-hours in this programme, but always failed to find clear anisotropies in stellar distribution. Kapteyn bowed to the seemingly inevitable and built a near-central position of the Sun and the Solar System into his model of the Galaxy (which was for him identical to the universe at large).[8] Kapteyn's universe was thus very small and at least approximately heliocentric: it was made of stars within a single stellar system of ellipsoidal or lenticular shape somewhat less than 17 kpc in diameter and about one-fifth thick. The Solar System was located only about 650 pc from the centre of the system, thus enjoying an essentially Aristotelian position in the centre. And outside of this stellar system there was nothing – just empty space, devoid of stars or any other visible sources. Before the advent of the expanding universe in 1929–1934, it was by far the most authoritative cosmological picture in existence, as testified by the fact that Einstein used it as a prototype of what relativistic cosmology has to explain.

Although Wallace did not use the precise Kapteyn model, his default cosmology was very similar, which is a reflection of the fact that even before Kapteyn's definitive publication of *The Structure of the Universe* in 1913, similar ideas had been widespread in the astronomical community, as Wallace's citation of John Herschel, Simon Newcomb and Norman Lockyer testifies.[9] Thus, effectively, the co-discoverer of natural selection used Kapteyn's model as a necessary background for his astrobiological speculation. If anything, the version of Wallace was still smaller: roughly spherical with a diameter of only about 1.1 kpc. The position of the Sun and the Earth in the central star cluster caused adverse commentaries, even among some of the contemporaries. Wallace compounded his error by arguing for a yet undiscovered reason why stellar radiation would be particularly conducive for biogenesis and the subsequent evolution of life; however, even such an error contained a grain of worth – this was the first speculation to the effect that there is *non-uniformity of habitable sites* – what we today call the Galactic Habitable Zone (henceforth GHZ).[10] For instance, Wallace writes:[11]

> Thus, it seems to me, the controlling force may be explained which has retained our sun in approximately the same orbit around the centre of gravity of this central cluster during the whole period of its existence as a sun and our existence as a planet; and has thus saved us from the possibility – perhaps even the certainty – of disastrous collisions or disruptive approaches to which suns, in or near the Milky Way, and to a less extent elsewhere, are or have been exposed. It seems quite probable that in that region of more rapid and less controlled motions and more crowded masses of matter, no star can remain in a nearly stable condition as regards temperature for sufficiently long periods to allow of a complete system of life-development on any planet it may possess.

The essential idea of the spatial and temporal correlations that lead to the non-uniform distribution of habitable planets behind GHZ is contained here in embryonic form. Wallace's conclusions went much further: he found not only that the Solar System and the Earth are unique – hence uniqueness of the terrestrial biosphere – but also that the evolution of intelligent observers cannot be explained without recourse to the 'disembodied mind' pervading the universe. After all, in a small universe with a still smaller central star cluster, the number of habitable sites must be small as well.

Stephen Jay Gould in his popular, but somewhat biased, essay on Wallace's version of the anthropic principle, calls him the 'first prominent exobiologist among evolutionists'[12] and the label seems fully justified. However, Gould simplifies the question too much when he brings into the discussion Wallace's spiritualism and connects it with the anthropic principle on the basis of the following citation from the concluding parts of *Man's Place in the Universe*:[13]

> Lastly, I submit that the whole of the evidence I have here brought together leads to the conclusion that our earth is almost certainly the only inhabited planet in our solar system; and, further, that there is no inconceivability – no improbability even – in the conception that, in order to produce a world that should be precisely adapted in every detail for the orderly development of organic life culminating in man, such a vast and complex universe as that which we know exists around us, may have been absolutely required.

From the subsequent description, it seems that Gould takes into account only old-fashioned teleological construals of anthropic reasoning, which are not taken seriously for quite some time; I shall discuss this central point in detail in Chapter 3. The implication is not that Wallace has understood the empirical basis of anthropic reasoning in the modern, disteleological sense. On the contrary, his motivation for undertaking his (proto)astrobiological project was teleological and this intrusion of teleology is one of two main reasons he reached his misleading conclusions about the uniqueness of humans. However, the main lesson of Wallace's pioneering work can be summarized as the importance of cosmology for the study of life in the general cosmic context, and the possibility of drawing *mistaken* conclusions about life from an inadequate cosmological background. The fact that nowadays we have a much firmer grasp on physical cosmology should not lead us to the mistaken conclusion that we necessarily understand life much better. (Another part of the problem which I cannot enter into here is that Wallace, as a good Baconian, insisted on inductive generalizations from the scientific knowledge of his day, which might be inappropriate for many sectors of the astrobiological endeavour; see Chapters 3–6 below.)

In addition, Wallace argued for *dis*continuity between life in general and intelligent life forms capable of tool making and civilization, thus being a precursor to the 'Rare Earth hypothesis' of present-day astrobiological research. I argue below (see, in particular, Chapter 6) that, following one of Wallace's most illustrious successors at the high table of evolutionary biology, J. B. S. Haldane, we should consider both inanimate matter, simple life forms, complex metazoans, intelligent beings – and possible post-biological intelligent entities – as parts of a single continuum, without sharp discontinuities.The reasons for this are not purely in the philosophical domain – nor were they entirely abstract and speculative in Haldane's time either (1920s–1940s), well before the advent of complexity theory, not to mention the astrobiological revolution of the 1990s.

Wallace's long-standing interest in astrobiological issues is expressed in his later 1907 booklet *Is Mars Habitable?*, again with a beautiful subtitle *A Critical Examination of Professor Percival Lowell's Book 'Mars and Its Canals,' With an Alternative Explanation.* In this sober criticism of Lowell's fanciful claims about Martian civilization, widely accepted in their day, he applied the method sketched in *Man's Place in the Universe* in more detail, showing that the physical conditions on our planetary neighbour are not conducive to biological evolution, at least not such as to enable the emergence of intelligent beings. Metaphysical and teleological aspects of Earth's (hence humanity's) uniqueness are again toned down and the treatise represents a reasonable precursor of later astrobiological research on the habitability of planets in the Solar System. As in so many other issues, Wallace has been regarded as a dissenter and heretic here as well, and has only recently attained a modicum of reappraisal. Even Gould – a sharp critic of teleology in all its guises – links Wallace with the anthropic principle and the ideas of a modern 'heretic' such as Freeman Dyson, thus offering a curious compliment to the actuality and relevance of the astrobiological thought of the co-discoverer of natural selection, almost a century later.[14]

Reappraisal also has many faces. Those who have hurried to make Wallace a precursor of modern 'Rare Earth' scepticism are making two different kinds of mistakes, apart from being anachronistic. The first is the claim that Wallace, having understood evolution by the action of natural selection, argued on that basis for the absence of intelligent life ('mind') on other planets. In reality, Wallace made all sorts of increasingly bizarre manoeuvres in order to force the two together; if anything, his strict adaptationism outside of the 'mind' domain hindered, rather than helped his monism. Hence, such amusing ideas like starlight being important for growing plants, etc. In fact, Wallace thought that if evolution were the only force acting throughout the universe, it would be very difficult to argue for the uniqueness of the terrestrial biosphere or the human *mind* (which was, obviously, his foremost concern).

The second mistake is assuming that Wallace's reasoning is easily 'transferable' from his tiny universe to the world of present-day cosmology. Cosmological background is, for Wallace, not a decoration – it is an important, active player. That is exactly why he was such an important precursor to astrobiology; there is no 'Whiggish' (mis)interpretation of the history of ideas in admitting that on this particular methodological issue he was right on target. For the same reason, erroneous conclusions reached through bad cosmological input cannot stay the same when the input changes. Therefore, we can hardly claim that Wallace's claim of Earth being unique in the universe was on the right track any more than, for example, we celebrate Thomas Wright's conclusion that Earth is located at the periphery of the Milky Way *on the basis of morality*. It is exactly the 'non-transferable' nature of Wallace's proto-astrobiology that makes his work and its lessons valuable for the researcher of today.

In the words of a modern – and sympathetic – historian of science, Wallace 'took a creative step of linking plurality [of life and intelligence] with the general question of the preconditions for organic evolution'.[15] It sounds practically a recipe for present-day research, doesn't it? And, although we discard teleology today, we often repeat Wallace's mistake in neglecting or misinterpreting the wider cosmological context. Clearly, Herschel's and Kapteyn's universe is outdated today; but what grand cosmological pictures substitute for it?

Six eras and the (New) standard cosmology

Modern physical cosmology 'officially' started with the Einstein 1917 paper, although a good case could be made for the Boltzmann–Zermelo debate about two decades earlier, at least on the basis of being the first debate that explicitly invoked the beginning of the universe within the discourse of physics which took place in major peer-reviewed research journals.[16] In any case, in almost a century since Einstein's first cosmological model, physical cosmology radically transformed our view of the world, perhaps the most radical such shift since the days of Copernicus and Galileo. Still, its history has been highly structured and for present purposes we can distinguish the following six major epochs from a cursory look at the record:[17]

?–1917: *Prehistory*: Among important early elements, one might mention Olbers' paradox and the so-called gravitational paradox originating with Newton. In the last part of this period, very interesting cosmological debate was engaged by Ludwig Boltzmann in his controversy with Ernst Zermelo (1895–96), regarding the origin of the thermodynamical disequilibrium of the universe – testified, as Boltzmann presciently argued, by the existence of us as intelligent observers.

1917–1929: *Static universe*: Two main early models, Einstein's static universe and de Sitter's empty universe emerged in 1917 as a consequence of the great theoretical breakthrough in formulating the first and still the best metric theory of gravity. Both models were characterized by contemporaries as static, although the label has subsequently only applied to the Einstein model. The dearth of empirical knowledge continued in this phase, and cosmology was regarded as a mathematical game rather than a description of physical reality. The seminal work of Alexander Friedmann, which took place in early 1920s, was rediscovered only later.

1929–1948: *Expanding universe*: Hubble's discovery of the expanding universe created conditions for a 'cosmological revolution'. Georges Lemaître and George Gamow, together with the belatedly understood contribution of Friedmann, laid the foundations for relativistic cosmological models of the expanding universe. What Lemaître called the 'primaeval atom' will become better known as the (hot) Big Bang. The work of those three, together with important input of Sir Arthur Eddington, Richard Tolman, and a few others who shyly started calling themselves cosmologists, helped physical cosmology attain a modicum of seriousness and authority. Nowadays, this epoch looks and is often described, especially in popular accounts, as heroic, but one should keep in mind that in it cosmology was still, in Fred Hoyle's pointed words, the science of 'two and a half facts'.[18]

1948–1965: '*The Great Controversy*' between the steady state theory and the family of relativistic world models (which during this time obtained the initially pejorative label of 'Big Bang' in a BBC radio programme hosted by Hoyle) was the key formative period in which most of the tenets of contemporary cosmology crystallized in serious, empirical form out of vague speculation. Crucial observational tests of world models were devised in order to resolve the controversy. A huge amount of philosophical debate took place in those years, as masterfully described by Helge Kragh, but this should not detract from the fact that some of the key observational tests of world models were conceived (and some even executed in practice) in this period.

1965–1998: *Hot Big Bang victorious*: Empirical discoveries of the cosmic microwave background (CMB) radiation and high-redshift objects like quasi-stellar objects (QSOs), coupled with increased theoretical understanding of processes like nucleosynthesis, recombination and growth of gravitational perturbations in an initially almost homogeneous medium, left no doubt that relativistic Hot Big Bang models point in the right direction. Cosmology entered its 'normal phase' in the Kuhnian sense, characterized by an improved accuracy of its models and the devising of new and better methods of testing predictions. New questions, especially the nature of dark matter and the theory of structure formation came to the fore, together with the computing revolution that enabled large-scale numerical simulations of cosmological processes.[19]

1998 – today: *New Standard model*: Although predicted by Steven Weinberg (and some other 'voices in the desert'), the discovery of dynamically large dark energy – usually thought of as Einstein's cosmological constant, though it can be misleading – in 1998 brought about a new mini-revolution in cosmology.[20] To this, other slower developments have been added, notably the successes of inflation in predicting the CMB spectrum, as verified first by COBE and then by WMAP satellite observatories. A whole bunch of new problems characterize this epoch, notably the origin of dark energy, as well as focusing attention on very early times (cosmological inflation, quantum cosmology, string cosmology, etc.) and on generality in the form of a multiverse. This period is occasionally called '*post*modern cosmology' in order to emphasize both its divergence from the Hot Big Bang roots and its often overly speculative character.

Even such a cartoonish sketch illustrates that not only the complexity of cosmological narrative and its historical emergence, but also the old-fashioned positivist view of gradual and linear accumulation of knowledge makes no sense. The two key ingredients of the modern cosmological worldview, namely the Hot Big Bang (and associated theories of nucleosynthesis and recombination) and the New Standard model, emerged in dramatically different ways. Since they are milestones in any meaningful discussion of cosmological knowledge, I shall briefly summarize them here.

Imagine travelling backwards in time. Matter and radiation in the universe gets hotter and hotter as we go back towards the initial quantum state, because it was compressed into a smaller volume. In this *Hot Big Bang* epoch in the early universe, we can use standard physical laws to examine the processes going on in the expanding mixture of matter and radiation. A key feature is that about 300 000 years after $t = 0$, the start of the Hot Big Bang epoch, nuclei and electrons combined to form atoms. At earlier times when the temperature was higher, atoms could not exist, because the radiation then had so much energy it disrupted any atoms that tried to form into their constituent parts (nuclei and electrons). Thus, at earlier times matter was ionized, consisting of negatively charged electrons moving independently of positively charged atomic nuclei. Under these conditions, the free electrons interacted strongly with radiation by Thomson scattering. Consequently, matter and radiation were tightly coupled in equilibrium at those times, and the universe was opaque to radiation. When the temperature dropped through the ionization temperature of about 4000 K, atoms formed from the nuclei and electrons, and this scattering ceased; the universe became very transparent (today we are able to see galaxies at enormous distances from us). The time when this transition took place is known as the *time of decoupling* – it was the time when matter and radiation ceased to be tightly coupled to each other. The sea of radiation will become what we today observe as CMB photons. After decoupling, matter formed large-scale structures through gravitational instability, which eventually led to the formation of the first generation of stars and is probably associated with the re-ionization of matter, which puzzles today's astrophysicists. At that time, planets could not form since there were no elements heavier than lithium present in the universe. The first stars aggregated matter together by gravitational attraction, the matter heating up as it became more and more concentrated, until its temperature exceeded the thermonuclear ignition point and nuclear reactions started burning hydrogen to form helium. Eventually, more complex nuclear reactions started in concentric spheres around the centre, leading to a build-up of heavy elements (carbon, nitrogen, oxygen, for example), up to iron. These elements can form in stars because there is a long time available (millions of years) for the reactions to take place. Massive stars burn relatively rapidly and eventually run out of nuclear fuel. The star becomes unstable, and its core rapidly collapses because of gravitational attraction. If the stellar mass is above a threshold, the consequent rise in temperature blows it apart in a giant explosion, during which time new reactions take place that generate elements heavier than iron; this explosion is seen by us as a *supernova* suddenly blazing in the sky, where previously there was just an ordinary star. Such explosions blow into space the heavy elements that had been accumulating in the star's interior, forming vast filaments of dust around the remnant of the star. It is this material that can later be accumulated, during the formation of second-generation stars, to form planetary systems around those stars. Thus the elements of which we are made (carbon, nitrogen, oxygen and iron nuclei, for example) were created in the extreme heat of stellar interiors, and made available for our use by supernova explosions (and, to a lesser extent, by the formation of planetary nebulae and mass loss from less-luminous stars). Astrochemistry, which logically precedes at least those sectors of astrobiology dealing with life *as we know it*, is deeply rooted in Hot Big Bang cosmological theory.

Over the past 14 years, a new, improved standard cosmology has been emerging. It incorporates the highly successful standard Hot Big Bang cosmology and extends our understanding of the universe to times as early as 10^{-32} s after $t = 0$, when the largest structures in the universe were still microscopic quantum fluctuations. This *New Standard Cosmology* is characterized by

- A globally flat, accelerating universe.
- An early period of rapid, exponential expansion (cosmological inflation).
- Density inhomogeneities produced from quantum fluctuations during inflation.
- Composition: $\sim$67% dark energy; $\sim$33% dark matter (both baryonic and non-baryonic); 0.5% bright stars; $\ll$1% neutrinos and radiation.

The New Standard Cosmology is certainly not as well established as the standard Hot Big Bang. However, the observational evidence is mounting from many sources, notably the CMB power spectrum and anisotropies, cosmological supernovae observations, precision abundance measurements of light nuclei, as well as precision measurements of large-scale structure from the Sloan Digital Sky Survey (SDSS) and other galaxy surveys.

The subsequent history of the universe is mainly the evolution of structure, dominated by local astrophysical processes, like star and planet formation. Finally, life on Earth emerged at –3.8 Gyr and until roughly –530 Myr stayed fairly simple. After that, an evolutionary pattern of greater complexity, punctuated by occasional mass extinction episodes, has been maintained until the origin of intelligent life, about –300 Kyr. This event and the emergence of technology heralds a new era in which life has gradually begun to dominate matter. There will be more on this later, but for the moment note that here the completely naturalistic and non-intentional narrative of cosmic evolution ends. This bookend is *necessary*, even in a laconic sketch of the cosmological story as given here. The failure to provide it in many accounts, both professional and popular, is exactly the reason why the tight relationship between cosmological and astrobiological narratives is so often obscured.

Infinite past(s) and the Davies–Tipler argument

Let us for the moment forget everything about the New Standard Cosmology and track back to an earlier epoch, before Einstein and the advent of physical cosmology. The simplest division of all cosmologies is into two broad classes:

those postulating the eternal universe and those that postulate some origin of the universe, or at least the part of it that cosmologists currently inhabit. This taxonomy dramatically predates modern physical cosmology. Aristotle and his pupils fiercely polemicized against the remnants of pre-Socratic teachings about the universe, infinite in space and time, teachings that, as we shall see, prefigure some of the modern ingredients of today's framing of astrobiological 'big questions'. In AD 1440, as both the Hundred Years War and the long history of Byzantium were drawing to their close, thus heralding a new age for Europe, the future cardinal Nicolas Cuzanus published his famous treatise 'On Learned Ignorance' (*De docta ignorantia*), sharply rejecting dominant Aristotlean scholasticism and arguing for an infinite and eternal universe, coexisting with the Christian deity.[21]

An alternative perspective had already been present. As the great historian of philosophy cogently noticed:[22]

> Here is a second major intellectual achievement, comparable both intrinsically and in its influence on later philosophy to the distinction between sensible and intelligible, namely the distinction between time and eternity, the recognition of *eternal* as a separate category from *everlasting*. To conceive of something as merely everlasting is to set it in time. One says that just as it is now, so it was thousands of years ago and will be in the future. But for the eternal 'was' and 'will be' have no meaning, and the time-sequence is abolished. Thus Plato taught that the physical universe was as old as time (it 'has been and is and shall be perpetually through all time'), but it is not eternal.

Both eternal and everlasting (in this Platonic sense) universes could aspire to adopt some sort of stationarity, a condition which is of singular importance in many branches of physics (inter alia because the law of energy conservation is closely connected with a translational symmetry of time), and which certainly simplifies the solving of specific problems. In the Enlightenment era, after the religious dogma about Creation in 4004 BC (or any other specific date) was abandoned, the universe has been considered eternal, although the great minds of classical physics, such as Newton or Boltzmann, began to perceive some of the difficulties associated with such a proposition.[23] The resistance to any opposing view (which eventually became what is today dubbed the standard cosmology) was exceedingly strong during most of the nineteenth and the early twentieth century. It is epitomized in Eddington's words, in his authoritative *The Nature of the Physical World*: 'As a scientist, I simply do not believe that the universe began with a bang.'[24] From the end of the Middle Ages until Hubble's observational revolution in the third decade of the twentieth century, the stationary worldview has been in one way or another the dominant one. This explains, among other issues, the dramatic reaction of most of the scientific community, including Lord Kelvin, Holmes, Eddington, Crookes, Jeans and others, to the discoveries of Helmholtz, Clausius, Boltzmann and other thermodynamicists, implying a unidirectional flow of time and physical change. Interestingly enough, even during this epoch, the idea – one of the least publicized triumphs of the scientific worldview – that the thermodynamical arrow of time originates in cosmology has occasionally surfaced.[25] The power of a stationary alternative to the evolutionary models of the universe has been reiterated in particularly colourful form during the great cosmological controversy in the 1950s and 1960s.[26] As is well known, the debate ceased when empirical arguments persuaded by far the largest part of the cosmological community that a universe of finite age is the only empirically acceptable concept. Thus, the position is now reversed compared with the situation in first decades of the twentieth century. The evolving universe with a definite beginning enjoys almost universal support, and the evolutionary paradigm has become unquestionably dominant.

How does it connect to astrobiology? A short notice by Paul C. W. Davies – a British physicist who only much later became well known as a popular-science writer, and still later became engaged in astrobiological research – appeared in *Nature* in June 1978.[27] In this succinct critique of the Ellis *et al.* static cosmological model,[28] Davies points out that:[29]

> [T]here is also the curious problem of why, if the Universe is infinitely old and life is concentrated in our particular corner of the cosmos, it is not inhabited by technological communities of unlimited age.

The same idea has been further developed and put on a mathematical footing by Frank Tipler.[30] As claimed by John Barrow and Tipler in their encyclopaedic monograph on anthropic principles, this is historically the first instance in which an anthropic argument has been used against cosmology containing the past temporal infinity.[31] This statement can be criticized on two grounds: (i) the 'anthropic' credentials of the argument are open to criticism, and (ii) some important precursors can be found in the history of ideas, which while not scientific in the Davies' sense, could still pass muster. However, the grounding fact remains: the Davies–Tipler (DT) narrative is an example of a surprisingly effective cosmological argument obviously based on assumptions about life and observers.

It is indeed strange that the same argument had not been considered earlier in the course of the twentieth century. The suprise is strengthened by the fact that cosmologies postulating an infinite past in scientific or half-scientific form have existed since the very dawn of science. In addition, since ancient times a belief in the existence of other *inhabited* worlds has also been present, in one form or another.[32] Today, the scepticism sometimes encountered against this mode of thinking is even stranger, when various SETI projects testify that there is no rational reason to a priori reject the possibility of the existence of technological civilizations other than the human one; I shall consider arguments for their *unlikelihood* in Chapter 7. The technological nature of the targets (the same one that produces the problem Davies wrote about) is a *sine qua non* of any sensible SETI enterprise. In spite of the huge volume of writings on the philosophical aspects of stationary cosmologies, this particular argument has not been discussed in detail so far.[33]

An ancient echo of this type of argument can be recognized in the surviving fragments of some of the most distinguished Hellenic philosophers of nature. From our point of view, especially interesting is the cyclic cosmology of Empedocles of Acragas (6–5th century BC), in which the universe is eternal,[34] consisting of the internally immutable four classic elements, as well as two opposing forces (Love and Strife, i.e., attractive and repulsive interactions). The cyclic motion of matter in the universe is governed by the change in relative intensities of the two interactions.[35] It is interesting to note that Empedocles' cosmology is *uniformitarian*, in the sense that all six basic constituents (four elements and two forces between them) are present in each instance of time in accordance with the eternal principles of mutual exchange. In some of the surviving fragments, Empedocles implies that although this uniformitarianism may seem counterintuitive, as we see things coming into being and vanishing, this is just our special perspective – *anthropocentrism* – and does not reflect the inherent state of nature.[36] This is similar to the uniformitarian notions present in the classical steady state theory.[37]

Probably the most lasting and controversial legacy of Empedoclean cosmology is his assertion that biological and even anthropological evolution are *inherent, necessary and inseparable* parts of global cosmological evolution.[38] Thus, speaking on the four elements, he states:[39]

> For out of these have sprung all things that were and are and shall be – trees and men and women, beasts and birds and the fishes that dwell in the waters, yea, and the gods that live long lives and are exalted in honor.
>
> For these things are what they are; but, running through one another, they take different shapes – so much does mixture change them.

However, if we accept this view that biological evolution and the appearance of consciousness and intelligence are contingent upon cosmological processes, which I shall call the 'Empedoclean picture', the eternal universe of Empedocles faces the same kind of problem as that of modern stationary cosmologies like the classical steady state theory or the stationary theory of Ellis *et al.* criticized by Davies. Why then, in the supposed infinity of time, are 'men and women, beasts and birds' of finite (and relatively small) age? In the eternally existing or recurrent cosmologies, there is a natural ambiguity surrounding the very concept of novelty, which has been poignantly expressed by the Ecclesiastes (probably influenced by Hellenic philosophy as well):[40]

> Is there a thing of which it is said, 'See, this is new'?
> It has been already,
> in the ages before us.
> There is no remembrance of former things,
> nor will there be any remembrance
> of later things yet to happen
> among those who come after.

So things might *look* new if there is no information about their previous existence ('remembrance'), be it finite or infinite. But why would such erasure of information occur?

Empedocles might have seen this himself and evaded the problem in the only natural way: by postulating two singular states in the beginning and in the middle of each of his great cycles. These singular states are moments (in absolute time!) of complete dominance of either Love (an ancient equivalent of the modern initial and/or final singularities) or Strife (no true equivalent, but similar to the modern version of heat death in the ever-expanding cosmological models[41]). In these states, life, with its complex organizational structure, is impossible and therefore they serve as *termini* for the duration of any individual history of life and intelligence. The maximal duration of any form of life and/or intelligence is determined exclusively by cosmological laws. Therefore, there are no arbitrarily old beings, time is *effectively* finite, and the anthropic argument of Davies and Tipler is inapplicable.[42]

Andrew Gregory, in a recent comprehensive study of ancient Greek cosmologies, defends the important idea that there are what he calls 'perennial problems' in the history of cosmological ideas:[43]

> A key argument of this book will be that there are perennial philosophical and scientific problems relating to cosmogony. This is not to say that these problems are insoluble, or that no progress has been made in relation to them. On the contrary, I believe we now have a much more sophisticated understanding of these problems and a greater and more sophisticated array of possible solution, even if we often lack definitive solutions. Once these problems have been recognised and formulated, they need to be dealt with by each subsequent thinker. ... While there may be perennial core problems in cosmogony, they manifest themselves in varying forms at different times and in different circumstances. The problems as perceived by the ancient Greeks are related to but not identical to the problems addressed by modern cosmology.

The scope of these perennial problems, tellingly, encompasses the domains of both cosmology and astrobiology. Gregory is careful to delineate the approach to these, rejecting a naive positivist view of 'anticipation' (whatever that might mean in the context of complex history of ideas), as well as postmodern relativism (claiming even the deepest theories of nature as non-external products of 'culture'). The synergy between cosmology and astrobiology is a fine example of the persistence of perennial questions and perennial explanatory approaches, and I shall return several times in this book to how well Gregory's project tallies with the extended mandate of astrobiology. In the specific case of Empedocles, he notices a theme of continued relevance:[44]

> If the fundamental constants are in some way reset, perhaps randomly, each time the universe contracts to a singularity, then there is a way of explaining the nature of our own universe. Over time, there are a multiplicity of universes with different values for the fundamental constants, so our universe with its apparently fortuitous values for the fundamental constants is no longer a surprise, but is one of many.
>
> Empedocles' problem is slightly different, in that he needs to explain how a *kosmos* can occur without intelligent guidance, both from a complete association and a complete dissociation of the elements. His answer is structurally the same as the modern answer though. If there is an element of chance, and if we have enough goes, then the occurrence of a *kosmos* can be explained as one among a multiplicity of accidents. Empedocles gives the same sort of answer in zoogony, and here too he finds a modern counterpart. Species evolve through a multiplicity of accidents leading to viable species. Whatever we may think of the details of the explanation that Empedocles gives us, the explanatory structure he uses is important.

Our acquaintance, Howard Lovecraft, would poetically describe the same perennial idea like this:[45]

> Then in the slow creeping course of eternity, the utmost cycle of the cosmos churned itself into another futile completion, and all things became as they were unreckoned kalpas before. Matter and light were born anew as space once had known them; and comets, suns and worlds sprang flaming into life, though nothing survived to tell that they had been and gone, been and gone, always and always, back to no first beginning.

Back to no first beginning. So, here we have an Empedoclean way of recycling of the universe, retaining the 'best of both worlds': no beginning of time and no arbitrarily old causal influences.

In the very first chapter of the immortal history of Thucydides, there is a famous statement that before his time – i.e., about 450 BC – nothing of importance (σ*υ* μεγαλα γενεσθαι) had happened in history.[46] This startling statement has been correctly called 'outrageous' by Oswald Spengler, and used to demonstrate the essentially mythological character of ancient Greek historiography.[47] It may indeed be outrageous from the modern perspective, but it does motivate a set of deeper questions, dealing ultimately with cosmology. The fact that Thucydides did not know (or did not care to know) previous historical events does not change the essential perception of *finiteness* of human history, inseparable from Greek thought. This property starkly conflicts with the notion of an *eternal, continuously existent material world*, as it was presented in both modern and ancient cultures.

Obviously, it is irrelevant which exact starting point we choose for unfolding historical events. In any case, the number of these events is finite, and the timespan considered is small even compared to specific astronomical timescales (some of which, like the precession period of the equinoxes, were known or prefigured in the classical antiquity, as is clear from the discussion of the 'Great Year' in Plato's *Timaeus*), not to mention anything about a past temporal *infinity*. Although there was no scientific archaeology in the ancient world, it was as natural then as it is now to expect hypothetical previous civilizations inhabiting Oikumene to leave some traces – in fact, an infinite number of traces for an eternally existent Oikumene! There are indications that pre-Socratic thinkers were aware of the incompatibility of this 'Thucydidean' finiteness of historical past with the eternal nature of the world. We have already mentioned the solution – periodic singular states – proposed by Empedocles himself. Even earlier, in the fragmentary accounts of the cosmology of Anaximandros, one may note that he proposed an evolutionary origin of humankind in some finite moment in the past, parallel with his basic postulate of separation of different worlds from *apeiron* and their subsequent returning to it.[48] In Anaxagoras' worldview, there is a famous tension between the eternity of the world's constituents and the finite duration of *movement* (and, therefore, the finite duration of relational time) in the world. At the same time, it seems certain that Anaxagoras, together with Anaximandros and Empedocles, was an early proponent of the evolutionary view, at least regarding zoogeny in general and the origin of humankind in particular.[49]

Of course, when Thucydides wrote that line, he did not necessarily imply the beginning of time, or even just the beginning of humanity. Conceivably, he could have thought about 'prehistory': a long interval of time in which humans lived in a primitive state, lacking any events a political historian might find interesting. In principle, this is compatible with even the *infinite* past existence of humans, their being co-terminous with the material world.[50] Although such a thought seems preposterously contrived, we need to investigate all options; after all, we today tend to regard as a thought experiment something that in ancient times could have been a concrete issue of history. Without going into detail, it seems safe to assume that, as a highly educated member of the Athenian upper classes, Thucydides was aware of the philosophical debates of sophists and pre-Socratics. He could even have been among the pupils of Anaxagoras, who taught in Athens before his exile about 434 BC. Again, one of the central tenets of Anaxagoras' system was the finite origin of motion, with the action of the Mind (*Noûs*) acting on the eternally existing material constituents. The action of the Mind might be plausibly regarded as the beginning of the *effective* passage of time. Therefore, for Thucydides (a rather practical and realistic writer, in any case) *effective* history started relatively recently. Of course, this prompts what could be called a cluster of 'origins' questions: what causal agency determined the beginning? How did it come into play? etc.

Finally, an almost modern formulation of the anthropic argument against the past temporal infinity was made in Roman times by Lucretius, who in Book V of his famous poem *De Rerum Natura* wrote the following intriguing verses:[51]

> Besides all this,
> If there had been no origin-in-birth
> Of lands and sky, and they had ever been
> The everlasting, why, ere Theban war
> And obsequies of Troy, have other bards
> Not also chanted other high affairs?
> Whither have sunk so oft so many deeds
> Of heroes? Why do those deeds live no more,
> Ingrafted in eternal monuments
> Of glory? Verily, I guess, because
> The Sun is new, and of a recent date
> The nature of our universe, and had
> Not long ago its own exordium.

For scientific-minded Lucretius, the shortness of human history (over which Thucydides passes without comment) *is* very strange in the face of a hypothesis of the eternal existence of the world. Although the references to 'eternal monuments' and 'other bards' may sound naive, it is clear that Lucretius had in mind any form of transmission of information from the past to the present; and an infinite amount of information from an infinite past. His empirical assessment of the surrounding world clearly shows the absence of such information. Therefore, an explanation is needed. The simplest explanation, as Lucretius was highly aware, is to treat the argument as *reductio ad absurdum* of the starting hypothesis (eternal nature of the world) and to assume that the world is of finite – and relatively short – age.

The depth of Lucretius' thought in this passage is amazing, especially when the historical blindness of subsequent generations to this same perspective is taken into account. Lucretius' argument applies to the classical Newtonian universe of infinite age, as well as to modern stationary alternatives to evolutionary cosmology. It emphasizes the *technological* nature of possible evidence ('ingrafted in... monuments'). This is exactly what modern cosmologists Davies and Tipler have had in mind when constructing the anthropic argument in order to refute the eternal cosmologies of our epoch. Lucretius' monuments play essentially the same role as Tipler's von Neumann probes sent by advanced intelligent communities.[52]

As in some fractal image of the history of ideas, we find the same theme when the history of our local habitat – the Earth – enters the realm of science. James

Hutton's solution to the problem of the duration of the past vs. limited information transmitted from those epochs is obviously motivated by the final causes. Hutton imagined a 'world machine': his mechanistic worldview found an excellent field of applicability in the geology of his day. Erosion of the soil is compensated by the uplifting of mountains; any other particular tendency is contrasted with an opposite one that is bound to return the world to one or more previous stages. Hutton's vision is a geological analogue of the Empedoclean cyclic universe; hence, the most famous passage of his, ending the 1788 short version of his *Theory of the Earth*:[53]

> If the succession of worlds is established in the system of nature, it is in vain to look for anything higher in the origin of the earth. The result, therefore, of our present enquiry is that we find no vestige of a beginning – no prospect of an end.

But, contrary to textbook history, portraying Hutton as the standard-bearer of a modern scientific outlook, this view of the 'world-machine' was not motivated so much by the desire to explain the observed phenomena, as by the metaphysical invocation of final causes. The final cause in question was nothing less than what modern astrobiologists would call *planetary habitability*: time and again, Hutton writes of 'mechanism of the globe, by which it is adapted to the purpose of being a habitable world'. The Earth was obviously constructed (at some *indefinite* epoch, not existing from eternity!) for the higher purpose of being a habitat for life and, eventually, for human domination. Hutton writes about 'a world contrived in consummate wisdom for the growth and habitation of a great diversity of plants and animals; and a world peculiarly adapted to the purpose of man, who inhabits all its climates, who measures its extent, and determines its productions at his pleasure.'[54]

Note that the age of the Earth (and perhaps the rest of the universe) was considered indefinite, but not infinite. Infinite age would conflict with Hutton's profound Christian religiosity, and he repeatedly implies that the ultimate questions of the beginning and the end of the world are not part of the scientific discourse. However, he subtly builds an insurance against the Lucretian eternal-monuments problem in his choice of words: not that there is no beginning – there are only no *vestiges* of the beginning! The cyclic nature of the world machine erases the relevant information from previous cycles and 'cleans the slate'. But teleology is still inconsistent: if the world is made for man, how come that the achievements of previous generations of humans are also erased? And if we accept pluralism about abodes of life,[55] then it is very difficult to conceive of an explanation for the failure of intelligent beings to overcome the slow processes of erosion and decay which erase information from previous cycles. Empedocles at least postulated catastrophic, singular events encompassing the entire universe; Hutton's world machine is much less efficient in this respect.

Among the precursors of the anthropic argument of Davies and Tipler, one may list the great British biologist, chemist, philosopher and author J. B. S. Haldane, whom we shall repeatedly encounter in the course of this book. His keen interest in cosmological issues has been characterized by his defence of Milne's two-timescales cosmological model in which (at least according to one timescale) the universe is of finite age and fundamental constants change with time. In the following interesting passage, through comparing the hypothesis of the origin of the universe in the finite past vs. the hypothesis of its eternal existence, he shows both his cosmological interests and appreciation for a melioristic and humanistic worldview:[56]

> On the first hypothesis, why was it not created better; on the second, *why has it not got better in the course of eternity*?... On neither theory have we very strong grounds for hoping that the world will be a better place a million, let alone a thousand, years hence, than it is today. But on Milne's theory the laws of nature change with time. The universe has a real history, not a series of cycles of evolution. Although, from one point of view, the past is infinite, life could not have started much before it did, or have got much further than it has at the present date. If this is so, human effort is worth while and human life has a meaning.

If we understand 'improvement' of the universe not in strictly ethical terms, but as an increase in its complexity brought about by evolution, the question posed by Haldane is the same as in the DT argument. Complexity may be achieved through either technologization or 'biologization' of the universe, and both lead to paradoxical consequences in an eternal universe.

For the sake of better understanding, let us consider a counterexample of a cosmological model involving past temporal infinity which the DT argument does not apply to. This is the Eddington–Lemaître universe, which was quite popular in the 1925–1935 period.[57] This model belongs to the class of relativistic models with a positive cosmological constant and without a singular beginning. Therefore, it was very appealing from the point of view of resolving the age discrepancies between cosmological models and various astrophysical and geological timescales.[58] Having appeared on the cosmological scene after the realization of the instability of the original Einstein static universe, this model[59]

> has an infinite past which was spent in the Einstein state. This has greatly attracted investigators since it seemingly permits an arbitrarily long timescale of evolution. The picture of the history of the universe derived from this model, then, was that for an infinite period in the distant past there was a completely homogeneous distribution of matter in equilibrium in the Einstein state until some event started off the expansion, which has been going on at an increasing pace ever since. The condensation of the galaxies and the stars from the primeval matter took place at the time the expansion began, but this development was stopped later by the decrease of average density due to the progress of the expansion.

This model is a good physical representation of the situation often considered in philosophical studies of the distinction between the relationist and absolutist theories of time: the situation in which an absolutely unchanged universe suddenly transforms into the changing world we observe.[60] From the formal point of view, in accordance with the Weyl postulate, the Eddington–Lemaître universe has an infinite past, i.e., the initial state is given by the formal limit $t \rightarrow -\infty$. However, this is a 'false' infinity, at least in the context of life, complexity and anthropic reasoning, because the period of time in which there are conditions enabling intelligent observers is necessarily finite.[61] In addition, this period is approximately equal to the time that has elapsed since the beginning of the expansion. In the Leibnitz–Berkeley–Machian relationist theories of time, time itself does not really exist before the onset of instabilities, i.e., the universal expansion. The period of complete homogeneity can be regarded as a state analogous to the epochs of complete dominance of Love or Strife in the cosmology of Empedocles or, even more accurately, to the time before the motion began in Anaxagoras' cosmology.[62] In both cosmologies it is necessary to invoke a state that prevents the propagation of information from an arbitrarily distant past to the present epoch. In both cases this goal is achieved by postulating states with a sufficiently high degree of symmetry.[63]

Obviously, in the case of the Eddington–Lemaître universe, the DT argument is inapplicable, since the *effective past* is finite. Intelligent observers (or von Neumann probes!) possess only a finite time for technologization of their cosmic environment. This is valid for the generic version of the Eddington–Lemaître model. Of course, the model pretending to describe the real universe is normalized to the present expansion rate, and therefore we conclude that this effective age is similar to the age of galaxies, or again of the order of H_0^{-1} (H_0 being the present-day measured value of the Hubble 'constant'). Therefore, the incompatibility argument in the core of the DT argument is lost and the problem reduces to the much weaker Fermi's paradox: why don't we perceive traces of older observers, allowed for by the age and size of our Galaxy?

Probably the more physical and meaningful way of restating the entire situation is to reject the notion of an infinite age of the Eddington–Lemaître model as a hollow formalism. A principle sometimes ascribed to Aristotle or St Augustine tells us that there is no time without a changeable world. The state of perfect equilibrium in the Eddington– Lemaître model in the $t \rightarrow -\infty$ limit is exactly such an unchangeable state, without means of determining either direction or the rate of passage of time. In the sense of a modal version of the Aristotle–Augustine principle, the temporal infinity in this model thus collapses into a purely formal notion. Newton-Smith's formulation of this principle:[64] 'There is a period of time between the events E1 and E2 if and only if relative to these events *it is possible* for some event or events to occur between them', explicitly points to the (macroscopic) indistinguishability of moments in the state of complete thermodynamical equilibrium. The same applies to the distant future of the universe in which, according to many models, a state of heat death is bound to occur. Barrow and Tipler, in a 1978 paper, suggest that a formally infinite future should be substituted with a finite interval, through an appropriate coordinate transformation.[65] A sort of counterexample, confirming the general thesis that the cosmic time established by the Weyl postulate should not be regarded as sacrosanct, is the diverging number of (possible) events in the finite temporal vicinity of either the initial or the final global singularity. In such a situation, a finite cosmic time may be less appropriate than an alternative infinite timescale.[66] This view received a poetic description in Lord Byron's *Cain* (1821): 'With us acts are exempt from time, and we / Can crowd eternity into an hour, / Or stretch an hour into eternity.' For instance, the ever-decreasing number of events in the world approaching future heat death (in the framework of some particular cosmological model) could well be described, in the relationist picture, with the finite time interval remaining; therefore, the time between the initial singularity and the final heat death could be represented by a $(-\infty, 0)$ interval.[67]

Finally, I shall briefly consider the DT argument in the most famous cosmology with past temporal infinity, the 1948 steady state theory of Hermann Bondi, Thomas Gold and Fred Hoyle.[68] Although there is some controversy whether the classical steady state cosmology represented a single entity or two disjointed theories (that of Bondi–Gold and the version of Hoyle), I shall refer to them as the classical steady state model, since particular technical differences between the two versions are irrelevant for the present argument.[69] While Hoyle's version is generally superior, being formulated in the language of classical field theory, for our purposes it is, in fact, the perfect cosmological principle (PCP) of Bondi and Gold that makes the important point most clearly. Its essentially non-mathematical character makes it even more transparent by providing the core formulation of uniformitarianism in cosmology.[70]

PCP can be simply expressed as the homogeneity of the universe in four-dimensional spacetime. This is just the generalization of the conventional cosmological principle that assumes homogeneity in space, but not necessarily in time.[71] Part of the appeal of the steady state concept can be found in the words of Sciama:[72]

> The steady-state theory opens up the exciting possibility that the laws of physics may indeed determine the contents of the universe through the requirement that all features of the universe be self-propagating . . . The requirement of self-propagation is thus a powerful new principle with whose aid we see for the first time the possibility of answering the question why things are as they are without merely saying: it is because they were as they were.

However, the germ of doom lies exactly in the concept of self-propagation, since it seems to be incapable of correctly accounting for a specific 'feature' of the universe, namely us. In other words, if we accept the Empedoclean picture (in which biological evolution is an inherent and necessary part of the cosmological evolution), then although it has to be self-propagating, the rise of intelligence at the same time *must not* be self-propagating.

The basic violations of perfect uniformity we notice empirically in the universe are galaxies. Newly created matter is *continuously* condensed in galaxies, and although the details of this process have remained controversial, mainly because insufficient theoretical work was devoted to it prior to the universal rejection of the steady state picture in the mid 1960s, the predictive power of the PCP is manifested here once again.[73] The answer offered by the theory to the question of the age distribution of galaxies on a sufficiently large scale is essentially independent of the physical details of galaxy formation. In the classical steady state model, the distribution function of galaxies' ages is simply $f(t) = \exp(-3Ht)$, where H is the Hubble constant, a true constant in contrast to the Big Bang models (in which the measurable, present-day value is traditionally denoted by H_0). The average age of galaxies is simply the average of this function between 0 and $+\infty$, namely $\langle t \rangle = (1/3)H^{-1}$. This illustrates a beautiful simplicity which the PCP imposes on the theory: the average age of galaxies is calculated without any reference to the complicated physics of galaxy formation (or unknown physics of matter creation!). However, the estimates of the Hubble constant relevant in the late 1940s were in gross violation of what we today know as the plausible interval for that quantity. For $H_0 \simeq 500$ km s^{-1}, which was the then reigning Hubble measurement, the Big Bang cosmology was in serious conflict with the age of the Earth and chemical elements.[74] At the time of formulation of the steady state theory, the mean age of galaxies was considered small ($\langle t \rangle \simeq 6 \times 10^8$ yr, due to the gross overestimate of the value of the Hubble constant), and the Milky Way was already an extraordinarily old galaxy, which certainly implied that surrounding galaxies were far less probable to achieve the same degree of chemical and biological evolution. Although this circumstance in fact does not resolve the DT argument, it certainly does have a psychological effect, making problems with technologization literally much more distant.

Similarly, the fraction $\delta(t)$ of galaxies older than the age t is given by

$$\delta(t) = \frac{1}{n} \int_t^\infty 3nH \exp(-3Hx)\, dx = \exp(-3Ht). \tag{2.1}$$

This is the mathematical root of the problem, reflecting the fact that the exponential function is everywhere finite. For instance, if we take $\langle t \rangle$ to be an order of magnitude higher, in accordance with the today's best knowledge of the magnitude of the

Hubble constant, the DT argument quoted above gains force. Since the fraction of galaxies that are older than age $t = 2 \times t_{MW} \approx 2 \times 12$ Gyr is given by Equation (2.1), it follows that there are almost 2% of all galaxies in any large enough comoving volume that are *twice* the currently known age of the Milky Way.[75] This subset would be quite numerous in any of the modern large galaxy surveys, and the lack of any anomalous structure or even detectable Kardashev Type 3 civilizations in them[76] would be hard to explain in the context of the steady state universe. Afficionados of 'alternative histories' may enjoy the thought of a counterfactual situation in which the 'Great Controversy' in cosmology is settled on the basis of a SETI observation.

The idea of stationarity on very large scales has been recently reanimated in the form of several similar inflationary scenarios. A typical example is the work of Andrei Linde and his collaborators, as well as Alexander Vilenkin.[77] In these models (known as 'chaotic' or 'eternal' inflation), bubbles are formed out of spacetime foam at Planck energy, each bubble evolving into an individual universe in its own right, with a specific topology, geometry, laws of nature, coupling constants, etc. The entire process of separation and inflation of these individual bubble-universes has no beginning or end, and therefore the entire ontological background of these processes (for which the appropriate name of *multiverse* is coined) is stationary. A significant similarity between classical C-field cosmology and inflationary scenarios in general has been noted.[78] The manifestation of that swing of the pendulum backwards from extreme evolutionism towards some form of stationarity can be seen in the very titles of several relevant papers, such as 'From the Big Bang Theory to the Theory of Stationary Universe'.[79] I shall return to this important topic in Chapter 4. It is enough to mention here that the application of the DT argument to these stationary multiverses is challenged by the existence of cosmological horizons and the physical nature of Planck energy substratum that makes any transfer of information hard to imagine, at best.

Time and chance: historical parallel of cosmology and astrobiology

In today's flood of both popular and research articles and news dealing with cosmology, it is hard to imagine those days when it was just a slow trickle, largely ignored by the mainstream of physical science; even when not ignored, it was treated like a mathematical game, devoid of any connection to reality. Philosophers heaped scorn on the nascent discipline, discarding it as a speculative fad (frustrated, perhaps, by this intrusion of 'mathematicians' on their hallowed domain of inquiry).[80] In spite of heavyweights like Einstein, Eddington, de Sitter and Dirac, who showed an active interest in those early days of cosmology, it was regarded as an eccentric pursuit, more appropriate for the elderly speculation of already distinguished scientists like Frederick Soddy or Walther Nernst, but hardly a hot topic for talented graduate students and incoming researchers, for example, not to mention fund-raisers.[81] In order to recognize an overarching analogy with astrobiology, one needs to perform a mental feat of historical 'time travel' and return to the second and third of the epochs of physical cosmology sketched above, when the future princess was still a 'Cinderella' among the sciences. First Eddington, and later Hoyle and Gamow, expended huge efforts in popularizing the new cosmological science, giving many talks, including using the new medium – the radio – and writing popular books and articles on relevant topics; however, this only in retrospect prepared ground for the renaissance of public outreach which followed Weinberg's *The First Three Minutes* booklet in late 1970s.[82]

When George Gamow coined the term *factual cosmology*, it was neither sarcastic, trivial nor tautological.[83] Quite the contrary, he tried bona fide to ground something that was too often regarded, by scientists and laymen alike, as fuzzy, unclear, ill-defined, too speculative and in general not worthy of serious scientific pursuit. He did not really regard all previous work in the field, including studies by Einstein, Eddington, etc., as speculative rubbish which could safely be ignored; on the contrary, as his published papers show, he was well acquainted with the cosmological literature of the time, and actively participated in some of the livelier debates. However, he felt the need to upgrade the discussion – and the time was ripe for that. In contrast, we do not yet have 'factual astrobiology' in the general Gamowian sense.[84]

Analogies are tricky, especially in historical sciences; they are also inescapable. The analogy between the development of physical cosmology and astrobiology has been noted several times, including in works by some of the practitioners of both, like Paul C. W. Davies. As mentioned above, the most detailed discussion framed from a historical point of view was given by Steven Dick; there is no need to repeat his arguments related to historical and thematical connections between cosmology and our views of life and intelligence in the cosmic context (i.e., astrobiological tropes). Instead, I shall adduce some further arguments for the same thesis, mainly from the methodological and philosophical domain.

As we have seen, Wallace's failure largely followed from his insufficient cosmological background – it was too early for the true emergence of astrobiology as a science. It *required* an adequate understanding of the architecture of the universe.[85] The emphasis had obviously shifted by the time astrobiology in its modern phase began with Joshua Lederberg, Carl Sagan, and perhaps also Iosif Shklovsky and Frank Drake in 1960s.[86] Cosmology had made huge strides by then, and there was a rather clear astrophysical background that could not be arbitrarily modified according to one's philosophical preferences, as was possible in Wallace's era. Of course, that does not mean that there are no such preferences and commitments in cosmology nowadays. Even the most strident positivists and 'anti-philosophers' or – to borrow a fine expression from Frank Wilczek[87] – 'philosophers in spite of themselves' cannot avoid (or, indeed, foresee) philosophical and methodological implications of their thoughts and results. This applies with equal force to both Eddington's speculations about the 'fundamental theory' and Weinberg's 'pointless universe', to take just two examples usually considered antithetical.

What has happened in the meantime? Obviously much, since it was still possible around the middle of the twentieth century for Sir Hermann Bondi to warn observers about the possibility of empirical mistakes, which was received with furore.[88] Today it is quite old news, at least in the domain of the physical sciences. The modern history of science – in which the history of cosmology plays a significant role – entirely accepts the idea that progress in science is driven indifferently by empirical and theoretical work, and that the empirical work is not in any way 'closer to the truth' than theoretical 'speculation'. In this process of abandoning the old-fashioned positivist and naive materialist views, the role of philosophical problems in science has been thoroughly re-evaluated. Note, for instance, several intriguing issues that have so far provided 'hot' topics for the philosophy of cosmology:

- 'The universe' is not uniquely defined, even as a working consensus. On the contrary, the evolution of cosmological theories dictated modifications, adjustments and re-evaluations of what we consider the basic, most fundamental concepts in cosmology.
- Cosmological theories have often been only weakly predictive and degeneracies are plentiful. Predictions have often been obviated by 'local peculiarities' or 'evolutionary uncertainties' (e.g., in the domain of evolution of galaxies).
- The distinction between dynamical laws and boundary conditions, all-important in most of the physical sciences, is blurred in cosmology. The reason is that we access only a single instantiation of allegedly general regularities (or 'laws') and even when other instances of 'a universe' are postulated, they are usually considered empirically inaccessible.
- While human cosmologists are undoubtedly part of the structure and evolution of the universe (in whatever loosely understood meaning), our existence is unambiguously linked to the subject matter. This leads, at best, to ubiquitous 'anthropic biases' which need to be corrected for.
- The universe seems to be evolving from a simple state toward a state of high complexity. However, there is no consensus as to how to conveniently measure or even express this growth of complexity. While broad outlines of the process are understood, the search for the dynamical mechanisms generating this huge complexity – which includes living and intelligent systems – is continuing with no end in sight.

All these (and some additional) points are present almost verbatim in the astrobiological context as well. We have already considered the problem in defining life and intelligence. The problem of (missing!) astrobiological dynamics will be addressed in Chapter 3 below, as well as the perennial issue of the observer and her biasing empirical results and working assumptions. I shall discuss to what extent Darwinian evolution on Earth could serve as a blueprint for the explanation of the increase in astrobiological complexity in Chapter 5. Thus, the parallels between philosophical problems and even scientific style in cosmology and astrobiology clearly exist – but how different is the spectrum of research activities in the two cases! While nowadays even tough chestnuts in the philosophy of cosmology could, at least in principle, be cracked by referring to the observed CMB power spectrum or constructing large-scale numerical simulations (or even building a Bayesian theory of observation-selection effects), the hope that something even remotely analogous could be done in astrobiology belongs to the future.

Such a mismatch between the 'deep' conceptual problems and empirical research programmes in the two fields can best be explained, I submit, by a historical mismatch between the development timelines. Astrobiology is currently in a state analogous to the one cosmology was in during the 1920s and 1930s, about the time of Hubble's great discoveries. A pivot, which the discovery of the expansion of the universe provided to physical cosmology, is provided for astrobiology by the discovery of a large number of extrasolar planets; both showed that it makes sense to discuss *dynamics* in a wider, non-trivial sense. If there were no systemic changes (like global Hubble expansion) in spacetime, it would still make sense to discuss local spacetime curvature and its geometrodynamics, like that manifested in Eddington's observation of the total solar eclipse. By analogy, even if the Solar System was the only planetary system in the Galaxy (as in the old Buffon–Jeans catastrophic theories), it would still have been possible to study the astrobiology of Earth or Mars. Nevertheless, in both these counterfactual cases the realm of the phenomena would have been quite poor.

As cosmology has had its many detractors, so has astrobiology. One could argue even further that there is a structural similarity between the classical criticisms of physical cosmology, such as those by Herbert Dingle or Stephen Toulmin on the one side, and the detractors of astrobiology, led by George Gaylord Simpson, Michael Hart, Frank Tipler and some contemporary philosophers, on the other side of the story. (I shall discuss Simpson's and some other criticisms in Chapter 7 below.) These are conservative criticisms, professing scepticism, emphasizing uncertainties, and deploring the 'overly speculative' nature of the endeavour considered. There is also a populist streak of comparing the research discourse in the disputed fields to extra-scientific discourse in both cases. For cosmology, especially the models with a finite past, it was religion; in the case of astrobiology, it is the discourse of science fiction. Both fields have been denigrated as extravagant, frivolous, or remote from daily reality. Both tend to attract amateur and fringe 'researchers', even if outright quacks are fewer than is usually assumed. Even the enthusiasm – or zeal – for the popularization of science, which has characterized distinguished practitioners, is a common trait of both cosmology and astrobiology. While the existence of the mentioned extra-scientific discourses certainly boosted the public appeal of both cosmology and astrobiology, part of the reasons for the continuing success of both fields as vehicles for public outreach seems to lie in strong intuition of researchers about wider relevance, and indeed importance, of their professional results.

The parallel between astrobiology and cosmology is so strong that it occasionally surfaces even in purely technical studies. In an interesting paper on the disputable subject of typicality (or otherwise) of the Solar System, we read, for instance, that[89]

> it is natural to ask whether the Solar system is special in some way compared with the majority of planetary systems to be found in the Galaxy . . . It is, of course, special in the sense that we are in it, and it is therefore no surprise that all the early models of the formation of planetary systems applied exclusively to the Solar system.

This is followed by the footnote which says: 'Just as almost all models in cosmology apply exclusively to the Universe'. We shall see (in Chapter 4) that this is not necessarily true any more in contemporary cosmology, although as a historical generalization it is certainly justified.[90]

The most important context in which the entire astrobiological drama unfolds is cosmic time. Nowadays, we know that the age of the universe is about 13.7 Gyr. This age is obtained in the New Standard cosmological model, after 1998, but it was not always so, especially in the 'dark days' when the age problem plagued relativistic cosmologies in general. For the current rate of expansion, given by the Hubble 'constant', the cosmological timescale is of similar magnitude as the lifetime of Sun-like stars, t_* (which gives, in a rough approximation, the duration of habitability of terrestrial planets). However, as shown by Dicke, the relation $H_0^{-1} \sim t_*$ is natural in the Big Bang cosmology, while it is pure coincidence in the steady state model![91] Thus, from the point of view of observation-selection, which I will expand upon in Chapter 3, finding ourselves at this epoch is another argument in favour of relativistic models and another instance of the entangled nature of cosmological and astrobiological enterprise.

Many personal experiences testify about the same deep link. Among the towering ones is that of Sir Fred Hoyle, one of the most original and inspiring figures in modern science.[92] While undoubtedly best known as a cosmologist and one of the fathers of the steady state theory – as a consequence of which he set foundations for the theory of stellar nucleosynthesis – at least from the late 1950s he was actively interested in the question of life in the cosmic context, and to make things more interesting, in both discoursive and fictional domains. Hoyle's first excursion into fiction writing was the celebrated *Black Cloud*, a novel about an alien intelligent life form embodied in a cloud of interstellar gas.[93] It was followed by a dozen other SF books, most of which deal with topics which would today clearly be labelled as astrobiological. *The Black Cloud* became somewhat of a common term for unorthodox forms of life, independent of terrestrial planets. Most of Hoyle's later writings were also devoted to astrobiological pursuits.[94] This is entirely consistent with the main strand of his thought, which had always revolved around cosmological 'big questions'. As one of the Cambridge mavericks who proposed the steady state theory in 1948, and as perhaps its last staunch defender (with Jayant Narlikar, in form of various conformally invariant cosmological theories), his belief in some form of eternally existent universe with an infinite past played a key part in his thinking about life as well. Similar to Cuzanus and some of the early thinkers, discussed by Crowe and Dick, Hoyle seemed to imply that life also has eternal duration, being coexistent with the universe. However, this presents us with many problems, as we have seen, not least being the Davies–Tipler argument considered above.

It is a pitiful misrepresentation of history that Hoyle's name in astrobiological circles is most often associated with the unfortunate 'junkyard Boeing-747' metaphor. This he used to mock (in his own peculiar style, similar to the cricket metaphors he used in fierce cosmological debates with Sir Martin Ryle) the ideas of spontaneous assembly of biochemical compounds into something similar to the level of biological complexity characterizing even the simplest living beings. Parts of the real history – not necessarily entirely unbiased, as always – can be glimpsed, for example, in his charmingly poetic autobiography (*Home Is Where the Wind Blows: Chapters from a Cosmologist's Life*), as well as in the long memoir of Chandra Wickramasinghe.[95] Hoyle clearly perceived his work in cosmology and astrobiology as tightly joined parts of the same underlying whole. His influence, not always justifiably, helped revive interest in some of the perennial astrobiological problems, notably panspermia, where viable models have been recently built, and which is undoubtedly in better shape today than it was, say, in the 1950s.[96]

Among contemporary astrobiologists, an excellent example of the same personal link is presented by Charles Lineweaver, the author of seminal works on the age distribution of habitable planets and the evolution of the Galactic Habitable Zone, who successfully maintains an extensive publication record in both cosmology and astrobiology. The focus of his cosmological work has been the CMB and the determination of cosmological parameters like the matter and the dark energy density Ω_m and Ω_λ.[97] Subsequently, he smoothly joined these research themes with astrobiological topics, again through ubiquitous observation-selection effects to which I shall return in detail in Chapters 3 and 4.[98] Lineweaver's very active career demonstrates, as an 'insider' in the full swing of the astrobiological revolution, how seemingly unrelated strands of thought are deeply intertwined to produce original research of exceedingly high quality – a feature, I wish to argue, characterizing not only the link between astrobiology and cosmology, but a wider intellectual consilience spearheaded by astrobiology.[99] If such consilience is indeed taking shape today, then there is a straightforward explanation for both the structural similarities of cosmological and astrobiological projects and the difference between their stages of development: they are complementary descriptions of the same underlying cosmic evolution. In a crude simplification, while physical cosmology, such as originated with Einstein and de Sitter, deals with the low-complexity sector of that same underlying process, astrobiology treats the high-complexity sector. Since the latter is logically contingent upon the former, it is not only clear why pre-physical-cosmology projects of treating 'life in the universe' topics, like Wallace's, were bound to fail, but also why there has to be a temporal offset between phases in our understanding of the two (allowing, of course, for many extra-scientific and societal factors influencing the magnitude of this offset).

3

Cosmology, life and selection effects

Treason doth never prosper: what's the reason?
Why if it prosper, none dare call it treason.

Sir John Harington

There can be no doubt that the Copernican revolution – and the wider period of 'scientific revolution' – played a crucial role in the birth of modern science and civilization. As already mentioned, Copernicanism (often misleadingly expressed in as the 'Principle of Mediocrity') in the narrow sense tells us that there is nothing special about the Earth or the Solar System or our Galaxy within large sets of similar objects throughout the universe. In a somewhat broader sense, it indicates that there is nothing particularly special about us as observers; our temporal or spatial location, or our location in other abstract spaces of physical, chemical, biological, etc. parameters are typical or close to typical. There are two crucial qualifications here, which are rarely explicated and discussed. First is that this does not mean that our locations in these spaces are *random*. The latter statement is obviously wrong, since a random location in configuration space is practically certain to be in intergalactic space, which fills 99.99... % of the volume of the universe. This is a longstanding confusion and the reason why Copernicanism is most fruitfully used in conjunction with some expression of the observation selection effects, usually known as the anthropic principle(s), a label which can also be misleading. While observation selection effects have been treated in detail in some recent sources,[1] I shall concentrate here on those aspects of relevance for astrobiology that have been neglected so far. The second qualification is that it is usually not, a priori, clear what defines the set in which typicality is subsequently assumed. While the intuitive picture of the situation generally relies on the images of the Copernican revolution historically – that the Earth or Sun are typical *in space* – the most interesting new applications of Copernicanism are glimpsed in *abstract parameter spaces*, not easily visualized or intuitively understood.

Non-controversial observation selection: extrasolar planets

Observation selection effects are often difficult to comprehend and accept. There is an unfortunate tendency to downplay these effects and minimize their impact on our conclusions – as if imposing limitations on our empirical efforts would somehow make us look less smart. Even in disciplines long accustomed to dealing with observation selection, there is often reluctance to face their tremendous impact. A truly astounding example comes right from the heart of astrobiology – from the domain of the search for extrasolar planets. In the entire pre-*Kepler* epoch (roughly 1995–2009), there was practically no description of the discovered planetary systems that did not emphasize the fact that all the planets are massive and all planetary configurations are strange and quite unlike the Solar System.[2] Of course, that was greeted with enthusiasm by the proponents of the 'Rare Earth hypothesis' (see Chapter 6 below) and the occasional creationism/'Intelligent' design fan. It has been commonplace for science journalism to conclude, 'astronomers have difficulty explaining why the vast majority of planetary systems are so unlike our own'. This historical fact – true for a limited span of dates and now slowly fading into the marginalia of the history of science – was even used to criticize Copernicanism: how dare those philosophers or naive dreamers assume that Earth and the Solar System are unremarkable, when 'hot Jupiters' are all the rave? The irony here is that the supposed explanandum is false – or at least very uncertain. The 'vast majority of *discovered* planetary systems' may have very little to do with the 'vast majority of existing planetary systems', in exactly the same way as the 'majority of voting US citizens having a phone number around the time of the 1948 presidential elections' might indeed have very little to do with the 'majority of voting US citizens around the time of the 1948 presidential elections'. Well, nobody felt that better than Thomas E. Dewey, the Republican hopeful. The famous photo of Harry Truman who holds a newspaper proclaiming the false (predicted) outcome of the elections clearly shows the ludicrous consequences of ignoring observation selection.[3]

The very first question a scientist, journalist or simply an interested reader should ask upon encountering a report on the discovery of a particular extrasolar planetary system X, allegedly unlike the Solar System, by using the concrete observational method M, is simply:

> If we, in a thought experiment, substitute the system X with a perfect copy of our Solar System, would M be able to discover it? (*)

This is the crucial test of fairness of any judgement involving typicality. And it really does not matter whether, instead of a single method M, we are dealing with a whole array of methods, $M_1, M_2, \ldots M_n$. Only if the answer to the question (*) is affirmative, should we dare to proclaim that the particular finding undermines (and even that, slowly, in a Bayesian manner) the typicality of our own Solar System. However, in most cases during the pre-*Kepler* era the answer to (*) was, lo and behold, *firmly negative* for practically all methods used. Why wasn't that fact on the front page of your popular science magazine? One of the reasons might be because explaining observation selection would be a rather hot potato for all but the most capable and sophisticated science journalists. Nevertheless, a part of the reason, I suspect, is in the implicit hostility of their sources – in this case astronomers – to the ambiguity in the seemingly straightforward interpretation introduced by observation selection effects.

In the particular case of extrasolar planets, observation selection effects are multiple and formidable, though they vary with the method used. The ubiquitous effect is the temporal scale of observations: most of the discoveries have been made based on observations lasting from a few nights to several years, at most. This should be contrasted with the most significant temporal scale in our Solar System – the period of Jupiter's revolution, $T_J = 11.86$ years. Clearly, not a single 'wobble' in solar spectral lines due to the solar motion around the common centre of mass with Jupiter (which is more massive than the rest of the planets combined, so it sets the scale) can be detected at much shorter temporal scales. There is no help for it, really, unless we can find a way, like a sort of anti-Tristam Shandy, to compress decades of observations into months in order to be acceptable to today's practical astronomy. Very similar reasoning applies to the occultation detection method: even if the line of sight to the Sun–Jupiter system is convenient for a hypothetical observer on a planet orbiting a distant star, she/he/it would observe just a single transit in $\sim$12 years. Luckily, the situation in this respect is improving rapidly and now transits of much smaller planets much closer to their parent star can be observed, providing motivation for the successful *Kepler* mission.

A whole cluster of observation selection effects is related to the metallicity of the parent star. The content of metals in the stellar atmosphere is the source of spectral lines, the study of which has been the source of most of the extrasolar planets discovered so far. The relationship is not simple; it depends on complicated and often poorly known parameters like the temperature of a particular atmospheric layer or the density of free electrons in plasma, but it definitely exists, and is modelled quite successfully with standard software packages like CLOUDY.[4] Such a relationship precludes the discovery of planets by spectroscopic means around stars where there are no metal lines at convenient positions in the spectrum or they are too weak. Therefore, a part of the real parameter space – the low-metallicity stars – is undersampled. And there are other complications. A fairly general class of cosmogonic models suggests that, with reasonable assumptions, high metallicity implies more efficient planetary formation, and even larger average masses of the planets formed out of the protoplanetary disc. (To the degree that metallicity of the protoplanetary disc is well represented by the metallicity of the parent star's atmosphere.) Thus, when we discover that there is a strong correlation between the positive signals of planet detection and the parent star's metallicity,[5] we cannot, at least not without a careful and complex study, say whether it is due to our increased chances of detecting them or to their increased inherent chance of formation; the latter is a physical effect, but the former is an observation selection effect.

All this is a far cry from accepting a nihilistic conclusion that we cannot say anything about the typicality of our Solar System (or otherwise) on the basis of extrasolar planets' research. In a recent study, for example, quite optimistic conclusions are reached, bolstering the entire front of astrobiological research:[6]

> The hypothesis that $\sim$100% of stars have planets is consistent with both the observed exoplanet data that probe only the high-mass, close-orbiting exoplanets and with the observed frequency of circumstellar disks in both single and binary stars. The observed fractions f that we have derived from current exoplanet data are lower limits that are consistent with a true fraction of stars with planets, f_t, in the range $0.25 \leq f_t \leq 1$. If the fraction of Sun-like stars that possess planets is representative of all stars, our result means that out of the $\sim$300 billion stars in our Galaxy, there are between $\sim$75 and $\sim$300 billion planetary systems.

Non-controversial observation selection: two examples

There are many well-studied instances of selection effects in numerous branches of science. Let me briefly describe two examples, originating in fields closely related to astrobiology.

Astronomy is an observational science, but it is no trade secret that most of the sources in the sky are below the limit of visibility in any particular observation. One of the selection effects related to that simple fact goes under the name of the *Malmquist bias*. It was described in detail by Swedish astronomer Gunnar Malmquist in 1920, although it seems clear that observational astronomers were aware of it earlier and it certainly played a role in attempts, mentioned in the previous chapter, starting with William Herschel in the second half of the eighteenth century, to determine the shape of our stellar system by counting the stars in small standardized patches of the sky. The Malmquist bias was generalized by Sir Arthur Eddington in 1940 (it is indeed sometimes referred to as the Eddington–Malmquist effect) and it is nowadays clear that we are dealing with a whole family of related biases.[7]

Suppose we are observing a population of celestial sources; for Malmquist those were stars of a specified spectral type, but it could in principle be any sources (stars, nebulae, galaxies, quasars, supernova remnants). If our observing equipment is such that it detects with certainty a source X if and only if X's flux (in whatever spectral band we are observing) exceeds some minimal value, the resulting sample of sources will be *flux-limited*. In most cases, though, astronomers are really interested in the *intrinsic* properties of sources, notably their luminosity, not in apparent flux – or apparent magnitude, in astronomical jargon – reaching our instruments on (or around) Earth. If they calculate luminosities of sources in a flux-limited sample, they will find a strange thing: *the average luminosity of sources increases with their distance*. This is, of course, because less-luminous sources at large distances will not be detected, since their flux will fall beneath the detection threshold of our sample. Thus, we are not dealing with a 'fair' sample of sources existing in nature, but with a biased sample.

A rather straightforward instance of the Malmquist bias is the set of stars visible with the naked eye. The human eye is a flux-limited detector, so the sample of stars visible by naked eye from any place on Earth is a flux-limited sample – so we will not be surprised to learn that most of the stars observable by naked eye, and consequently, most of those which received historical or, in astronomical parlance, trivial names, are either main-sequence giants/supergiants (Rigel, Deneb, Polaris, Hadar, all of the Pleaides) or red giants/asymptotic giants (Antares, Betelgeuse, Arcturus, Aldebaraan). These are stars with huge intrinsic luminosities – but are they really so numerically predominant in the Galaxy? Of course not! Quite to the contrary, a complete sample of stars within 50 parsecs of the Sun – obtained with modern telescopes, to be sure – clearly shows, by far, that *red dwarf stars* of extremely low intrinsic luminosity are numerically predominant. A classroom example of the latter is the closest star to the Sun, Proxima Centauri, a distant third component of the Alpha Centauri star system. This is an extreme red dwarf of spectral class M5.5, whose luminosity is only 0.0017 solar,[8] so it requires a powerful telescope to be seen, even though it is so close to us (1.3 pc). A survey of all stars within 5 parsecs of the Sun – something that is called a *volume-limited sample*, in contrast to a flux-limited sample – reveals that 49 out of 62 stars are red dwarfs (79%). In general, it is usually assumed that about four-fifths of all stars in the Milky Way are red dwarfs, and not a single one of them is visible with the naked eye. Parenthetically, this example also shows that human cultural predilections are by default framed by selection effects – the trivial names of stars are given in old catalogues of stars visible to the naked eye, products of Hellenistic and medieval Arabic science. Only in the last couple of centuries have astronomers started to rectify the 'injustice' towards the huge population of red dwarfs by giving some of them trivial names – examples include Barnard's Star and Luyten's Star.

This example hints at a solution for correcting the Malmquist bias: we need to use a sample that is not flux-limited (e.g., one that is volume-limited). The antidote is far from being easy to obtain, especially in extragalactic astronomy, which suffers from the severest form of this selection effect. The determination of extragalactic distances usually depends on assumptions about intrinsic luminosities of galaxies or clusters, and thus is sensitive to the same selection effects as in the case of stars. Suppose we wish to determine the distance to a galaxy that lies at true distance r_0 by using the estimator $\hat{r}$, which can be a function of the observable quantities, like apparent flux, spectral line widths, etc. Let $p(\hat{r}|r_0.)$ be the probability distribution function of the estimator $\hat{r}$ conditional upon the true distance of the galaxy. By definition, $\hat{r}$ is unbiased if the expectation value of $\hat{r}$ is equal to the true distance: $E(\hat{r}|r_0) = r_0$. In the realistic case, the bias $B(\hat{r}, r_0)$ of $\hat{r}$ is defined as:

$$B(\hat{r}, r_0) = E(\hat{r}\,|r_0) - r_0 = \int p\,(\hat{r}\,|r_0)\,\hat{r}\,d\hat{r} - r_0. \tag{3.1}$$

It is clear that the bias of the estimator is a function of the unknown true distance, r_0. Analogous relations apply to estimators of *any* parameter in astronomy; astronomers have been plagued with observation selection effects for a long time. Malmquist bias in the general sense is any such bias involving estimators based on apparent fluxes (or magnitudes).

The situation is further complicated, since in extragalactic astronomy we deal with distances large enough that look-back times allow for the *intrinsic physical evolution* of the sources (this is usually not the case for stars and other objects within the Milky Way, which we observe effectively at their present-day states). Distant sources could be intrinsically more luminous than their nearby analogues, in which case the effect would be magnified by the Malmquist bias. Or they could be intrinsically fainter in comparison to nearby analogous ones, which could be dampened, or even completely cancelled by the bias. A sizeable literature has been devoted to disentangling this conundrum and finding unbiased estimators of various quantities.[9] If we are dealing with objects at cosmological distances, the evolutionary corrections might indeed be such that the objects at high redshift – hence younger in terms of their evolutionary status – are intrinsically brighter; such is the case for QSOs and most other active galactic nuclei.

One precious piece of wisdom to be drawn from the very description of bias in Equation (3.1) is that observation selection effects are more effectively written and understood in the Bayesian framework. After all, the distinction between 'true' a-priori and 'observed' or 'inferred' a-posteriori values of a parameter is essential for the Bayesian approach to data analysis and inference. We shall encounter other Bayesian formulations when we consider anthropic reasoning and its relevance for astrobiology in some detail below.[10]

The second example of an observation selection effect comes from Earth sciences. In palaeontology, one of the basic unknowns in the reconstruction of life's history is the duration of mass extinction episodes. Were those mass extinctions documented in the fossil record sudden events, or did they unfold over a geologically relevant time (several Myr, for example)? In many cases, palaeontologists – especially those accepting the long-reigning gradualist mode of thinking – subscribed to the latter view, arguing, for instance, that ammonites and dinosaurs were in decline for millions of years before their final disappearance at the end of the Cretaceous, 65 Myr ago. This formed one of the basic tenets of the strong opposition to the Alvarezes' impact hypothesis, proposed in 1980, for an explanation of the K/T extinction. It is unnecessary to postulate a speculative event like an asteroidal or cometary impact to explain the extinction of ammonites and dinosaurs, sceptics argued, when the fossil record tells us that their numbers (individuals, species and higher taxa) were decreasing for a long time before the exact K/T boundary. Then, in 1982, Philip Signor and Jere Lipps, two geologists working at the time at the University of California, Davis, noted a glaring omission in the sceptics' reasoning. How do we determine the epoch of a species' extinction in the fossil record? Obviously, by finding the *latest* fossil of that particular species and dating the corresponding stratum. The difference between the first and the last fossil remnant of a species is called the species (stratigraphic) range. However, since the fossil record is always sparse, even for the most numerous species and higher taxa, the end of the stratigraphic range is simply the stratum in which the last individual was fossilized – and found by human observers. The species could have really ended after that epoch, but certainly not before it. Thus, the *density* of fossil record comes into play. Although the fossil record is always rarefied and the absolute difference in the number of fossils of, say, cynodonts and archaeopteryxes, are not that large, even a modest relative difference can have important consequences for our understanding of the dynamics of extinction.

Phylogenetically very close species may vastly differ in population size; today, rats are orders of magnitude more numerous than beavers, and the huge population of humans dwarfs our mammal relatives, giraffes. Suppose that a sudden catastrophe, like a nuclear winter or supervolcanic eruption, causes the extinction of both humans and giraffes at some future time t_X. Even if the fossilized fraction is constant over morphological differences, this would mean that – for hypothetical palaeontologists of the far future, long after t_X – the probability of finding a human fossil remnant near the end of the present geological stratum would be much higher than the probability of finding a giraffe fossil at the same location. Instead, the giraffe fossils would be much more likely to be located near the centre of the stratum and the range of giraffe fossils would be artificially truncated at some epoch t_G much earlier than t_X (see Figure 3.1). Now, we might imagine a scientific meeting of those far-future researchers on a general topic of 'The Extinction of Mammals', where a gradualist faction would argue that the data point to a prolonged extinction pattern, lasting at least $\Delta t = t_X - t_G$. In principle, and especially since we humans are, unfortunately, reducing the giraffe population through

Figure 3.1. The Signor–Lipps effect in a simplified scheme. I use giraffes here for the same reasons Georges Cuvier used elephants in his famous speech of the 1st Pluviose in year 4 of the French Republic (in the unliberated parts of Europe known as 21 January 1796), before a public session of the National Institute of Sciences and Arts in Paris, in which he decisively demonstrated the fact that species could become extinct. Simply, these large and lovely mammals are easy to spot, even in a very incomplete fossil record. (Courtesy of S. Popović.)

hunting and habitat destruction, the duration Δt could be fairly large. But we know that it is an entirely artificial effect of the sampling, since we have set up the thought experiment in this way! Postulated extinction was instantaneous – in geological terms, at least – and yet the prima facie data would not have indicated that. Of course, this example for two species can be generalized to a greater number of species comprising higher biological taxa and their extinction.

The lesson is clear, and quite general. *What is in fact sudden will, in retrospect, tend to look prolonged*, since both the natural process of fossilization (sporadic) and the human sampling of fossils (incomplete) act to select that part of the parameter space consistent with prolonged extinction. 'Evidence contradicting a catastrophic mass extinction might not be derivable from fossil range data alone.'[11] This has been dubbed the Signor–Lipps effect; it is a selection effect of paramount importance in Earth sciences. (In fact, Signor and Lipps have discussed another sampling bias in their seminal 1982 paper: the artificial supression of diversity of higher taxa, like families or phylae – understood as the number of species alive at a given time – due to small sampling sizes at particular epochs. However, it is the artificial range truncation that usually goes by the name of the Signor–Lipps effect.)

The Signor–Lipps effect provides another even more universal lesson for epistemology. Selection effects cause biases, which, if left uncorrected, are likely to reinforce prejudice and support comfortable scenarios against the more original, radical and disturbing views. This is especially applicable to observation selection effects in cosmology, known as the anthropic principle or anthropic bias. Just as prolonged extinctions seemed to tally well with Lyellian gradualism, which ruled geosciences dogmatically for almost a century and a half, so empirical observations that the universe seems fine-tuned for life is too often taken as the evidence for comforting teleological or theological views, especially those compressed into the Design hypothesis (see pp. 67–69 below). Such doctrines are occasionally insidious and subtle, creeping in as a part of 'a standard picture', and it is easy to miss important selection effects if we uncritically accept such standard pictures.

When considered in the context of the recent history of science, the Signor–Lipps effect may occasionally provoke a sort of Huxleian reaction: 'How extremely stupid of me not to have thought of that!' (Thomas Henry Huxley's response upon encountering Darwin's theory of natural selection). Such was the reaction of several astronomers I informally polled about it. Indeed, the reasoning behind it is so elegant in its deep, incisive simplicity, that one might have expected it to have been embedded in our way of thinking about the patterns of history much earlier. It is hard to avoid the impression that the 'provocation' caused by the paper of the Alvarezes and collaborators was immensely healthy and beneficial – one of the fruits of the controversy being the discovery of Signor and Lipps. Conversely, one might use this effect as an argument that gradualist dogma has indeed stifled imagination and originality – if sudden events are of small or no importance whatsoever, then the impact of this selection effect would be minuscule. To be sure, the bias would still operate for each particular species in the fossil record, but it would not tell us anything particularly interesting or important at the higher level – about the evolution of higher taxa or about the overall patterns of life's history. Only in the (neo)catastrophic context does the Signor–Lipps effect have a significant impact on our reconstruction of the past. Again, it is hard to avoid a Popperian conclusion that the context of problem-situation is crucial for both the emergence of novelty and its acceptance in science.

Somewhat controversial observation selection: fine-tuning arguments

On 26 October 1996, a remarkable football match took place at The Dell in Southampton, the smallest arena in the Premier League at the time.[12] The reigning champions, Manchester United, managed by Sir Alex Ferguson and led by legendary striker Eric Cantona and midfielder David Beckham, which went on to retain the championship, played with a home team that was threatened by relegation... and was soundly defeated. Southampton won 6–3 in the highest scoring game of all season in the Premiership.

The result was certainly unexpected, as could be amply testified by the bookies, but do we need an explanation? In spite of the excessively complex nature of a football game, some simple explanations are certainly possible. Perhaps the match was fixed or the referee bribed; perhaps a new and traceless performance-boosting substance was used by the players of Southampton. All these – and many other – simplistic hypotheses are unlikely, and most of them can be rejected based on existing evidence. However, is this a sign that we should seek a deeper and more complex explanation – or accept the unexpected event as a fluke (at least regarding the form of the Manchester United players)? Should we be *less* eager to search for a particular explanation of the Southampton game if we knew that only six days before the team from Manchester was thrashed by the main rival for the title, Newcastle United, 0–5? Does obtaining the additional information make an observed fact – the Southampton result – less surprising and more explicable as a part of the prolonged downward fluctuation ('slump') in the otherwise superb playing level of the Manchester team?

The first difficulty in any problem-situation is to establish what requires explanation – what is 'unexpected' or 'surprising' in our observations. While there is no universal 'suprise-meter', there are some quite reasonable procedures to follow. Notably, an important question is whether further observations – in the same, admittedly vaguely defined, field – increase or decrease the degree of surprise? If we find a particular cosmological parameter X has an atypical value, how are our approaches to finding an explanation influenced by the subsequent information that a fundamental physics constant Y also has an unexpected, peculiar value, or a range of values?

So-called anthropic fine-tunings of various constants of physics and cosmological parameters of our universe have been known for quite some time. They are usually expressed in the form of counterfactuals: *if the parameter A was outside of a small interval around its actual value, the universe would not be habitable for intelligent observers.* This is an empirical matter, with ample support from both observations and theoretical models.[13] Our universe is much more complex than most universes with the same laws but different values of the parameters of those laws. In particular, it has a complex astrophysics, including galaxies and long-lived stars, and a complex chemistry, including carbon chemistry. These necessary conditions for life as we know it are present in our universe because of its complexity, which in turn is made possible by the special values of the parameters. For example, we observe that the cosmological density of matter at the present time ($\Omega_{m0} \approx 0.3$) and the dark energy density ($\Omega_{\lambda 0} \approx 0.7$) are of the same order of magnitude. There is no causal explanation of that observation. However, a recent study clearly demonstrates that the distribution of ages of habitable planets is such that observers are most probably in the cosmological epoch in which $\Omega_{m0} \sim \Omega_{\lambda 0}$ – that is, the fine-tuning 'coincidence' is an observational selection effect.[14]

In what is historically the best-studied example of an 'anthropic' argument winning over the 'dynamical' explanation of fine-tuning, Paul A. M. Dirac in 1937 pointed out the near-equality of several fundamental large dimensionless numbers of the order 10^{40}. One of these large numbers varied with time, since it depended on the age of the universe. Thus, there was a limited time during which this near-equality would hold. Under the assumption that observers could exist at any time during the history of the universe, this large-number coincidence could not be explained in the standard cosmology. This problem motivated Dirac (and subsequently Pascual Jordan) to construct an ad hoc new cosmology with modified dynamics, whose one famous consequence was secular change in the Newtonian gravitational constant G.[15] Alternatively, Robert Dicke proposed that our observations of the universe could only be made during a time interval after carbon had been produced in the universe and before the last stars stop shining.[16] Intelligent observers require billions of years of evolution on the surface of a planet near a stable long-lived star, and Dicke was able to argue that the physics of stars implies that these conditions would only hold during an era in the universe where Dirac's large-number coincidence would also hold. He concluded that this *temporal observation selection effect* – even one so loosely delimited – could explain the coincidence without invoking a new cosmology. This episode in the history of science is particularly illuminating, since predictions of Dirac's alternative dynamics could be checked directly with improved techniques a couple of decades later, both in the Solar System and in the famous binary pulsar – and they have been decisively falsified.[17]

Four different kinds of explanation of the empirical facts of fine-tuning have been proposed. Whether the list is really exhaustive is unclear, but the long tradition of searching for solutions of this 'deep' problem suggests that even if some additional family of solutions exists, it must be quite exotic. According to those views, the observed fine-tunings are either:

1. **Brute facts** (an explanation of the deepest properties of the physical world is neither possible, nor should be sought), or
2. **Co-emergent with observers** (the physical world emerges with intelligent observers and the act of their observing determines its properties), or
3. **Intentional design** (the physical world is created in the unlikely state by an intentional designing agency), or
4. **Observer-selected local facts** (the physical world is embedded in the wider ensemble, the multiverse, and is selected by its habitability).

Clearly, this is a heterogeneous list. On most metrics, the explanatory nihilism of option 1 is the least satisfactory. The history of science clearly testifies against it, since each time somebody cried *ignorabimus!* it turned out that hoisting the white flag was premature. Remember that even for 'mundane' problems like the origin of biological homochirality we could postulate a brute-fact 'explanation'. Interestingly enough, the history of that particular problem reflects well how the brute-fact approach is unfruitful.[18] In addition, this sort of answer suffers from a glaring inherent problem, known since the time of Zeno's arguments against plurality and motion. Proponents of this view can give us a satisfactory causal account of many important questions in both science and everyday life; thus, they encounter no problem in giving an intelligible causal answer to the question, *Why is the Earth round?* In the answer, they will not be able to avoid using some properties of the gravitational force (namely that it is a long-range central force). However, when we go just one step further and ask for an explanation of these properties (or ask why is gravity so weak in comparison to the other forces or why is the sign of G positive?) – since they play an important role in at least some anthropic fine-tunings – proponents of option 1 simply shrug them off as brute facts.[19]

The situation is even worse in the general case, since gravity is rather well known and intuitively taken for granted, at least in physicists' minds. According to option 1, we are *always* facing a slippery slope whenever we wish to engage in explanatory projects. In general, we cannot know in advance around which corner we shall run straight into a wall of brute fact(s). This might occur at the third or at the hundred-and-third step of the explanatory ladder. And indeed, how many brute facts are there in the universe? It is not only incumbent on supporters of this view to answer this question, which they usually do not provide, but it seems that any possible answer must be rather awkward. If the number is large, this suggests that our explanatory powers are rather limited; if it is small, then one can never be sure why the process of reducing some of the previous brute facts to 'true' brute facts must stop exactly there. Of course, option 1 always stays with us – its doubtful virtue is that we can always return there if everything else fails. But, *pace* some distinguished philosophers,[20] there is no reason to run there any sooner.

Option 2 is perhaps the least-examined proposal for resolving the fine-tuning problem.[21] Consequently, it is poorly defined, although it does seem a logically coherent possibility. The closest to this option comes Wheeler's idea of the 'participatory anthropic principle'. In brief, it suggests that the universe is like an improvised 'twelve questions game'. In the normal game, a selected person imagines an entity and the participants need to determine what has been imagined based on up to 12 yes/no questions. 'Is it alive?' 'Is it white?' 'Is it a person?' In Wheeler's version, there is no fixed entity, but the protagonist accommodates the imagined entity to each new question (requiring, consequently, more and more thinking, i.e., time for each subsequent answer). So, when at last one of the participants asks 'Is it a cloud?' and the answer is 'Yes!', it is not the case that 'cloud' was present from the beginning – rather, it *emerged as a consequence of questioning*. The participatory anthropic principle suggests eponymously that observers participate in the emergence of the physical properties of the universe they inhabit. The best reason for discarding, at least for now, such an 'emergent universe' explanation is that it has never been formulated as a clear-cut hypothesis for explaining *particular* instances of empirical fine-tuning. While it can be speculated that the reality created by 'cooperation' with observers should have properties conducive to the emergence and evolution of further intelligent observers (of our particular kind), this remains on a very speculative level. Like option 1, there is no *inherent* answer as to which features of the observed universe are explainable in a conventional manner and which are due to this mystical form of 'emergence'.

Most accounts of fine-tuning simply ignore options 1 and 2 and jump straight to the two remaining hypotheses, perhaps fascinated by the alleged (and always newsworthy) conflict of religion and science. Explanatory option 3 is the classical design hypothesis, though it need not be necessarily supernaturalistic. There has been some discussion lately about naturalistic design, or 'lesser designer', or 'basement universe' scenarios, which do not involve supernatural elements (although, on the internal perspective – those of created or simulated observers – the capacities of the designer could be even *greater* than the capacities usually attributed to deities).[22] Even these represent a council of despair, and that still more forcefully applies to the creationist versions of the design hypothesis. Like option 1, we can always return to design if everything else fails.

Thus, we come to what most cosmologists regard as the only plausible and scientific way to account for the empirical fine-tunings of physical and cosmological parameters – embedding our atypical case in a wider ensemble. Option 4 is usually known as the *multiverse hypothesis*. In brief (details can be found in many already classic references[23]), if there is an ensemble of worlds with sufficient variation in local values of fundamental constants, cosmological parameters and whatever else is under scrutiny, a particular atypical set of values – or, more precisely, a set of intervals for these values – is to be found somewhere in that ensemble. From the 'external' perspective of the ensemble itself, there is nothing atypical, surprising or unexpected in such a situation. From the 'internal' perspective of any individual world in the ensemble, the key difference is that some sets of values will support life and observers, while the others will not. Therefore, the 'internal' question – which is, in our case, the question about fine-tunings of *our universe* – is posed only where the conditions and prerequisites for the existence of observers are satisfied. The remaining question is, what is the measure of the subset of parameters satisfying the prerequisites for observers (i.e., entailing *a habitable universe*) in the total set of parameters? If we have logical, empirical or computational reasons to believe that this measure is small, all the reasons for surprise or unexpectedness disappear. Of course, all this depends on the existence of mechanisms generating the required variety of the ensemble. (Strictly speaking, one could insist that the world ensemble characterized by the required variety is a brute fact; this way we return to the desperation of option 1 above.) The support for the multiverse hypothesis in recent years has been fuelled exactly by the fact that modern physics and cosmology, as we shall see in the next chapter, provide such mechanisms.[24]

Thus, the 'mystique' of anthropic reasoning is nothing more than observation selection. Just as the Malmquist bias and Signor–Lipps effect carry the lesson of reality being distorted by our ways of observing, a deeper *anthropic bias*[25] influences all our other observations. It does not influence any conclusions that are indifferent to the number and other properties of observers themselves, but it has to be taken into account in cases that are not indifferent. This immediately tells us why this is important in fundamental physics and cosmology, but also why it plays a role in fields such as planetary sciences, evolutionary biology or risk analysis; all these, which constitute a significant part of the astrobiological enterprise, influence the number and properties of observers. For example, anthropic reasoning tells us that the chance of finding an impact crater larger than 100 km in diameter and younger than 5 Myr anywhere on Earth is *exactly zero* – and without the effort and expenses of a field trip! If such a destructive catastrophe occurred so recently in the history of our planet, there would have been no observers on the planet today.[26] This is an extreme case of the wider constraint that the presence of observers on today's Earth imposes on the complete set of admissible histories of our planet and its biosphere. Mutatis mutandis, this applies to any astrobiological observation involving the properties of observers; in applying our particulars to general astrobiological questions we need to take precautions against introducing our anthropocentric biases. (As I will show in the next chapter, these particulars apply not only to humans, terrestrial metazoans, the Earth, Solar System and Milky Way, but also to our universe as a whole.) Anthropic reasoning is the antitoxin for this long-standing disease of human intellectual history.

Ironically, the central methodological position of anthropic reasoning for the study of life in the widest context is rarely appreciated. Whenever and however properties specific to terrestrial life and us, as observers, are invoked, explicitly or implicitly, we need to use anthropic reasoning to cleanse our thinking of the selection biases. This simple conclusion is usually overlooked by both astrobiologists and their critics. Cohen and Stewart thus write in purported criticism of astrobiology:[27]

> Our aim is simpler: to map out some of the main areas that must go into xenoscience. Of course one major area has to be 'Earthlike environments and Earthlike life', and it is this area that astrobiology concentrates on. There's no harm in that as long as we appreciate that it's just one major area, and as long as we have a proper grasp of the true diversity of Earthly life. In this book we will also devote plenty of space to Earthlike scenarios, but we will do that because it is in this area that we can be most specific, not because we think that nothing else can happen. When it comes to evolution, for instance, we view Earth's story as just one sample from what is probably a far broader range of possibilities. And we can ask two things. One is: just how broad *is* the range? The other is: are there any general, unifying principles?

Apparently contrary to the authors' intentions, there is nothing here that an open-minded astrobiologist would not subscribe to enthusiastically. A unifying principle would be the continuity between living and non-living matter, as well as between living and 'complex living'/intelligent matter; I shall discuss various forms of the continuity thesis in Chapter 6. The other question posed by Cohen and Stewart cannot be answered without a satisfactory account of observation selection effects, i.e., anthropic reasoning. They do not only constrain what could be expected to be empirically accessible, but also – in conjunction with some sort of more general theory – give us a Bayesian probability measure of any set of actual observations; in the next chapter we will encounter an example of such usage in the framework of more general theory.[28]

A widespread misunderstanding of anthropic reasoning is based upon the charge of naive Panglossianism: everything that exists must be for the best. In a more sophisticated form, critics deplore the 'curious reversal of causality' (the phrase used by Stephen Jay Gould to describe Freeman Dyson's teleological views[29]) seemingly implied by anthropic principles. Life and observers – meaning here *known*, i.e., terrestrial life and humans – could not fit into a different universe described by different forces, fundamental constants, etc. All empirical fine-tuning observations can be listed here. Therefore, according to this Panglossian construal, the universe 'in some sense must have known that we were coming'. From this, it is a tiny step to concluding that the universe must have been constructed or designed for our benefit.

This profoundly anti-Copernican view is justifiably condemned by critics, but it has nothing to do with anthropic thinking. The latter deals with observation selection effects and does not reflect any supernatural action, or mystical holism, or any such nonsense. Thus, if we use the locution that our universe was *selected* for its life-bearing properties, this simply reflects the inadequacy of human language – the motive we repeatedly encounter in the astrobiological context, finely explicated in *The Colour Out of Space*. Selection is just the consistency requirement, nothing more.

In Lee Smolin's criticism of anthropic reasoning, this is reiterated together with some other misconceptions (for instance, a dogmatic insistence that we need anthropic explanations only in fundamental physics and cosmology, while they are entirely acceptable in many other fields, like planetary sciences, ecology, economics or risk analysis). In particular, when Smolin concludes that Hoyle's type of argument for the fine-tuning of nuclear resonances in the ^{12}C nucleus is not really anthropic, he makes a semantic error and misses the wider point:[30]

Hoyle's argument:

1. X is necessary for life to exist.
2. In fact X is true about our universe.
3. Using the laws of physics, as presently understood, together with perhaps other observed facts, Y, we deduce that if X is true of our universe so is Z.
4. We therefore predict that Z is true.

In Hoyle's case, X is that the universe is full of carbon, Y is the claim that it could only be made in stars, and Z is the existence of a certain resonance in carbon. We see clearly that the prediction of Z in no way depends on step 1. The argument has the same force if step 1 is removed . . . There are other examples of this kind of mistaken reasoning, in which an argument promoted as "anthropic" actually has nothing to do with the existence of life, but is instead a straightforward deduction from observed facts.

To see what the problem is with this sort of misconstrual, imagine that, instead of removing Smolin's step 1, we remove step 2. The argument obviously has the same force. What is the main difference between the two cases? Clearly, we can infer at least some of the requirements for life without understanding the properties of the universe as a whole; we could imagine Earth's atmosphere being as opaque as that of Venus, or even us being a species of intelligent fish (as Leonard Susskind suggests in his oft-cited parable[31]). Of course, we would have understood the need for carbon – Hoyle's original argument – but we would *not* have been able to understand, for instance, the need for galaxies or other large-scale structures in the universe. There certainly are many other prerequisites for life and observers, which are not as simple as the need for carbon; the same sort of argument should be applicable for all of them – they are also informative tautologies that illuminate the complexity and continuity of evolution. One such non-obvious requirement is the abovementioned lack of a young $D > 100$ km impact crater on our planet. In addition, Smolin's protestation against the use of the 'anthropic' label is misplaced as well – to say that the argument for the existence of carbon 'has nothing to do with the existence of life' is disingenuous if we understand both the existence of carbon and the existence of life as parts of universal, cosmic evolution. While I shall return repeatedly to the issue of continuity and the universality of evolution, it is enough to emphasize here that anthropic reasoning pertains to any link in the excessively complex chain of events determining the number and capacities of observers. Further specification seems more like a semantic issue rather than one of substantial content. (Occasionally, extreme specification of X is an intentional strategy, usually used by detractors of X, as will be discussed in detail in Chapter 7 concerning SETI projects.)

Now I shall show how a particular philosophical puzzle can arise from systematic thinking about properties of intelligent observers in the cosmological context. Besides demonstrating time and again the deep connection between astrobiology and cosmology, this puzzle and a proposed explanation may serve as a blueprint for similar paradoxes we shall encounter in further chapters.

Controversial observation selection: Olum's problem

In an elegant and thought-provoking paper, physicist and philosopher Ken Olum[32] argues that 'a straightforward application of anthropic reasoning and reasonable assumptions about the capabilities of other civilizations predicts that we should be part of a large civilization spanning our galaxy.' Starting from the assumption of an infinite universe (following from the inflationary paradigm), Olum conjectures that somewhere there are civilizations much larger than our own (the latter consisting of about 10^{10} observers at present). The spatial extent and amount of resources at the disposal of such large civilizations would lead, in principle, to a much larger number of observers (for example, 10^{19} observers).[33] Even if 90% of all existing civilizations are small, similar to our own, anthropic reasoning suggests the overwhelming probabilistic prediction is that we live in a large civilization. Since this prediction is spectacularly unsuccessful on empirical grounds, with a probability of such failure being about 10^{-8}, something is clearly wrong.

I shall refer to this alleged incompatibility of anthropic reasoning with observations as 'Olum's problem'. In a less-refined manner, it has been foreseen by J. Richard Gott (in a founding paper on the 'Doomsday Argument'):[34]

> In the limit where (cosmology permitting) a supercivilization is able to accumulate an infinite amount of elapsed conscious time and an infinite number of intelligent observers . . . , the fraction of ordinary civilizations such as ours that will develop into such a supercivilization must go to zero so that the set of observers born on the original home planet is not an infinitesimal minority of all intelligent observers.

The cosmological background was not so clear at the time of Gott's paper, a finite recollapsing universe was still a widely accepted option. In contrast, recent results have strongly confirmed predictions of the inflationary paradigm, which is generically eternal and faces us with a strong form of Olum's problem.[35] The same applies even more forcefully to cosmologies following from M-theory,[36] where there are one or more extra dimensions in which our universe is just an embedded 'slice'; similar to the way Edwin A. Abbott's classical 'Flatland' is embedded into our usual three-dimensional space. In all these cases, we face an infinite universe with all the philosophical conundrums that infinities have traditionally presented us with; in addition, new problems at the interface of cosmology, astrobiology and philosophy, such as Olum's, are bound to arise.

Olum gives us several possible solutions to the problem of why we do not find ourselves members of a large civilization:

1. anthropic reasoning does not work;
2. anthropic reasoning should use civilizations instead of individuals;
3. one should consider observers who live at any time;
4. selection biases;
5. infinitesimally few civilizations become large;
6. the universe is not infinitely large;
7. colonization of the Galaxy is impossible;
8. we are a 'lost colony' of a large civilization;
9. the idea of an individual will be different in the future;

admitting that none of them are very likely, but maybe

10. some combination of them might work to alleviate the problem.

Recently, I have advanced a proposal for a general solution which escapes Olum's analysis, although it has some connection with several of his proposals.[37] This in an interesting testbed for studying the philosophical implications of recent developments in both cosmology and in our astrobiological understanding of habitability and evolution.[38]

The devil hides in the details. When Olum writes, 'Something must be wrong with our understanding of how civilizations *evolve* if only one in a billion can survive to colonize its galaxy' (emphasis added), he is on the right track. Unfortunately, he does not offer a glimpse of what that 'wrong' might be, so let me try to fill in the gap here. The title of the relevant section of Olum's paper is 'Infinitesimally few civilizations become large'. This is an instance of sliding into the Parmenidian timeless view so dear to some philosophers.[39] The correct title would be 'Infinitesimally few civilizations have become large *so far*'. There is no inconsistency here. The universe, be it infinite or finite, evolves: it changes with cosmic time.[40] What has been a sufficient condition for X at epoch t_1, need not be sufficient at epoch t_2.

'The problem will exist even if we confine ourselves to those observers who exist presently.'[41] This reasonable stance implies that we have *indexical knowledge* about the epoch we are living in. This is an important piece of additional data that has to be taken into account in anthropic reasoning. The quoted sentence of Olum, for instance, could not be uttered if 'presently' were to apply to epochs incompatible with the existence of intelligent observers, e.g., the times before galaxy formation, or the epoch in our distant future when all protons decay. Although it sounds tame, this simple constraint may, in fact, undermine any reasoning that presumes our understanding of the necessary conditions for the existence of observers and civilizations.

There are strong empirical reasons to conclude that the universe has been less hospitable to life earlier in its history. One instance of such behaviour is related to the chemical enrichment of matter; fewer elements heavier than helium mean a smaller probability for the formation of terrestrial planets, and perhaps a smaller probability of biochemical processes leading to life, intelligence and observers. This has been recently spectacularly quantified by Lineweaver and collaborators who showed that, for instance, the median age of Earth-like planets in the Milky Way is 6.4 ± 0.9 billion years; before then, conditions were far less favourable for the formation of possible life-bearing sites.[42] Another, effect, possibly crucial, are catastrophes capable of disrupting the evolutionary sequence leading from the simplest prokaryotes to complex life, to animals, intelligent beings, and to civilizations, small and large. Notably, in the last two decades astronomers have confirmed that γ-ray bursts, detected by satellites in Earth's orbit at a rate of about one per day, are traces of explosions occurring in *any* galaxy within our cosmological horizon.[43] Their occurrence in any one particular galaxy (say ours), means a catastrophic event capable of destroying life forms in a large part of, or the entire, GHZ.[44] Fortunately enough, we also know from the cosmological research that their frequency *decreases with cosmic time*.

Let us call the cumulative effect of all these occurrences the *hostility parameter*. The hostility parameter obviously has its spatial and temporal distributions, and these act as Bayesian constraints on any a-posteriori estimates of the probability of finding ourselves in a large or a small civilization. Observe a representative volume of space and count N_0 habitable sites where life and intelligence can develop.[45] Let us assume that the probability of a small civilization becoming a large one evolves over time as:

$$p_l(t) = p_{l0}[1 - \exp(-t/\tau)], \tag{3.2}$$

where t is the cosmic time (measured from some relevant moment, say the galaxy formation), τ is the hostility parameter expressed as a characteristic timescale (say for γ-ray burst rarefaction), and p_{l0} is the asymptotic 'standard' probability, *ceteris paribus*, of a small civilization making the transition to a large one. One can think about Equation (3.2) as representing recuperation or recovery from a large adverse perturbation. A simple model of traffic flow gives the probability of bus arrival per unit time at your stop, after a major traffic collapse, by an equation of the same form, p_{l0} playing the role of 'regular' probability based on bus statistics over many days. In the astrobiological case, of course, no previous equilibrium existed (except, trivially, the dead galaxy before planet formation began). Thus, we can understand Equation (3.2) as the approach to an equilibrium state in which perturbations from past large-scale physical processes (like nucleosynthesis and γ-ray bursts) will cease to play a significant role, and the only parameter describing the transition between small and large civilizations is the asymptotic probability, p_{l0}. In Olum's study, this probability is at least 0.1 (since his 'timeless' argument assumes that 90% of the currently existing civilizations are small ones).

In this toy model, the fraction of observers living in a large civilization is, clearly

$$\begin{aligned} f_{\text{large}}(t) &= \frac{\text{no. of observers in large civilizations}}{\text{total no. of observers}} \\ &= \frac{\text{no. of observers in large civilizations}}{\text{no. of observers in large civilizations} + \text{no. of observers in small civilizations}} \\ &= \frac{N_0 p_l(t) \langle n_l \rangle}{N_0 p_l(t) \langle n_l \rangle + N_0 [1 - p_l(t)] \langle n_s \rangle}. \end{aligned} \tag{3.3}$$

Here $\langle n_l \rangle = 10^{19}$ is the average number of observers in a large civilization, and $\langle n_s \rangle = 10^{10}$ is the average number of observers in a small civilization (using the very same numbers as in Olum's study). All numbers of observers are taken to be time-dependent, evolving quantities. Now, from Equations (3.2) and (3.3), after trivial algebraic transformations, we obtain

$$f_{\text{large}}(t) = \frac{p_{l0}\langle n_l\rangle}{p_{l0}\langle n_l\rangle + \left[\frac{1}{1-\exp(-t/\tau)} - p_{l0}\right]\langle n_s\rangle}$$
$$\approx \frac{\langle n_l\rangle}{\langle n_l\rangle + \frac{1}{p_{l0}}\frac{1}{[1-\exp(-t/\tau)]}\langle n_s\rangle}. \tag{3.4}$$

(We neglect here the $p_{l0}\langle n_s\rangle$ term as arguably insignificant.) Upon inspection of the last expression in Equation (3.4) we conclude that *at least one* of the following propositions must be true: (A) $f_{\text{large}}(t) \approx 1$; (B) $p_{l0} \approx 0$; or (C) $\exp(-t/\tau) \approx 1$.

Proposition (A) is what Olum finds paradoxical (and corresponds perhaps to his solution 8, claiming that we are in fact part of a large civilization without being aware of that fact[46]). Proposition (B) corresponds to the idea that infinitesimally few civilizations *ever* become large ones, presumably because of some inherent problem like self-destruction or relinquishing interstellar travel. This might be valid for *some* small civilizations, but is rather exclusive; and we should prefer non-exclusive explanations, if available.[47] But (C) is something entirely new. Here we have a *global evolutionary effect* acting to impede the formation of large civilizations. *It does not clash with any observation, as* (A) *does, nor does it imply something about (arguably unknowable) alien sociology, as does* (B).

Of course, the toy model is certainly not realistic. One way of improving it would be to recognize that there might be many 'critical steps' in evolving towards a large civilization.[48] Each critical step takes some time. For instance, we may envisage the transition from simple life to complex life as a step having its own hostility timescale (suppressing and postponing 'Cambrian explosion analogues' everywhere), and the same for the transition between complex life and intelligence, intelligence and small technological civilization, etc. In the simplest analogy with the toy model above, we could substitute a single term in the denominator of Equation (3.4) with something like

$$\prod_{i=1}^{n} \frac{1}{1-\exp(-t/\tau_i)}. \tag{3.5}$$

Clearly, in this case, we would not need to worry about Olum's problem as long as the proposition p: $(\exists i)\ \tau_i \gg t$ is true. It is obvious that p is not true for all times, so at some particular epoch we would have to face the problem again. Fortunately, that epoch most likely lies in the distant cosmological future, so we need not worry about it now.

In brief, by taking into account the physical evolution of the universe and the underlying requirements for civilizations (either large or small), we can help to resolve the problem, pointed out by Olum, in a manner quite in accordance with current tenets of empirical astrobiology, Copernicanism and anthropic reasoning. The solution presented here should not be misconstrued as the simplistic assertion that a seemingly paradoxical situation *has not occurred yet*, in the same sense as a lottery player can assert that a particular number has not come up yet. The current astrobiological state of affairs of our Galaxy is not a sort of fortuitous state of affairs randomly pulled out of a lottery. Our intention is to emphasize systematic and *on the average* deterministic evolutionary processes in physical reality. Such processes have precluded (in the probabilistic sense) 'paradoxical' states of affairs from arising so far. The qualification 'on the average' is necessary here, since astronomers cannot predict, for example, exactly where and when a particular supernova or a γ-ray burst will occur. Nevertheless, the general trend of decreasing frequencies can be fairly uncontroversially inferred based on large observational surveys.

As we have seen, with increasing time elapsing from the galaxy-formation epoch, the chance of finding a large civilization gradually increases. In that sense – which is only a trivial application of uniformitarianism – we can assert that the universe becomes more hospitable to life and intelligence. The problem of conflict between anthropic reasoning and observations will become more acute – assuming that *both* anthropic reasoning *and* our observations do not change in future. However, it is reasonable to conjecture that our observations regarding this issue are going to change, either by the discovery of a large civilization in our past light cone, or by *becoming* a large civilization ourselves. However, even under the most optimistic timescales, this is not a very pressing concern.

Of course, other solutions are possible. Deng Ho and Bradley Monton have argued that Olum's problem arises because anthropic reasoning is misunderstood.[49] According to this view, when measuring one of the large-scale parameters of the universe, any observer should expect to get a result that was antecedently considered improbable, from the standpoint of anthropic reasoning. This is due to overspecification – we do not expect the value of the fine-structure constant of $(137.03599074)^{-1}$ to be typical *with such precision*. The real inference is, according to Ho and Monton, that we should be one of the observers whose observations are to be considered improbable, making anthropic reasoning compatible with observations. I find this suggestion misleading, since anthropic reasoning deals with intervals compatible with the existence (or number) of observers, not with narrow values. Even if the interval of some constant compatible with the existence of observers were large, it would still have been a legitimate target for an anthropic explanation. I shall return to the problems stemming from overspecification in later chapters.[50]

I conclude that it is *too early* (on the cosmological timescale) in the history of the universe for a situation to arise which contains Olum's problem. Given the present situation, it is *too early* (on a human timescale, and in the slightly different sense of being premature) to conclude that either astrobiology or our understanding of observation selection conflicts with observation. The missing ingredient here is a *dynamical mechanism* or a set of mechanisms explaining the number of observers under the widest possible range of environmental parameters. This key ingredient is what I shall consider next.

The 'real thing': the astrobiological landscape

A casual survey of ongoing astrobiological research reveals a very heterogeneous, fragmented realm. How can astrobiologists persuade anyone of the unique span and significance of their mission when they 'speak in so many tongues'? Herein, I shall argue that the *astrobiological landscape* – an abstract landscape-like structure in the space of astrobiological parameters – is a concept capable of unifying different strands of thought and research. By arguing that the astrobiological landscape is a working concept, useful for actual research and not just a metaphor, I shall invoke several lines of reasoning, some to be briefly explained in this section, and most gradually emerging through discussions in the rest of this book.

We need solid foundations on which to base any universality. In the oft-cited final words of *On the Origin of Species*, Charles Darwin contrasted the simple motion of our planet under the law of gravity (with its implied regularity, analytical and short description) with an 'entangled bank' as a metaphor for the biological complexity attained by evolution:[51]

> It is interesting to contemplate an entangled bank, clothed with many plants of many kinds, with birds singing in the bushes, with various insects flitting about, and with worms crawling through the damp earth, and to reflect that these elaborately constructed forms, so different from each other, and dependent on each other in so complex a manner, have all been produced by laws acting around us. . . . There is grandeur in this view of life, with its several powers, having been originally breathed into a few forms or into one; and that, whilst this planet has gone cycling on according to the fixed law of gravity, from so simple a beginning endless forms most beautiful and most wonderful have been, and are being, evolved.

I shall return to this important image in Chapter 5, when dealing with those aspects of evolutionary theory crucial for astrobiology. For now, I wish to use the image as an example of a spectrum of things any concept of life in the cosmic context must encompass.

'The entangled bank' here embodies characteristics such as the *complexity* of a *particular kind* of (eco)system, distributed over a particular region of *space* in a particular epoch of *time*. Complexity plays the key role – Darwin chooses this image to convey the impression of a huge number of interrelated and intertwined biological processes.[52] However, the complexity of an entangled bank is different from the complexity of a scrub plain or the complexity of a savannah. The difference is not just in geography. In other words, there is a *differentia specifica* of the entangled bank in comparison with other ecosystems, even if they are very closely located in physical space. There is a set of parameters required to describe the entangled bank that are different from the set of parameters required to describe scrub plain or savannah. Let us call these parameters *dynamical parameters*; I shall briefly explain the reasons for this label. Dynamical parameters might be a black box to us – we may feel that it is easy to distinguish between a river bank ecosystem and a rainforest ecosystem, but what we really wish to have are a few – the less the better – underlying principles which include all the differences, and not a full encyclopaedic list of differences. The principles might be unknown, but it is both epistemologically and practically prudent to assume that they exist.

Thus, in our mental image, we bundle together four things: (1) the complexity of life, (2) the space in which we observe this complexity, (3) the time at which we observe this complexity, and (4) the parameters specifying the kind of environment in which we observe this complexity. What does here point exclusively to Earth and the specific biosphere we currently observe? Nothing, it seems. Darwin's words could resonate – to the best of our knowledge and imagination – with a non-human naturalist on the fourth planet of ζ Reticuli planetary system when allowance for different kinds of ecosystems is made. Any other situation can be analogously accommodated in the landscape framework.

How do we visualize the astrobiological landscape? Of course, since it is an abstract multidimensional space, we cannot hope for a literal visualization, but one graphical approach is shown schematically in Figure 3.2. While changes in location and epoch can obviously be compressed into one axis, the condensation of – yet unknown – astrobiological dynamics into a single axis may look controversial. Since the theory dictates our working concepts, we cannot yet have a complete list of those parameters we compressed under the banner of 'dynamical variables'. However, as long as we stick to naturalism, there is no reason to doubt their *existence*; and as long as we stick to Copernicanism, there is no reason to doubt that by local research – say within the Solar System – we shall be able to discern them. Equally reasonably, one expects that there is a universal zero point of astrobiological complexity – the lack of all life, what is denoted by 'dead space' in the figure. In addition, the property of all complex systems is that complexity – irrespective of the exact metric – depends on the spatio-temporal region considered and the values of at least some dynamical parameters of the system.

Alternatively, we can understand the astrobiological landscape as a *set of viable evolutionary histories of life* in a particular region of space. In view of the highly clustered cosmological distribution of matter, it is only natural to focus on our

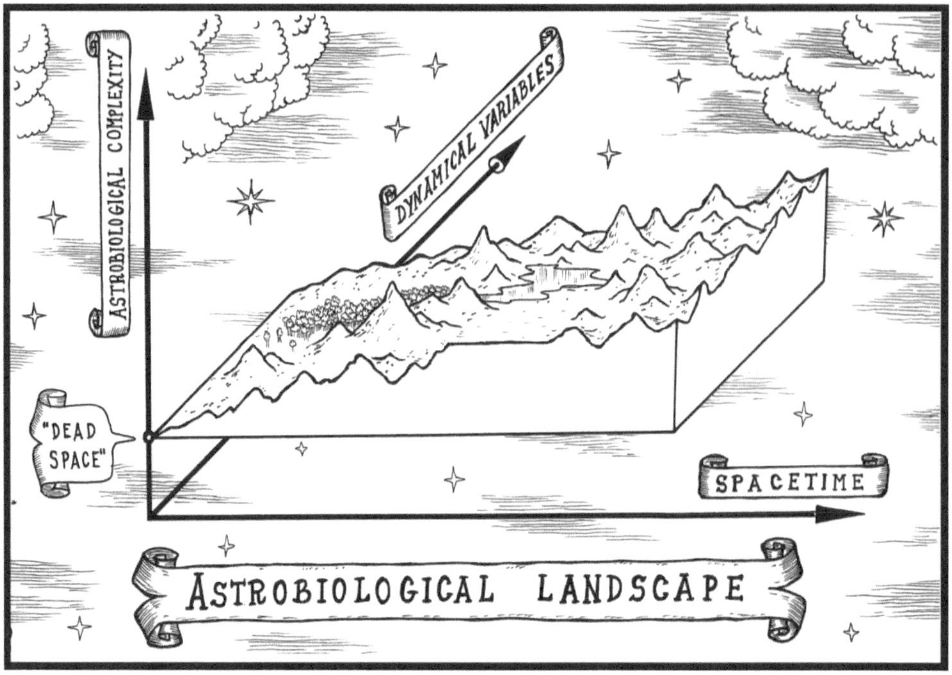

Figure 3.2. A schematic representation of the astrobiological landscape. Since we are dealing with multidimensional parameter space, it is impossible to visualize it directly, so this simplified sketch condenses space and time coordinates into one axis, and all astrobiological dynamical parameters into another. Thus, the landscape links astrobiological complexity with the yet poorly understood set of dynamical variables and spatio-temporal region under consideration (both could be defined in many contexts, from a local planetary to the global cosmological). (Courtesy of S. Popović.)

own Galaxy. Consider, for instance, the following astrobiological scenarios (or 'histories') for the Milky Way:

1. Dead Space: The Galaxy is entirely dead and it stays that way in all epochs.
2. Sporadic Life: Life forms emerge here and there, without any particular correlation either in space or in time.
3. Rare Earth: While simple life is ubiquitous in the Galaxy, complex biospheres, like the terrestrial one, are very rare due to the exceptional combination of many improbable requirements.
4. False Precision: the number of inhabited planets in the Galaxy behaves like $A \ln(t/t_{MW}) + B$, where t_{MW} is the present age of the Milky Way, and A, B are given constants.

5. Red Dwarf Kingdom: While simple life forms are ubiquitous in the Galaxy, peaks of complexity could be found only around M-class red dwarf stars, which create the most stable conditions over their huge lifetimes.
6. Galactic Club: Life, intelligence and civilizations evolved independently long ago in various places, and many older civilizations are aware of one another and are actively communicating and/or collaborating.
7. Extinct Galactic Club: The same as in the Galactic Club, except that some unspecified natural or intentional cataclysm has destroyed most or all advanced civilizations at some moment in the past (rather close to the present in comparison to t_{MW}).
8. Black Clouds with a Vengeance: Life on planets is a rare exception and most astrobiological complexity lies within giant molecular clouds and their low-density, low-temperature ecosystems.

Obviously, there is a huge number of conceivable scenarios; some are specific, some are more general, including many others within them. Each broader category includes many related scenarios differing only in the numerical values of their parameters (explicit or implicit). What exactly – and how tightly – constrains such scenarios? This question is central not only for understanding the entire concept of the landscape, but also for outlining useful directions for future astrobiological research. This especially applies to the theoretical domain, where, as sketched in Chapter 1, there is still so much work to be done. We can still infer many important constraints from our limited knowledge. It is obvious that the Dead Space does not work – we are here. Sporadic life is falsified by the recent work on the Galactic Habitable Zone, which indicates strong correlations in both the spatial and temporal distribution of habitable sites in the Milky Way.[53] While it would seem that False Precision could be fitted by some future database and by particular choices of the constants, it is completely arbitrary and, lacking any theoretical motivation, should not be taken seriously. Further, it seems that the more extravagant versions of the Galactic Club did not work either, at least not as it was assumed in the period of wild SETI optimism in the 1960s and early 1970s; otherwise, our search efforts would have been crowned with success long ago.[54] Rare Earth is a serious and well-developed astrobiological hypothesis and I shall discuss it in detail later in this book.

Clearly, a part of the constraints comes from our understanding of the overall physical structure of our Galaxy. For example, we cannot imagine spherically symmetric panspermia over scales of kiloparsecs in the Milky Way, since the distribution of possible habitats is clearly non-spherical over such spatial scales, the vast majority of habitable stars being concentrated in the thin-disc component of our

Galaxy. This example immediately shows that with better astrophysical knowledge we can always do much better: with the information on precise densities of habitable stars (which has become available recently), we can put other, tighter constraints on the pace and probabilities of panspermia. For instance, since we know that the density of stars in the disc varies almost exponentially with galactocentric distance, we may conclude that this will enhance the probability of successful transport of biotic or prebiotic material between the nearest stellar neighbours in the inner regions of the Milky Way in comparison to the outer regions. While this is almost trivial, it is important to keep in mind that any generalizations quickly become highly complex. The same applies to temporal scales: that the age of the thin disc is somewhat below 12 Gyr forces us to reject those astrobiological histories in which life forms have had, say, a petayear (10^{15} years) to evolve at *any* habitable site. (It also excludes the notions of 'eternal life', popular in the nineteenth century and briefly discussed in Chapter 2.) Thus, some constraints on the landscape come from cosmological considerations, others from more local astrophysical, astrochemical, planetological, etc. findings.

The expansion of our astrobiological knowledge may be construed as obtaining a sharper and more focused view on specific areas of the landscape. For instance, natural further steps include a more precise understanding of GHZ and its evolution – acceptable histories have to be in agreement with the results of Lineweaver and collaborators mentioned in the previous section. From the empirical point of view, acceptable histories will necessarily be fuzzy. Recall the Areoparadise Lost scenario introduced in Chapter 1, in which microfossils are found on Mars, morphologically similar to terrestrial microfossils. Such a discovery would strongly constrain possible astrobiological histories of our Solar System, but since we would be unable to determine – without additional information – whether we have discovered an independent biogenesis or a case of interplanetary panspermia, fuzziness will remain.

Another part of the constraints comes from what we understand by habitability or, more precisely, what we consider a reasonable extension and generalization of *capacities* of life. This is usually only tacitly assumed and rarely discussed or questioned, but it is of paramount importance, since biotic feedback plays an important role in shaping the global properties of our planet, starting with the obvious (atmospheric chemistry), up to complex local ecological details. The ongoing search for biomarkers in the set of extrasolar planets relies exactly on such a methodological basis.[55] Unknown astrobiological dynamics will most likely need to incorporate this non-linear feedback in any complete description. For example, no matter how quick biogenesis has been on Earth, it is still quick in geological and astronomical terms, in 'deep time'. It is unreasonable to assume – though perhaps cannot yet be decisively refuted – that somewhere else in the Milky Way, in a

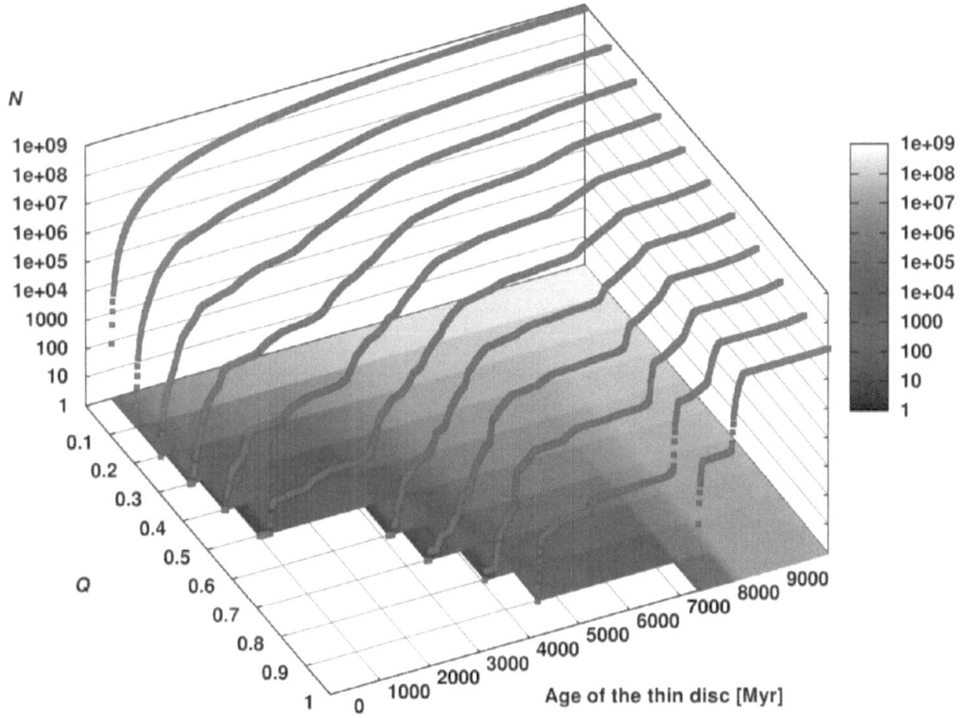

Figure 3.3. A toy model of the astrobiological landscape in the context of GHZ. Presented is the number of planets that have achieved noogenesis at least once (cumulative plot), as a function of the age of the Milky Way thin-disc stellar population and the mean extinction probability Q per catastrophe.

similar type of habitat, biogenesis could be acomplished in a dramatically shorter interval, say on the scale of years or months. Perhaps such astrobiological histories can be excluded from the landscape as well. Further insight into prebiotic chemistry and biogenesis will enable clearer delineation and sharpening of our view of this important sector of the astrobiological landscape.

Finally, the most interesting constraints come from insight into the specific problems of astrobiological research; this notably applies to Fermi's paradox, which makes the original optimistic Galactic Club scenario unacceptable. This obviously demonstrates the role of such constraints as *boundary conditions* to yet unknown astrobiological dynamics.[56] Of course, the problem is that insufficient research on Fermi's paradox has not yet enabled us to make precise quantitative models. As an extremely simplified illustration, a toy model of a very special type of astrobiological landscape built in order to study Fermi's paradox is shown in Figure 3.3.[57] A major motivation has been the possibility of generalizing Stephen Jay Gould's 'third tier' to the Galactic Habitable Zone: '. . . mass extinctions are sufficiently frequent, intense, and different in impact to undo and reset any pattern that might accumulate during normal times.'[58] Here, astrobiological complexity is represented only by a step function: 1 for each site containing an extraterrestrial civilization, 0 otherwise (and the cumulative plot is shown). It is assumed that all habitable sites are equal and homogeneously distributed in the GHZ. Since the model has been constructed to test whether evolution dominated by global Galactic catastrophes can explain Fermi's paradox (in the specific case of γ-ray bursts, as mentioned in the discussion of Olum's problem above), the only dynamics of interest is represented by the severity parameter Q, describing the probability of extinction of life under the effect of a Galactic supernova or a γ-ray burst. (Implicitly present as parts of the relevant dynamics are the random distribution of burst events and a log-uniform distribution of biological timescales.) Initial conditions include the distribution of ages of planets, according to Lineweaver's results.

Results show that the system exhibits a systematic shift of behaviour as we move from small values of Q (gradualism) to large values of Q (catastrophism). At large Q, we have a step-like succession of astrobiological regimes, governed by external timescale forcing.[59] In each regime, it is obvious that the ages of inhabited planets are not independent and uncorrelated; just the contrary, as we expected from the considerations above. While these results are *very* tentative, they not only demonstrate that quantitative models of specific astrobiological problems are possible (in that study, Fermi's paradox), but also that by highlighting specific aspects of the astrobiological landscape we can detect and, at least in principle, test new explanatory hypotheses. I shall provide further examples in the course of this book that the concept of astrobiological landscape can work as a heuristics for difficult problems.

The astrobiological landscape is large – and necessarily so, since it potentially contains all useful (and much useless) astrobiological information, just as phase space contains all useful (and much useless) information in statistical physics. It is also spatially additive, in the sense that we may imagine it for a Galaxy, or for a group of galaxies, or just a small part of a Galaxy (like the Solar System or any other extrasolar planetary system). In other words, it can be zoomed in and out, revealing structure that reflects the physical distribution of matter in the universe. Although we may think that parts of it will be similar (for instance, the landscapes of the Milky Way and our nearest giant spiral neighbour, M31, at the same epoch), this is essentially an empirical hypothesis to be tested – in the long term, at least – by observation. Of course, in practical terms it is the Galactic astrobiological landscape that is of foremost interest for research.[60] In theory, we can imagine zooming out so much that we include galaxies located at cosmological distances, which are impossible to observe except in the very early stages in their evolution; this is, of course, purely an abstract point, since nobody really expects to find anything but the Dead space zero level at cosmologically early epochs. But it serves another important epistemic purpose: to remind us that the distribution of matter – or even more, *kosmos* in the ancient Greek meaning of the word – neither stops at the faintest visible galaxies nor at our cosmological horizon. Nor does it stop with the domain ruled by our familiar, effective and low-energy laws of physics. In the next chapter, I shall try to show that it is meaningful to discuss life playing a role on a grand stage; indeed, the very grandest of them all.

4

Cosmology, life and the Archipelago

The idea of one Universe . . . is preposterous.

John A. Wheeler

Even the largest things are occasionally hard to find. In 1541, the Spanish conquistador Francisco de Orellana was, almost against his will, chosen to lead an expedition along the Coca and Napo rivers in eastern Ecuador, in search of the fabulous 'Land of Cinnamon'. Completely unexpectedly, Orellana and his companions found themselves in the sprawling basin of a huge and mighty river, descending thousands of kilometres on an improvised, barely seaworthy ship. On 26 August 1542, they finally reached the Atlantic Ocean. What the Spanish serendipitously discovered was the Amazon, the second longest and by far the most powerful river in the world, hitherto unknown, not only to Europeans, but even to the great pre-Columbian civilizations of Meso-America and the Peruvian altiplano.[1]

In 2005, Princeton astrophysicists J. Richard Gott and Mario Jurić announced the discovery of the 'Sloan Great Wall' (named after the Sloan Digital Sky Survey, the most comprehensive galaxy catalogue so far); it was the largest known structure in existence, measuring about 420 Mpc along the longest axis and made of millions of galaxies.[2] The Sloan Great Wall is, in fact, so huge that larger structures are impossible according to our best cosmological theory; with it, we reach 'The End of Greatness', i.e., the spatial scales on which the distribution of matter in the universe is indeed homogeneous, in accordance with the cosmological principle of Eddington and Milne. It could be argued that the Sloan Great Wall is the largest empirically accessible 'thing', apart from the universe within our cosmological horizon itself.[3] Ironically, it took decades of observations of galaxies and tremendous advances in data reduction and interpretation to discover such a huge aggregate of matter – something we would intuitively expect to associate with difficulties in finding the smallest constituents of matter in particle physics (vibrating strings? membranes?)

or the smallest biochemical systems capable of some form of self-reproduction (prions?). It seems that size is not (always) important in the complex parameter space of science itself, which explains why we have only recently discovered the ultimate extension of physical reality – the multiverse.[4]

Multiverse: a universal solvent

The multiverse concept, in all its multiple guises,[5] has provoked a lot of criticism lately. This is in part a conservative reaction to the rapid expansion of the use of the concept in several different disciplines; on the other hand, some critics are motivated by old-fashioned and naive empiricism. The opponents of atomic theory at the end of the nineteenth century, headed by the illustrious Ernst Mach, have claimed that since postulated entities (= atoms) are beyond the empirical domain, the theory is just a well-concocted fiction; the term 'science fiction' did not exist in Mach's time. The lesson of Mach's defeat in his debate with Boltzmann and other atomists did not become widely accepted, however. That much of the later-day criticism of the multiverse is not serious, is attested by the fact that fierce opponents of one kind of the multiverse (the one following from the string theory landscape), like Lee Smolin, readily admit to a need for the multiverse 'by other means' (e.g., through the cosmological natural selection).[6]

Here and elsewhere in the book I fully endorse the old adage, *the problem is not that we take our best theories too seriously – the problem is that we don't take them seriously enough.* In its many versions, it can be ascribed to Ludwig Boltzmann, Paul A. M. Dirac, Kurt Gödel, Hermann Bondi, Richard P. Feynman, John A. Wheeler, Linus Pauling, Leonard Susskind, among many distinguished scientists, and it has been particularly forcefully argued recently by David Deutsch.[7] Accordingly, I shall not engage in fruitless 'ontological' debates about the reality (whatever) of the multiverse and instead assume that it indeed presents the best explanatory concept of our best theories in fundamental physics and cosmology. The obvious task at hand is, then, to explore what consequences, if any, there are for astrobiology and its foundations.[8]

As discussed in the previous chapter, the multiverse explanation functions only in conjunction with observation selection effects, which explains why we perceive the values of fundamental physical and cosmological parameters as belonging to a small and a-priori atypical subspace of the entire parameter space. Since 1961 and Dicke's rejoinder to Dirac, we have understood a great deal about the prerequisites for delineating this subspace, but a huge amount of work is still facing us, most of it in the astrobiological domain. Four developments that took place in the last ten to twenty years, in particular, mark the intellectual convergence necessary for discussing multiple worlds in the physical context:

(I) The emergence of the 'new standard' cosmological model of a flat universe dominated by dark energy (after 1998), connected with the predominance of the inflationary paradigm, especially chaotic inflation, seen today as the generic form of the process.
(II) The rise of string theory, and generalization in the form of M-theory, as the best candidate for the 'Theory of Everything' and realization that it has a huge number of low-energy sectors (or vacua) forming a populated 'landscape'.
(III) The elucidation of anthropic principle(s) as observation selection effects, leading to rejection of the old-fashioned teleological (mis)interpretations.
(IV) The rise of an 'information paradigm' in many sciences, from biology to fundamental physics to computer science.[9] In parallel, the whole PC-revolution in practical computing has literally brought home the possibility of simulating multiple outcomes of a single event, even in very complex dynamical systems.

In general (though the particulars may vary), the multiverse[10] is a set of large, topologically connected or disconnected, causally connected or disconnected, cosmological domains – conventional universes. The multiplication of theories involving this concept has occurred over a wide spectrum of disciplines, from cosmology and mathematics to quantum information theory to philosophy. The major point here is that ontological enlargement is *required* by many different theories in many independent disciplines. While I have argued in the previous chapter that the multiverse is the only really satisfactory explanation of cosmological fine-tuning, and that we have reasons to accept it similar to Boltzmann's nothing-more-practical-than-a-good-theory acceptance of atomic theory, one may also opt for a more modest approach. In such an approach, the key reason for acceptance of the multiverse would be the fact that a well-established cosmological theory such as inflation requires it (again, over and above the fact that cosmological horizons divide even a single Friedmann universe into an effective ensemble of causally disconnected sub-universes, which constitute a Level 1 multiverse according to Tegmark's classification). It was realized in the mid 1980s, first by Andrei Linde and subsequently by almost all cosmologists, that the generic form of inflation is 'chaotic' or 'eternal inflation', which implies an infinite fractal ensemble of universes, a subset of which undergoes inflation at each particular time. As hinted at in Chapter 2, the whole is thus more akin to the old steady state theory (not accidentally, sharing the same global de Sitter metric) than to the conventional picture of an evolving universe. The conventional picture applies to our particular 'bubble universe', while the entire multiverse is in a steady state equilibrium.

A related form of multiverse is offered by string theory, which is, in the form of M-theory, the best current candidate for the 'theory of everything'.[11] In the rest of this chapter, I shall use the anthropic landscape of string theory as a prototype physical realization of the multiverse, while strongly emphasizing that it serves as just a placeholder and that no conclusion is crucially dependent on the choice of ultraviolet completion of unified field theories. In brief, the string theory multiverse as formulated by Leonard Susskind (who prefers the term 'megaverse', consisting of 'pocket universes') is the physical realization of the string theory landscape. While the landscape is a mathematical construct (a hypersurface in the parameter space of string theory), the resulting multiverse is a physical realization of a set of low-energy ('vacuum') solutions, corresponding to different low-energy breakings of the underlying high-energy symmetry.

More than a single book would be necessary to give a fair treatment to all hypotheses involving some form of multiverse. My goal here is only to argue that (i) the concept of a multiverse is the product of an important intellectual movement on a very wide front of contemporary research, and (ii) it offers a new and richer perspective on the prerequisites for life and observers, whose determination is, with only seeming *contradictio in adjecto*, one of the perennial questions of astrobiology. To see this, we need a further generalization of the concept usually regarded as a given – our habitable universe.

The Archipelago of Habitability

The idea of the 'Archipelago of Habitability' was introduced by Max Tegmark in his intriguing paper on the relationship between the mathematical and physical worlds.[12] Conceptually:

> **The Archipelago of Habitability**: a set of regions in parameter space describing those parts of the multiverse that are hospitable to life and intelligent observers of any kind.

Physically, the Archipelago is a subset of the populated landscape of either string theory or any other overarching 'Theory of Everything' with multiple low-energy solutions ('vacua'). Thus, the Archipelago is part of the abstract space defined by whatever physics determines the structure of the multiverse. It is both logically and physically contingent on the reality of the multiverse; but that is still not saying much, since as Tegmark emphasizes, simple variants of the multiverse are completely uncontroversial and legitimate consequences of our firmly established cosmological theories. Whether the multiverse is infinite, as in the currently popular cosmological theory of eternal inflation, or finite, as in some construals of the string theory landscape, reflects directly on the structure of the Archipelago as well.

What is an island?[13] It is a set of parameters describing habitable universes that are close in parameter space; whether we can specify the meaning of 'closeness' beyond simple intuition depends crucially on the structure of the multiverse itself. In other words, the multiverse imposes a natural – or at least a convenient – metric upon the Archipelago. By definition, there is at least one island in the Archipelago – our home island. If the multiverse contains arbitrarily small variations in the constants of nature or cosmological parameters or even the mathematical shape of physical laws, then it is reasonable to conclude that the Archipelago is as dense as the remainder of the multiverse. For instance, it is clear that the change in the coupling constants of fundamental forces of one part in 10^{20} in comparison to those actually existing in our universe (not those actually measured, since we are unable to measure them with such precision yet!), will not change anything in the habitability status of our universe. If such minuscule variations are actually realized within the multiverse (for instance, through eternal inflation or through Smolin's self-reproducing universes), a universe otherwise described by identical laws and parameters to ours also belongs to our habitable region = our home island. It is easy to see that various constants and parameters of the multiverse play the role of geographical coordinates in maps of terrestrial archipelagos. If we start on a land point in, say, Sumatra, and continuously (or in sufficiently small steps) change either longitude or latitude along any chosen direction, we are bound to end up in the ocean. Similarly, if we start with a universe like ours and continuously change some parameter – say the strengths of forces, or the baryon-to-photon ratio, or the cosmological constant Λ – we shall inevitably end up in a universe lacking the pre-requisites for life and observers. However, in the same manner as it is possible to start in Sumatra and by a non-continuous increase in longitude end up on some other island, for example, Borneo or New Guinea, it is possible that after a large interval of non-habitability, our parameter again enters an interval which (with appropriate changes in other parameters) enables the existence of life and observers. This situation is shown in Figure 4.1. Thus, even if we did not know anything about possible different habitable universes, it would still be both rational and prudent to allow for the existence of other islands beyond our home island in the Archipelago.

The prediction of the existence of a small positive cosmological constant in the anthropic manner more than a decade before its actual observational discovery is often attributed to Steven Weinberg.[14] Irrespective of whether one regards this was a prediction of something that has been occasionally invoked for 70 years, it was an important result and a theoretical success that opened the way for a dramatic expansion of anthropic arguments in cosmology. It was also a first step towards cartography of the Archipelago. In the current cottage industry of multiverse predictions in theoretical physics (we shall see further examples below), a special

Figure 4.1. A 'cross-section' of a piece of the Archipelago of Habitability. Only a single parameter X is shown, instead of multiple 'fundamental' constants and cosmological parameters characterizing low-energy ('vacuum') states. Two possible peaks are shown and the observed value in our – obviously habitable – universe is denoted by X_0. Contrary to misleading criticisms of anthropic reasoning, X_0 need not correspond to maximum habitability (this particular region of the multiverse is symbolically labelled as 'paradise'). Some possible reasons why the multiverse is non-habitable outside of the Archipelago are also indicated, like the domains ('universes') where all matter is in the form of black holes, or those where there has been no structure formation at all. (Courtesy of S. Popović.)

role is reserved for the equation giving the probability $p(X)$ that some observer anywhere in the multiverse measures a feature X:[15]

$$p(X) = \frac{\sum_n \sigma_n(X)\, V_n\, \rho_n^{\text{obs}}}{\sum_n V_n\, \rho_n^{\text{obs}}}, \tag{4.1}$$

where the index n labels all possible vacuum states (all different low-energy '*physicses*' or different universes in the multiverse). In current versions of string theory there is a finite number of such states, although it is huge (10^{500} or so),[16] but in principle it could be infinite. The latter case poses some interesting problems in the theory of probability,[17] but in general, it will not preclude the usage of Equation (4.1) with appropriate weightings. V_n is the spacetime volume belonging to the universe n, ρ_n^{obs} is the density of observers in the same universe, and

$$\sigma_n = \begin{cases} 1, & \text{if universe } n \text{ has property } X, \\ 0, & \text{otherwise.} \end{cases} \tag{4.2}$$

In principle, V_n is calculable from our understanding of cosmological physics, although in weird enough universes it might be impossible to calculate in practice (or at least in the finite time available to human cosmologists!). It is also likely to be infinite in some or most of the universes, so an appropriate weighting procedure is certainly necessary. But, of course, the biggest uncertainty comes from the quantity ρ_n^{obs}, the density of intelligent observers. It is usually assumed to be proportional to the density of galaxies,

$$\rho_n^{\text{obs}} \propto \langle \rho_{\text{gal}} \rangle_n, \tag{4.3}$$

the latter being the main features of the structure of our own universe. There are two kinds of uncertainty linked with this assumption:

(i) We assume that the density of galaxies is a well-defined function of the constants of nature (including coupling constants of all fundamental interactions) and a number of cosmological parameters, usually thought of as the initial conditions of a particular universe. However, even if we fix the low-energy physics of a universe – as we do in standard astrophysical cosmology – we still fail to derive unequivocally the density of galaxies without additional assumptions dealing with the process of structure formation. In other words, our understanding of the astrophysics of galaxy formation is still insufficient to give anything like an accurate prediction of the number of galaxies per unit comoving volume. It is to be hoped that the increase in quantity and quality of simulations of cosmological structure formation will remedy this problem in the near future.[18]

(ii) There are 'internal' uncertainties as to the degree galaxies are regarded as preconditions for life. Even if we discard nit-picking qualms about what exactly can be regarded as a galaxy (some observed astrophysical objects, especially in the low-mass range keep defying simple classification schemes), we are still left with controversial issues such as: Which types of galaxies are conducive to the evolution of observers? How to treat possibly extremely non-uniform temporal distributions of habitable sites in different types of galaxies? Is it physically possible – and it is certainly conceivable – that a stellar population with habitable planets arises independently of galaxies? How do details of galactic dynamics on a large scale influence the formation of habitable planets in any given locale? All these questions are extremely difficult to tackle at present, since we know very little about the astrophysical preconditions for observers under the known laws of physics – not to mention the vast generalization that is required to answer these questions in the general multiverse context.

Obviously, more sophisticated research on the conditions for habitability, i.e., in astrobiology, is needed here; I shall return to this point in the next section.

What Weinberg did in 1987 was a very instructive example of a new kind of research result based on observation selection effects. His anthropic bound on Λ showed clearly that crucial explanatory tasks in cosmology could be tackled by embedding a particular case in a larger ensemble. For $\Lambda < 0$, the situation is rather clear, since those universes must end in a Big Crunch (irrespective of their global topology), and the obvious condition is that the lifetime of the universe is greater than the lifetime of those stars whose systems could harbour habitable planets. If the latter timescale is denoted t_*, and the recollapse is driven by the vacuum energy, this reduces to (in $c = 1$ units)

$$|\Lambda|^{-\frac{1}{2}} \geq t_*. \tag{4.4}$$

However, the case of $\Lambda > 0$ (vacuum energy manifested as a repulsive force) is much more interesting and – as it turned out 11 years later – realistic. The habitability condition here is that the introduction of the cosmological constant ought not to deform the power spectrum of density perturbations in such a way that it prevents the formation of galaxies. In order to calculate it to the first-order perturbation around a homogeneous and isotropic Friedmann universe, Weinberg used the approximate equation

$$\left(\frac{da}{dt}\right)^2 + \Delta k = \frac{8\pi G}{3} a^2 (\rho + \Delta\rho + \rho_\Lambda), \tag{4.5}$$

where a is the cosmological scale factor in the Friedmannian case, $\Delta k > 0$ is the positive curvature constant, ρ is the matter density and ρ_Λ is the vaccuum energy density. In terms of the cosmological density fraction, this cosmological constant energy density corresponds to

$$\Omega_\Lambda \equiv \frac{8\pi G}{3H_0^2} \rho_\Lambda. \tag{4.6}$$

Solving Equation (4.5) for the simplest case of a spherical density perturbation obtains the limit on the cosmological constant, which can be expressed as:

$$\rho_\Lambda < \frac{500}{729} \tilde{\rho}, \tag{4.7}$$

where $\tilde{\rho}$ is now the minimal matter density inside the perturbation that leads to the perturbation growth (and, presumably, to the galaxy formation out of the growing perturbation). The calculation of this quantity depends on the details of the perturbation growth – and the amount of dark matter, nowadays thought to be

Figure 4.2. A prototypical archipelago – Earthsea, a fictional realm originally created by Ursula K. LeGuin for her eponymous cycle of novels. (Courtesy of U. LeGuin.)

non-baryonic CDM – but the results available to Weinberg gave the upper limit as

$$\frac{\rho_\Lambda}{\rho} < 550. \tag{4.8}$$

This was not a very strict limit and Weinberg acknowledged that much.[19] However, it was extremely important for several reasons. First, it still very effectively limited the cosmological constant in comparison to the infinite interval of values it could a priori take. Second, Weinberg's result was obtained when the vast majority of physicists believed that there was a cancellation mechanism for driving ρ_Λ to exact zero. Finally, it provided a blueprint for further studies linking a broad distribution of a cosmological parameter or a fundamental constant to the issue of habitability of a universe.

A joint venture of fundamental physics and astrobiology

How to assess whether a particular set of physical laws and initial conditions is compatible with the existence of life and observers? If the assessment is positive, how to determine the key factor ρ_n^{obs}? The answer is given, in principle, by astrobiology.

Habitability determines the altitude profile of an island in the Archipelago: how high it will be, whether it will have steep or shallow slopes, great plains or even depressions inside it, and so on. We all agree that our universe is hospitable to at least some forms of life and there is very strong empirical evidence that significant changes in values of some or all parameters will lead to an inhospitable universe (the subject matter of classical anthropic fine-tuning arguments). Thus, the problem of the determination of ρ_n^{obs} is recast as a problem of surveying.

There are many ways to survey the Archipelago, just as in terrestrial geography one can either use satellite imaging, or search for water, or observe in the infrared, or prospect for minerals, and so on. It is impossible here to give more than a cartoon sketch of some of the directions currently being actively pursued (although not always, or even mainly, recognized as belonging to the astrobiological domain; but such is often the historical pathway of young interdisciplinary fields). Research dealing with the preconditions for life and observers as we know them in a physical, chemical or geological sense is part of the mapping endeavour. It tackles the central issue of contention between proponents and opponents of the use of the anthropic principle in string theory. As described by Susskind, the usual complaint (which is repeated verbatim by critics of astrobiology in general) is that anthropic arguments deal with only the kind of life we know about, with the implication that some other forms of life will require different chemical, physical and cosmological prerequisites. As correctly emphasized:[20]

> "How do we know that life can't exist in radically different environments?" . . . I think that correct response to this criticism is that there is a hidden assumption that is an integral part of the Anthropic Principle, namely: *the existence of life is extremely delicate and requires very exceptional conditions*. This is not something that I can prove. It is simply part of the hypothesis that gives the Anthropic Principle its explanatory power. Perhaps we should turn the argument upside down and say that the success of Weinberg's prediction supports the hypothesis that robust intelligent life requires galaxies, or at least stars and planets . . . The populated Landscape plays the same role for physics and cosmology as Darwinian evolution does for the life sciences.

So – we have it all here: although Susskind does not explain it, there is clearly a job for astrobiological research. It is exactly the task of astrobiology – and part and parcel of what I call its *extended mandate* – to expand our ability to recognize life even in its strangest forms (i.e., those most removed from terrestrial experience). But astrobiologists could aspire to even more: since the results of their research are sought by *both sides* in the debate on anthropic explanations. Contrary to the arbitrary assertions of some critics, the tradition of speculative thinking about alternative forms of life is as old as astrobiological thinking itself.[21]

Speculations – now supplemented by computer simulations – concerning whether life can be based on silicon, fluoride compounds, solid-state physics, nuclear forces, or even universal scalar fields belong here, and are likely to increase in number and quality in the near future. The prevailing tendency even in the most conservative astrobiological circles has been to acknowledge an *increase* in the robustness of life (e.g., in discoveries of various extremophiles), thus justifying the heightening of our expectations for the possibility of life elsewhere in the multiverse.

A particular form of surveying is embodied in research on *artificial life* (A-life). From a rather eccentric game of computer scientists (albeit one with august ancestry, with von Neumann), A-life has become an important mainstream research discipline with wide-ranging influence and an increasing number of applications within fields such as evolutionary computation, cellular automata, genetic algorithms and even evolutionary art.[22] It is often stated that the goal of A-life is to *recreate* biological systems in a digital environment, but there are some distinguished researchers, such as ecologist and computer scientist Thomas Ray, who claim that such experiments are actually *creating* a new kind of life.[23] Clearly, if this strong A-life thesis turns out to be correct, it will have important epistemological implications for cosmology as well, since numerical experiments of Ray's type would make the seashores of the Archipelago amenable to much more precise probing – something akin to the revolution of Gerardus Mercator and Abraham Ortelius in cartography in the 1560s.

Most pertinently to our goal, we can ask, with Tegmark, what are the minimal conditions for a physical environment containing intelligent observers? Tegmark lists three basic conditions, namely (1) complexity, (2) predictability, and (3) stability. (He imposes them on mathematical structures describing arbitrary physical laws, in accordance with his 'Platonist' position on the relationship between mathematical and physical worlds; but the use of these criteria does not really hinge on this philosophical viewpoint.)[24] The first necessary condition is almost self-evident: even the most ardent physicalists are unlikely to deny that an excessively high level of complexity is the reason why, among other things, we have not yet managed to create living or intelligent beings in the laboratory. The other two criteria are meaningful if we consider the operation of measurement by which knowledge is received by observers, which requires both (i) that the results do not fluctuate too much in the past or the future to preclude the use of inference from past experience, and (ii) the existence of habitable conditions at least during measurements. Of course, these are necessary, but not sufficient criteria for habitability. The search for sufficient criteria, and thus the outlines of the Archipelago, continues.

All this (and, arguably, much else) constitutes an 'extended mandate' of astrobiology – and one immediately answering all the conventional criticisms. Almost

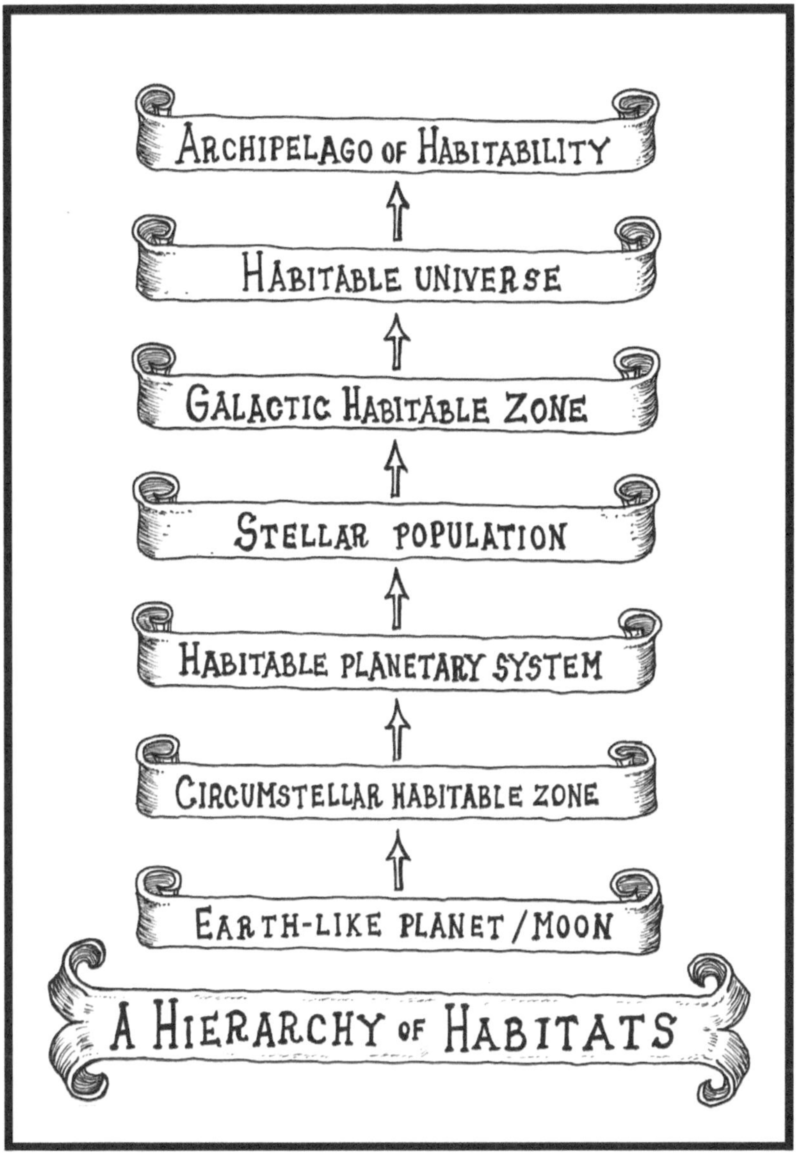

Figure 4.3. Discrete and hierarchical distribution of matter in the universe gives rise to the discrete hierarchy of habitats for life in the sense of astrobiology's extended mandate. Here, habitats are ordered by size with the smallest at the bottom and the largest on the top. The hierarchy shown here is 'conservative' in the sense that it is grounded in the concept of life similar to the terrestrial one and includes levels of organization of matter characteristic for our low-energy laws of physics. Astrobiology in an extended sense deals with the entire spectrum of habitats, from the bottom to the top. (Courtesy of S. Popović.)

by definition, it looks outward to the general, rather than inward to the particular; it easily avoids the charge of being 'by its nature too narrow-minded and too unimaginative to tackle the really big questions about aliens'.[25]

Geography and cartography would be unintelligible without an appropriate metric; in the terrestrial case, the units of length (such as kilometres and nautical miles) combined with the angular coordinates (such as longitude and latitude), provide us with a hugely successful description of any explored piece of landmass or ocean. The extension of our analogy to the cartography of the Archipelago of Habitability emphasizes the analogous role of changing various parameters of physics (including the very form of low-energy physical laws). Let us forget, for the moment, the issue of altitude of points on an island (i.e., the degree of habitability) and suppose that we are dealing with a single-contour two-dimensional map. Classical anthropic arguments from fine-tuning have amassed significant empirical evidence that we cannot change the parameters that describe our universe *very much*, at least as long as we keep other parameters constant.[26] For example, the initial low entropy of our Big Bang seems excessively fine-tuned,[27] as does the related amplitude of primordial density perturbations seen as anisotropies in the cosmic microwave background.[28] This would mean that our island is a rather small one, since even small changes in any single coordinate (with other coordinates kept constant) would lead us quickly into the ocean. In such a situation, it seems natural that early anthropic thinkers concluded that our island is the only one in existence. They were wrong, but not because of some ulterior and heinous agenda; it was perfectly reasonable for them to think so – we might compare their rationality to that of a hypothetical ancient philosopher of Easter Island, pondering the huge ocean surrounding his/her home.

But the contemporary explorers of the Archipelago have been able to refute this prejucide. In an often cited and almost as often poorly understood study, Anthony Aguirre found a 'cold' Big Bang model – a counterfactual one with low entropy-per-baryon in comparison to our actual Big Bang – which was habitable even according to the very high standards imposed by a *terrestrial* kind of life.[29] This type of universe is also capable of producing structure, such as galaxies and stars, as well as the chemical building blocks of life; and there is sufficient time for chemical and biological evolution in such an astrophysical environment, so drastically different in other respects. Some opponents of anthropic reasoning hailed this result as a long-awaited refutation of the alleged 'anthropocentrism' of such reasoning, whereas what Aguirre does, in fact, is to discover another habitable region in parameter space – another island of habitability!

While this is a seminal result, it is a far cry from being a 'refutation' or 'disproof' of anthropic reasoning. In fact, the supporters of anthropic reasoning have stronger reasons to cheer on Aguirre's bold exploratory voyage. It actually demonstrates that such exploration is meaningful, and that the question of the probabilities of observing various parameter values can, at least in principle, be answered quantitatively and operationally. The role of astrobiology sketched above finds some expression in the endnote of Aguirre's paper, which states:

> I have assumed that the number of independent measurements in a subuniverse is proportional to the number of suitable stars, [and] neglected such effects as, for example, high metallicity leading to more planets upon which multiple civilizations might arise in a single solar system. I have also neglected factors like extinction-causing impacts, radioactivity-induced plate tectonics, etc., which are possibly but not clearly necessary for the evolution of observers.

All these mentioned (and many other) factors clearly belong to the domain of astrobiological research; the reason why they are emphasized here by Aguirre is related to the Rare Earth hypothesis, which I shall consider in Chapter 6. However, the underlying lesson is clear: any elaboration of classical anthropic arguments, no matter whether supportive or undermining, straightforwardly necessitates the use of astrobiological knowledge. Moreover, Aguirre shows that we need such knowledge from *our universe* and our experience, even when dealing with other, disconnected parts of the Archipelago.

But Aguirre's is not the only island different from our home island discovered so far. In fact, the island discovered by Harnik, Kribs and Perez is both larger and more remote from our home universe: they investigated nothing less than a set of conceivable universes without weak interaction ('weakless' universes).[30] Before engaging in the analysis, they explain their methodology:

> We do not engage in discussion of the likelihood of doing simultaneous tunings of parameters nor the outcome of statistical ensembles of parameters. These questions are left up to the ultraviolet completion, such as the string landscape, which is outside of the scope of effective field theory. Instead, we are interested in 'running the universe forward' from a time after inflation and baryogenesis through billions of years of evolution. We will exploit the knowledge of our Universe as far as possible, adjusting standard model and cosmological parameters so that the relevant micro- and macro-physical outcomes match as closely as possible. We emphasize that this is really a practical matter, not one of principle, since *any significant relaxation of the 'follow our Universe' program would be faced with horrendously complicated calculations*. Put another way, there is probably a wide range of habitable universes with parameters and structures that look nothing like our Universe. For us, it is enough to find one habitable weakless universe about which we can make the most concrete statements, hence matching our Universe as closely as possible.
>
> *[emphasis by M.M.Ć.]*

This truly looks like a bold sailors' manifesto! The difficulties bound to be encountered by any stray navigator in the dangerous waters ('horrendously complicated calculations') are explained, and the independence of the basic approach to the form of the multiverse (i.e., string landscape or any other) explicated. This makes the results of the study still more impressive, since they find at least a single piece of rock above the water in the midst of the ocean – an apparently habitable universe missing so crucial an ingredient of our world as one of the four fundamental low-energy interactions. Whether this is just a rock protruding from the waters or the tip of a much larger island is unclear at present; Harnik *et al.* honestly admit that much further work could be fruitfully invested in researching this question.

Why, then, do these and other authors repeatedly perceive themselves as refuting or undermining anthropic arguments? The reasons are at least twofold. First, most anthropic arguments in the literature (like the one setting the electroweak scale, criticized by Harnik *et al.*) are narrowly or even badly formulated. They are based on narrow-minded calculations of the effects of small (one order of magnitude or so) changes in a small number of parameters, and not on a true insight into the underlying physics, which would indicate the way in which multiple parameters could be varied – at least in principle – whilst retaining habitability. Here is a clear parallel with the Rare Earth arguments in astrobiology of the form: 'If there were no Jupiter in the Solar System, the Earth would have been catastrophically bombarded by large comets and complex life would have been impossible.' Such arguments are deeply misguided, as will be elucidated in a later chapter. Instead of bad counterfactuals, we need a more refined and detailed *physical insight* into the evolution of planetary systems. The same applies to the habitability of universes in the string theory landscape or any other form of the multiverse.

The second reason why studies like those of Aguirre's or Harnik *et al.*'s are often misconstrued is that the resulting conclusions are based on a misconstrual of anthropic reasoning itself. The widespread perception of anthropic principle(s) as teleological naturally leads to the erroneous conclusion that habitability is, if not identical with the criteria for evolution of *Homo sapiens* (such an extremist position is relegated to the 'Intelligent' Design proponents), then it is at least *centred* on it. According to this way of thinking, it is enough to show that changing a parameter by a factor of a few or an order of magnitude is incompatible with some prerequisite for life as we know it (read: *Homo sapiens*) is to argue that this parameter must be anthropically selected. In the language of my central metaphor, if you can walk a hundred steps in a particular (single!) direction and get your feet wet, that is enough to prove that you live on an island. This logic is anthropocentric and flawed; it has nothing to do with the real meaning of anthropic reasoning. It may well be true that we are indeed living on an island (after all, even Eurasia can be regarded as a rather large island!), but the justification for it is wrong.

In a more general sense, studies such as that of Chavanis on white dwarf stars in an arbitrary number of dimensions, or Adams on stellar evolution with unconventional parameters, are part of the research helping us elucidate the Archipelago.[31] We do not find any direct connection between us and white dwarfs at present, but the elegance and simplicity of the problem of such physical objects in D dimensions gives us a feel for the approaches and ways of thinking necessary to sketch out the cartographical task. Getting as detailed a physical picture of other universes as possible is imperative on the way to determining their habitability status. Of course, interest in this sort of research has increased significantly in the last couple of years, so we might be on the verge of a true age of multiverse exploration!

How can the concept of the Archipelago of Habitability help us elucidate Equation (4.1)? Even very cursory thinking points towards the way, which I shall dub 'altitude weighting', indicated in Figure 4.1. Obviously, even islands of identical shape on the map can have vastly different habitabilities – the universes in two groups could have a vastly different density of observers. We have thus far discussed only surface coordinates of islands in the Archipelago, but their essential property (which makes them islands) is their altitude profile. Very high points (analogues of the volcano peaks of the Hawaii islands or *Islas Canarias*) represent sets of parameters describing extremely bio-friendly universes with many habitable planets and other life-supporting sites, in contrast to bare rocks protruding from the waters, where life forms are possible but unlikely or very rare. The same pertains to intelligent observers: regions of high altitude will contain a predominant fraction of the entire set of observers in the multiverse. These regions correspond to high values of ρ_n^{obs} in such regions of parameter space.

Thus, in order to make anthropic arguments cogent against the background of the Archipelago, we need to weight universes by their 'altitude', i.e., the measure of habitability. Clearly, this is a task for astrobiology in the widest sense. As discussed above, the very definition of the Archipelago is contingent upon resolving the issue of criteria for life to arise in a naturalistic manner. However, in order to obtain the altitude profile, we need much more effort – we need at least some hold on the emergence and dynamics of intelligent observers, traditionally the province of SETI studies (and the recreational discourse of SF books and movies). On the other hand, there is no discontinuity here; by studying already 'traditional' astrobiological themes, like the Galactic Habitable Zone, we expect to survey viable SETI targets and assess the density of extraterrestrial civilizations, even if some specific modifications are made when we consider intelligent observers rather than just complex life forms. Obviously, much further research needs to be done in that direction. The encouraging news, though, is that several approaches have been developed recently which encapsulate elements of the future whole picture, notably the counting of observer-moments discussed by Bostrom in philosophical terms,[32] and the counting of the number of possible observations in a particular universe devised by Starkman and Trotta in the physical domain.[33] There seems to be an epistemological similarity between the proposal of weighting by astrobiological description ('altitude' in terms of my central metaphor) and Trotta's suggestion that a 'fully Bayesian approach' could resolve the difficulties with anthropic arguments,[34] especially the lack of universal weight criterion for ρ_n^{obs}. In practice, this task is enormous; perhaps generations of astrobiologists will pass before we get close to quantitative precision, but the size of the task should not detract from the fact that it is a well-defined problem from the start.

This is a key point of this book, in a thematic, as well as a metaphoric sense. Through the concept of the Archipelago of Habitability, astrobiology gains its widest logically possible embedding, as well as its most general mandate. This expansion of our view in both physical and methodological spaces has, however, to be accompanied by a corresponding focus on particulars and developing new methods for studying 'traditional' astrobiological topics. In both directions, observation selection plays the crucial role. Before I discuss specific applications in the fields of evolutionary biology, planetary science, and SETI studies, I shall briefly, and in a relaxed format, address some refuted, but still repeated ad nauseam, criticisms of the anthropic reasoning.

On two views of anthropic reasoning: a dialogue

Simplicio: I admit to having all sorts of prejudices against the anthropic principle. Hasn't it been used to support Intelligent Design and similar quasi-scientific doctrines?

Salviati: Indeed it has been, in the same way fire has been used to kill people and books have been used to incite hatred. I would prefer to speak about *misuse* in such cases, my dear Simplicio. The other evening I watched a fine documentary about the Enron Corporation bankruptcy scandal, and I was flabbergasted to learn that the CEO of Enron claimed to be inspired in his management practices by a book by Richard Dawkins, *The Selfish Gene*.[35] Talk about bizarre misuse... Luckily, the times for such teleological nonsense concerning anthropic reasoning are bygones. Today, we know much better, on both philosophical and cosmological sides of the story.

Simplicio: How so?

Salviati: We've talked enough about the importance of the multiverse concept for present-day cosmology, so I'll just repeat that since the realization that generic inflation is of a 'chaotic' or 'eternal' type, and especially following Weinberg's successful prediction of a small positive cosmological constant on the basis of an anthropic argument, the share price of anthropic reasoning has soared. As to the philosophical side of the story, nothing can be preferable to reading the book of

Nick Bostrom, entitled *Anthropic Bias*,[36] a perfect antidote to the misconstrual of anthropic thinking as anthropocentric or 'cozy' or teleological. The line of 'argument' goes like this: anthropic principles point to peculiar properties required by the (part of the) universe we inhabit. Since these are of measure zero in the set of all possible properties (or values of physical and cosmological parameters), and since we obviously exist, our domain has been intentionally created (or fine-tuned) for the purpose of our existence. All sorts of teleological and theological inferences have been drawn from that point onward. Bostrom clearly and decisively demonstrates that this is not the case; the line of reasoning above is a misconstrual. The prevalence of the confusion and mispresentation of anthropic reasoning is attested to by the fact that some theologically-minded *and* some materialistically-minded philosophers and scientists have insisted on this modern version of the classical design argument. Yet, as Bostrom amply shows, anthropic reasoning may and should be interpreted not just disteleologically, but rather *anti-teleologically*. That is, instead of pointing out that nothing is really special about our cosmological domain and its physics, it emphasizes that any special feature we might observe is an illusion, a consequence of our restricted viewpoint. Thanks to this important clarification, we are in a position to develop deeper Bayesian theory accounting for any – at least in principle – observational bias in any observation we make. In this manner, anthropic reasoning embodies Copernicanism.

Simplicio: What? Are you now saying that we need to use anthropic reasoning to explain *everything*?

Salviati: No, although strictly speaking there is a grain of truth there as well. First, we need to establish whether an explanation is required at all. Since I happen to know you are a football fan, I'll remind you of the Southampton win over Manchester United we discussed the other day. In fields as complex as football (and lesser social activities), it is often a tough call to judge to what extent an observation requires explanation. But then, all observations known to us are made by human observers, so the compatibility of their outcomes with cosmological, physical, chemical, geological, etc. preconditions for observership must be satisfied (in this case human, but in general some other indexical information fills the slot). These constitute a sort of filter.[37] *Any* observation whatsoever in the world should be put through the filter, but it won't make any difference in most cases. Since most observations are indifferent towards the number of observers, they will remain the same after the procedure. However, a small subset of observations, mainly in the domains of fundamental physics, cosmology, evolutionary biology and astrobiology are logically connected to the number of observers (in the appropriate reference class). The observation that 'there are N oaks in forest X' is indifferent towards the number of observers; the observation that 'there have been M global catastrophes causing mass extinctions of species in the last T years' is not. The underlying lesson is clear: to approach problems dealing implicitly or explicitly with the properties of observers (e.g., problems in quantum physics or cosmology or astrobiology) in the same manner as, for instance, classical physicists studied the motion of a pendulum or the planet Uranus is not just epistemologically naive – it is plain wrong.

Simplicio: Aren't the shores of the Archipelago of Habitability very poorly defined even for our island? According to many authors, classical fine-tuning arguments are often imprecise, meaning that they are not, well, that fine-tuned . . . [38]

Salviati: True enough, but it is not obvious what one should make of it. To me, it sounds much like a 'Whiggish' charge that Christopher Columbus, Pedro Álvares Cabral or Fernando de Magallanes did not use precise maps; their errors in navigation were often an order of magnitude or more larger than what we in the GPS age know about the *correct* geography. Are their discoveries any the less appreciable for that undeniable fact? Do we mock them – or, in contrast, celebrate their achievements, while emphasizing the simple means they had at their disposal? Understanding that we are at the very beginning – an overstatement, if anything – of the exploration of the Archipelago should lead us naturally to expect large errors in our positioning of its shores – how could it be otherwise?

Simplicio: OK, let me try to understand a puzzling issue in which anthropic reasoning gives an outrageous prediction. That is the infamous problem of Boltzmann brains. You will remember that it was explicitly posed in the 'disturbing' study of Dyson, Kleban and Susskind[39] as a consequence of the New Standard Cosmology dominated by a positive cosmological constant. Such a cosmology leads asymptotically to a future eternal de Sitter spacetime. In the eschatological limit of future eternity, it is equivalent to a thermal state with an extremely low characteristic temperature of $\sim 10^{-29}$ K (roughly corresponding to thermal radiation with a wavelength of the size of our cosmological event horizon). At first glance, such a low temperature coupled with the extreme dilution of matter at such late epochs is unlikely to yield anything interesting, but in fact, like any thermal system, such a late vacuum will fluctuate at a non-zero rate per unit spacetime volume. And, out of such fluctuations, across large eschatological time spans, complicated systems, e.g., intelligent observers, can arise, again at a non-zero rate. Since de Sitter spacetime is eternal, such 'freak observers' will be infinite in number (in fact, their production rate will exponentially increase with time) and will thus completely overwhelm 'normal' observers, like us, who emerged through physical, chemical and biological evolution. Why, then, per anthropic reasoning, are *we* not freak observers?

Salviati: A few days ago we discussed a nice little problem formulated by Ken Olum, dealing with the puzzle that we did not find ourselves a member of a large civilization, containing zillions of observers. The Boltzmann brains problem is a close relative, perhaps a bit more unpleasant, but not really the type one can't get rid of; if you look in the literature, you'll see that the threat has been taken seriously. In particular, Page argues that the simplest answer to the puzzle is that the future is finite and that our universe will vanish before Boltzmann brains start to dominate the total observers' census.[40] On a more optimistic note (at least from the selfish point of view of evolved observers), Carlip shows that the time variation of fundamental constants can lead to a small measure of freak observers,[41] and Vilenkin demonstrates that the same effect is achieved by the bubble nucleation in a de Sitter vacuum under some fairly reasonable assumptions.[42] From the present point of view, the problem of freak observers is easily generalized to a multiverse, since de Sitter spacetime is usually regarded as the source of an inflationary self-reproducing fractal. A rather uncontroversial aspect of anthropic reasoning is that we need to suppress the measure of freak observers in the entire Archipelago in order to explain our evolution and our ordered observations (and, clearly, Page's recipe will not work for a multitude of universes). Although this problem will require further work, the good news is that there are ways out, such as those shown by studies of Carlip and Vilenkin. As the titan of science after whom those freaks were named put it in another context: 'It may be objected that the above is nothing more than a series of imperfectly proved hypotheses. But granting its improbability, it suffices that this explanation is not impossible. For then I have shown that the problem is not insoluble, and nature will have found a better solution than mine.'[43]

Of course, in a sufficiently large universe (not to mention the multiverse), there are likely to be Boltzmann brains of all kinds, including the deluded ones thinking that they are you or me having exactly this conversation. While this may raise some interesting issues in ontology and the theory of identity, I fail to see why anyone would consider it paradoxical or detrimental for anthropic reasoning. We have already talked about how in the classic Friedmann open and flat universes (without a cosmological constant or any other form of dark energy) there are infinitely many galaxies, which contain infinitely many copies of you and me.[44] If the existence of doubles doesn't disturb us too much – and nobody has argued against relativistic cosmology on the basis of *that*! – why would a phenomenologically weirder, but essentially similar situation with Boltzmann brains cause us any sleepless night?

Simplicio: But people are stubborn in their view that the anthropic explanation represents a defeat of science, or a capitulation, etc. Thus David Gross, one of the best minds of today's physics, likes to quote Churchill and his 'Never give up' speech and Michael Turner mentions the anthropic possibility with a 'God forbid' disqualifier. Even many supporters are accepting anthropics as the last resort and admit being unhappy with it. Isn't that a sign of the bankruptcy of an idea when even the supporters are so demoralized?

Salviati: This has to do more with human psychology and the history of Western science than with nature itself. If anthropic principle is cleansed from all teleological sediment, it is not just luxury to apply it when we run out of other explanations. It is our *duty* to apply it first, and subsequently explain what is left after observation selection is filtered out. There is nothing to give up whatsoever. The problem is that some hoary, tired recipes as to what should be legitimate science have been handed down from the time of Francis Bacon on, and practical people in the sciences seldom question these. These dusty prescriptions contain a dangerous presumption that it is always possible – and desirable – to find an 'Archimedean point', i.e., 'investigate phenomena with an observant but empty mind'.[45] Roughly, this Baconian view alleges that we should rationally explore nature *only* as seen from that completely neutral and immobile standpoint – not as real, physical beings evolved through various processes of physical, chemical and biological evolution, but as disembodied abstract observers. Only the latter viewpoint, suggest these researchers, yields true causal explanations and satisfies the overall goals of science. Anthropic reasoning does not conform to this model and, consequently, 'has no place in physics or cosmology'.[46] Today's followers invoke all sorts of heavy rhetorical artillery ('betrayal of the Enlightenment', 'a defeat of scientific spirit', etc.) just for the lowly expedient of avoiding accounting for themselves – in the literal sense! To me, this anti-anthropic attitude looks incredibly arrogant and hubristic. Are observers natural? It is somewhat surprising to see that the same critics of anthropic reasoning tend to believe in physicalist explanations of mental phenomena, including observership itself. This makes the criticism of anthropic reasoning on the basis of perceived lack of causal explanation a case study in hypocrisy. If observers are part of the nature to be studied, how can one justify ignoring their properties? Should they not be regarded as part of the wholeness of reality that is to be rationally and empirically explained?

This is the way to understand the frustration felt by Gross, Smolin and some other scientists mired too much in the Baconian – and possibly some crypto-positivist – views of the nature of scientific endeavour. Those views, which were in place for studying the preservation of chicken meat by cold, are certainly inappropriate for studying processes influencing the number of observers in a universe. (Come to think of that, is it just a coincidence that Bacon himself was sceptical of the then ongoing Copernican revolution?) At least they did not stoop to bizarre charges that 'anthropic principle' is a misnomer, in contrast to some philosophers.[47]

Simplicio: What happens if physicists somehow prove that there is only one universe?

Salviati: Well, what happens if they discover that the world indeed originated in Æsir finding places for the Sun, the Moon, and the stars, as told by the Old Norse *Völuspá*? Or that it is turtles all the way down, as Stephen Hawking has recently reminded us? There are a great many funny questions one might ask; as Quine and other modern philosophers of science taught us, any hypothesis barring impossibility or logical incoherence can be 'salvaged' through a series of skilful reformulation manoeuvres. But it is not incumbent upon us to prove the untruth of low-probability hypotheses. We are searching, which is too often overlooked, for the best explanation of *specific* data, in a *specific* moment in history, in *specific* circumstances. Thus, one who, as I do, considers the probability of the multiverse to be very high, could answer your question in the spirit of the old Greek saying, *hic Rhodes, hic salta!* In other words, you should first prove the impossibility of the multiverse, and then we shall discuss the issue.

If you still feel dissatisfied, I have yet another answer for you, which states that the domain of anthropic reasoning would be severely restricted, but it would still be as valid as usual within that restricted domain. In this regard, it puzzles me as well that even defenders of anthropic reasoning are so timid and defensive. All applications within a single universe would still be completely legitimate, starting with Dicke's rejoinder to Dirac's 'large-number hypothesis', up to Olum's problem we talked about earlier, applications in probability theory, ethics, and the 'anthropic shadow' in risk analysis.[48] In particular, applications in astrobiology as it is conventionally understood (i.e., study of life forms sharing the same set of low-energy laws of physics with us, or even more conservatively, sharing the same biochemical basis) will be unaffected. We will still be fully entitled to ask questions such as 'Does the Rapid Appearance of Life on Earth Suggest that Life Is Common in the Universe?' or 'What does the life span of the terrestrial biosphere tell us about the inherent probability of evolution of intelligence?' or even, with Fontenelle, 'Why are the observed orbits of comets highly inclined?'[49] Observation selection effects simply work whenever we deal with processes impacting on the number of observers or the processes historically leading to the emergence of observers. This remains true irrespective of our position on the alleged 'mystique' of the multiverse.

Simplicio: That does sound reasonable. But still, how do you explain the fact that so many people in sciences and philosophy developed an allergic response to the whole field of anthropic reasoning?

Salviati: One thing, my friend, has been well known to historians of science for a long time: scientists and philosophers are not really the rational beings they usually fancy themselves to be. Lacking deeper (evolutionary psychological?) explanation of *that*, you would do better to accept it as a *brute fact*.

5

Astrobiology as a natural extension of Darwinism

> Is evolution a theory, a system, or a hypothesis? It is much more: it is a general condition to which all theories, all hypotheses, all systems must bow and which they must satisfy henceforward if they are to be thinkable and true. Evolution is a light illuminating all facts, a curve that all lines must follow.
>
> *Teilhard de Chardin*

Our picture has been zoomed out to its logical maximum, encompassing the greatest conceivable strangeness of the Archipelago of Habitability. Its very scale, coupled with the abstract nature of the parameter space, may cause some depressing thoughts. Copernicanism is, as noted many times in the history of ideas, from Cardinal Bellarmine to Freud to Dennett, a depressing doctrine; huge scales and the sheer severity of observation selection just strengthen that feeling. One may speculate that there indeed exists an evolutionary reason for such gut feelings – it might not be advantageous to accept that we are infinitesimally small specks within a huge hierarchy of cosmological 'wholes'. This would help explain why our oldest, most primal cultural entities, mythologies and religions, are with a few exceptions so unashamedly anthropocentric. That feeling, however, should not persist. As I have attempted to show thus far, we have strong reasons for optimistic belief in the capacity and power of our best theories of present and near future to outline the bizarre shores of the Archipelago. To do that, we need to zoom in our view again, very hard, in order to capture at least a glimpse of 'astrobiological dynamics', the common foundations underlying each particular instance of the appearance of life and observers. From this point on, I shall focus on pieces of this overarching construction we can reasonably hope to discern. Both by virtue of our provisional definitions (see Chapter 1) and by virtue of Occam's razor, we expect the known Darwinian evolution of the terrestrial biosphere to be a major supporting column of this edifice.

Galactic Darwinism?

The end of one journey often hides the starting point for many others. Consider again Darwin's famous closing words in *On the Origin of Species*, quoted on p. 78 above. While the 'entangled bank' picture presents a powerful metaphor of the astrobiological landscape, there are some further gems to be found in these words, which have been cited, revered, celebrated and subjected to all sorts of exegesis during the last century and a half. Therefore, I shall briefly mention two points, which may be significant for the future task of achieving a fuller understanding of the philosophical underpinnings of astrobiology.

1. *The 'Entangled bank'* vs. *'this planet'*. Clearly, Darwin performed a skillful rhetorical manoeuvre by choosing a particularly rich and diverse ecosystem as his final example. Such diversity is likely to make a strong impression on any observer. By embedding it firmly in the global context of our planet, he underlined the search for a general – and rather simple – explanation of very complex particulars. Whatever the 'true' origin of biological forms is, it must be able to account for both simple and excessively complex systems. The scenery of the 'entangled bank' is evocative and powerful for its complex and colourful ecology; in principle, Darwin could have chosen a desert shore or an Alpine crag ecosystem as an equally valid scientific example, but the image would have lost some of its power. On the other hand, both the entangled bank and all the other ecosystems are small parts of 'this planet'; by contrasting them, Darwin also achieves another purpose, namely to stress the *effective* nature of our explanatory statements, regardless whether we call them laws or not.
2. *Truly universal laws*. The parallel between Newtonian gravity and natural selection (and whichever other evolutionary mechanisms are at work, since Darwin was, like the majority of biologists nowadays, a pluralist) has often been discussed. How universal the laws of both physics and biology were in Darwin's day and how universal they are when considered today is quite another story. In Darwin's day, not much was known about the architecture of the universe outside of our Solar System; as I have discussed in Chapter 2, almost half a century later Wallace and Kapteyn seriously entertained world models where the Earth was placed near the centre of the known universe. Newtonian mechanics was at its zenith, confirmed in its majesty by the relatively recent and spectacular discovery of a new planet, Neptune, its uncannily precise position for the pre-computer age predicted by the French astronomer, Urbain Jean Le Verrier in 1846. Thus, the regularity of Earth's orbital motion mentioned by Darwin in this passage presented a truly high standard of comparison in 1859.

Today, the rise of advanced computational techniques has enabled easy prediction and retrodiction of events in celestial mechanics, including even tiny relativistic corrections. For instance, we know (i.e., can retrodict) that there indeed was a total solar eclipse on 14 January, AD 484, observed by one of the last classical philosophers, Proclus of Athens. Is there a better proof that celestial mechanics, based on Newtonian laws (with relativistic corrections where necessary), gives an excellent explanation of such perennially puzzling phenomena? But wait! As explained by Charles Misner, Kip Thorne and John A. Wheeler in their monumental *Gravitation*, the eclipse could *not* be seen from Athens without plausible assumptions about the change in the rate of Earth's rotation – the totality strip being quite different and passing far from Proclus' home.[1] In other words, it has eminently *not* been enough to crank up our planetarium and turn it back to the fall of the Western Roman Empire. But the change of Earth's rotation due mainly to tidal slowdown because of the influence of the Moon encompasses much more than simple Newtonian laws – we need additional assumptions about the mass distribution of the Earth and the Moon, the influences of other planets, the influences of impacts and plate tectonics, etc. – in fact, so many more assumptions that one is justified in regarding it as effectively a 'historical' change.

Therefore, even laws are not really what they used to be. There has been much discussion on the topic of biological laws in the literature on the evolution, history and philosophy of biology. There is almost universal agreement that the classical 'laws' suggested from time to time in evolutionary biology are not real laws in the sense adopted in the physical sciences, but trends with many exceptions, or not even that (e.g., 'Cope's Rule', 'Dollo's Law', Haeckel's 'law' of ontogenetic recapitulation). Yet, there is still much discussion whether there are other laws of evolution, and, for example, to what extent the operation of natural selection is a law in itself. This has been coupled occasionally with more general and involved issues, such as whether scientific laws generally need to be empirical or not, or whether talk about laws is epistemologically useful at all.[2]

Perhaps the most illustrious participant in these discussions was Sir Karl Popper, who famously argued – at least for most of his long career – for the non-existence of laws in historical sciences, including evolutionary biology.[3] To support this conjecture he invoked an explicitly astrobiological scenario, namely, the possible diversity of life (or the lack of it) on Mars:

> For assume that we find life on Mars consisting of exactly three species of bacteria with a genetic outfit similar to that of terrestrial species. Is Darwinism refuted? By no means. We shall say that these three species were the only forms among the many mutants which were sufficiently well adjusted to survive. And we shall say the same if there is only one species (or none). Thus, Darwinism does not really *predict* the evolution of variety.[4]

Alfred Russel Wallace would have been satisfied to read these lines, though he might not condone Popper's conclusion that Darwinism does not offer a real explanation of diversity. Interestingly enough, in an early rejoinder, the philosopher of biology Michael Ruse argued that empirical considerations make Popper's scenario unfair to evolutionists.[5] Namely, Ruse claimed that we know with sufficient certainty that Mars is at best hostile to life and that this presupposes the low diversity of hypothetical life forms. In other words, Ruse seems to claim, the external physical constraints on the Martian environment (low temperature and pressure, lack of liquid water, high incident flux of harmful ionizing radiation, etc) preclude the application of at least some of the law-like machinery of evolution which would, *ceteris paribus*, lead to high biodiversity. On the other hand, when Popper was writing, the issue of life on Mars had already had a long and complex history and the example might have been chosen for its rhetorical impact as well; the issue had become part of scientific debate mainly through the gradual discrediting of fanciful ideas about a highly developed technological civilization on the red planet . It seems clear that Popper's passage would sound less persuasive if the thought experiment pertained to, say, the fourth planet of ζ Reticuli, which would be epistemically as valid.[6] Of course, whether the *ceteris paribus* clause implicit in Ruse's rejoinder could be sustained is an entirely different issue; more on this in the next chapter. In near future, we will be facing the exciting likelihood of discovering empirically whether Mars is or was inhabited by microorganisms of native or terrestrial origin.[7] Whatever the outcome, even for the participants in that early debate – from the point of view of philosophy of biology – it was rather clear that astrobiology has much to say about evolution.

Occasionally, the talk about universal biological laws is substituted by talk about universal biological concepts, which are, supposedly, present in any situation involving living beings. The examples commonly cited are mutation, adaptation, speciation, inter- and intraspecific competition, sexual reproduction, predation, parasitism, symbiosis and others. Elucidation of all these concepts, together with an explanation of their mutual interaction would, according to this view, lead to a theory expressible in nomological ('law-like') statements, which will be truly universal, valid on Earth, or on Mars (if inhabited now or in the past) or the fourth planet of ζ Reticuli . They will be equally so – though it seems more challenging – for hypothetical nuclear-force-based inhabitants of a neutron star, or a population of Hoylean 'black clouds', or other even stranger, yet unconceived possibilities like the ones constituting the outlying parts of the Archipelago of Habitability. Having general laws governing the evolution of life is an old idea – but it is hard to see how it could be completely meaningful without exploring other biospheres, i.e., without some help from astrobiology.[8]

While the existence of laws is controversial, it is an underlying assumption conventionally used in evolutionary biology and many derivative disciplines, like sociobiology or evolutionary medicine, that the *processes* of Darwinian evolution – be they monist or pluralist – are truly universal. Even in treatments explicitly disinterested in life beyond Earth, such as Shanahan's *The Evolution of Darwinism*, it is claimed that the operation of natural selection (and, one may safely assume, other evolutionary mechanisms like genetic drift) should be supposed to operate at any place and any time, including places outside Earth.[9] This is one point on which there is agreement between ultra-adaptationists like Richard Dawkins and pluralists, with regard to evolutionary mechanisms, such as Stephen Jay Gould.[10]

There is a cultural predilection for this universalism as well. We have seen that H. G. Wells has implicitly regarded a Martian invasion as a consequence of 'survival of the fittest' (although this misconstrual of Darwinism is due to the influence of Herbert Spencer and contemporary social Darwinism, rather than Darwin and Wells' teacher Huxley). Stanislaw Lem offered one of the most carefully constructed scenarios of this universalism in *The Invincible*, a short 1964 novel about the eponymous spaceship discovering a strange non-biochemical ecosystem on a distant planet in the constellation Lyra.[11]

Regis III: A Mars-sized planet where native life occurs in oceans, but not on the land any more, although it once did. The reason for the dramatic change is that several million years ago, some machines capable of self-reproduction and modification have arrived on it by accident. A machine evolution ensued, in which for different machine populations there were, apart from abiotic pressures (long-term cooling of the planet), two types of biotic pressure: one from the native lifeforms and another from its fellow von Neumann automata. The outcome – when human observers arrive and try to explain the situation – is the predominant population of small insect-like machines capable of aggregating in large clouds if any challenges appear in the environment. These 'black clouds' have all but eradicated any other form of land life, either of native biological or machine origin. In particular, huge, complex machines capable of sophisticated planning and acting have some time before lost the battle for scarce resources (mainly energy) and their 'fossil' remnants are mistaken by humans for long-abandoned ancient cities. Still, apart from a fantastic degree of versatility, the dominant machine-insectoid forms are utterly lacking in intelligence, or indeed *any other complex features*. Human intruders in this strange ecosystem have had a hard time, as expected, understanding and accepting that the 'pinnacle of evolution' on a planet might be so weirdly different from the terrestrial case.

Keep in mind that Lem's novel predates Popper's published thoughts on biological evolution by about a decade (and NASA's definition of life by two decades more!). It is no wonder that this scenario is so detailed, since Lem was a deep scholar of evolution, as testified, inter alia, by his *Cyberiad* stories, as well as his monumental *Summa Technologiae*, a truly brilliant non-fiction book which still strangely refuses to age.[12] Observe that overall, the evolutionary trend on Regis III is *passive* (as in Gould's 'drunkard walk' model of the increase of biological complexity[13]), and in fact, it leads to the *decrease* in diversity. However, the explanation given by Lem's protagonist is not only perfectly compatible with Darwinian evolution; it is an *affirmation* of Darwinism. It is a biologist, Lauda, explicitly described as very old (i.e., presumed conservative), who first among the human crew of *The Invincible* comes up with the best evolutionary explanation of the puzzles they had previously encountered on Regis III. Lem toys with the developmental constraints (*der Bauplan*) in evolution of his fictional biosphere as well; when human scientists perform detailed analysis of collected samples, it turns out that each insectoid element of 'clouds' is capable of linking with exactly three others via extended 'Y' connections and one other via a central port. This is not a frivolous detail, since it limits the quantity and functional role of the information stored in any individual element. While it could be adaptationally advantageous to be immediately connected with 400 other elements instead of four, it is clearly prohibited by the *constraints of form.* Lem's clear message is that evolutionary theory is capable of giving a truly universal explanation; it is only our parochialism and anthropocentrism that undermines our trust in the validity of such explanations, even when they sound wildly speculative. As in the multiverse case, the problem is not that we take our best theories too seriously – but that we do not take them seriously enough. The flow goes in the other direction as well: it is the degree of strangeness of *life as we do not know it* (to quote the apt title of a recent popular book on astrobiology[14]), which offers both the motivation and the ultimate test of mettle for the Darwinian explanation. The central irony of Lem's book – reflected, as some of its critics readily pointed out, in the very bitterness of its title – is that time and again, in a highly advanced technological future, as well as in the time of T. H. Huxley and bishop Wilberforce, such bold explanatory hypotheses face not only the usual epistemological sieve, but deep emotional resistance as well.

Replaying the replay

As we have seen in preceding chapters, cosmology traditionally blurs the difference between dynamical laws and boundary conditions. However, something similar is at work in evolutionary (paleo)biology due to its contingent nature and dependence on unknowable (at least in practical terms) details of initial and boundary conditions. In both cases, the blurring follows the ambiguity whether we study unique cases – one universe, one biosphere – or just a selected sample from a wider whole. In the biological case, 'initial conditions' apply not only to the process of biogenesis and prebiotic chemical evolution, but also to speciation events leading to the origins of individual species and taxa. 'Boundaries' could be construed as changes in the physical environment, which are also effectively unknowable on sufficiently long timescales. As an example, while the pattern of ice ages, obviously very important events for any biota, is reasonably well known in the late Tertiary and Quaternary periods (whence our understanding of Milanković's cycles comes), our knowledge of ice ages in the Carboniferous is quite speculative. This, in turn, makes our understanding of the fitness landscape in the relevant part of Carboniferous fauna incomplete at best. This is one of the perennial problems in the study of complex nonlinear systems: its effective unpredictability, originating in both sensitivity to small perturbations (in the beginning or in the subsequent course of dynamical evolution), and finite precision with which we, as imperfect observers, can determine any kind of boundary conditions.

So, even if biological evolution were predictable *in principle*, and in well-defined simplified situations where both biotic and abiotic pressures were properly understood, it still does not follow that it would have been predictable in practice. The weather is, after all, also predictable in principle and in well-defined simplified conditions, but we all know the quality of weather forecasts drops very rapidly with time in generic 'real' situations. Moreover, the drop in predictability is eminently not a consequence of introducing any new mechanisms, neither in the meteorological nor in the evolutionary case. It is not just that any specific product of biological evolution is context sensitive, as is sometimes erroneously used as an argument for the absence of laws in biology.[15] All the specific products of the operation of physical and chemical etc. laws are context sensitive as well. To explain our observation that water boils at roughly 373 K, we need to take into account not only parameters determining the very general phase diagram of water (ultimately the causal outcome of quantum statistical mechanics applied to molecules of H_2O), but also the context, such as the altitude at which our observation was performed, the concentration of various compounds dissolved in water, etc. The problem is not context sensitivity per se, but sensitivity to the unknown context (in practical terms, often *unknowable*).

Remember the problem of the chirality of terrestrial life sketched in Chapter 1 as an illustration of an astrobiologist's 'working hours'. There necessarily was an initial asymmetry, possibly in entirely and indisputably non-living systems – cosmic radiation in the Galaxy, or chiral clay surfaces, or something such – which interacted with prebiotic biochemistry. However, the specific sign of such asymmetry has to be 'put by hand' into our theory explaining chirality – it is contingently unknowable from a deeper theory (say the one explaining the origin, polarization and transport of interstellar radiation in the Milky Way at a particular epoch, even if such a deeper theory were available). Maybe a particular burst of UV radiation from a nearby supernova or magnetic neutron star ('magnetar') has been polarized in such a way that, when absorbed by key prebiotic molecules, decisively skewed the distribution of stereoisomers – but we can hardly hope to ascertain that after billions of years.[16] In fact, it is quite probable that we shall *never* be able to retrodict such astrophysical phenomena precisely, in much the same way that we shall never, completely regardless of our exponentially increasing computing (and, hence, modelling) capacities, be able to predict weather with arbitrarily small uncertainty.

If all these constraints apply to the terrestrial biosphere, whose physical, chemical and geological environment is rather well known, they will a fortiori apply with full force to any extraterrestrial biosphere. Consider the following scenario:

TwinEarth: astrobiologists discover life with the same biochemical basis (proteins and nucleic acids, water as solvent, etc.), emerging on an Earthlike habitable planet orbiting a Sun-like Main Sequence star, with relevant timescales – of 'crucial steps' – similar to the terrestrial case. The distance to the planet is large enough so that we can be sure that no interstellar panspermia has occurred between the Earth and its twin.

TwinEarth is arguably the most conservative scenario of 'grand breakthrough' one might expect (or hope for) in astrobiology; without discussing whether it is probable or not – it would imply a strong form of convergence in space of possible ways of biogenesis, among other things – we could ask the question of how predictable would the unfolding of biological processes on TwinEarth be. I shall discuss the continuity thesis – the idea that biological evolution is a continuous extension of physical and chemical evolutions of matter – in a subsequent chapter, but we can assume it in rough form here and suppose that biogenesis and subsequent complexification, including the 'Cambrian explosion analogue', proceeded on TwinEarth similarly to those on Earth. The ecological details are, of course, quite different on Earth and TwinEarth. What could we say in this thought experiment about the outcome of the evolutionary processes on TwinEarth approximately 3.8 Gyr after biogenesis?

In essence, TwinEarth is the same as Gould's thought experiment of 'rewinding the tape of life'. In his bestselling *Wonderful Life*, Stephen Jay Gould has strongly advocated a view of biological evolution as a primarily contingent affair, with innumerable branchings and crossroads where it could have taken a different turn. Using the Cambrian fossils of Burgess Shale as the paramount example, Gould suggests that we imagine travelling back in time to the mid-Cambrian and watch the 'replay' of evolution, trying to determine which representatives of the wondrously diverse Burgess Shale fauna will survive, flourish and ultimately be honoured as ancestors of the present-day dominant taxa:[17]

I call this experiment "replaying the life's tape." You press the rewind button and, making sure you thoroughly erase everything that actually happened, go back to any time and place in the past – say, to the seas of the Burgess Shale. Then let the tape run again and see if the repetition looks at all like the original. If each replay strongly resembles life's actual pathway, then we must conclude that what really happened pretty much had to occur. But suppose that the experimental versions all yield sensible results strikingly different from the actual history of life? What could we then say about the predictability of self-conscious intelligence? or of mammals? or of vertebrates? or of life on land? or simply of multicellular persistence for 600 million difficult years?

If you find the last rhetorical questions too extravagant, compare Gould's with the Regis III thought experiment above. Clearly, entirely Darwinian evolution can bring about the simplification of morphologies, not only their increase in complexity. As Gould argued in *The Structure of Evolutionary Theory*, this particular sort of contingency is rooted in the underlying functionalism of evolutionary mechanics, and thus is, in principle, applicable not only to natural selection, but also to Lamarckian inheritance, horizontal gene transfer, etc.[18]

This 'rewinding of the tape of life' *Gedankenexperiment* has provoked a lot of thinking and debates in recent years, with arguments focusing on where exactly the origin of contingency lies (initial conditions or subsequent 'external' or 'internal' fluctuations), as well as its relative significance.[19] It was even claimed, notably by John Beatty, that this has become at least partially an empirical issue, on either a micro- or macroevolutionary level, although it has been immediately disputed. My intention here is not, of course, to try to settle this difficult dispute, but to point out something that should have been obvious in the first place, but which still lies mostly 'below the radar' amongst life scientists: that both main contending issues – the origin and relative significance of contingency – are strongly related to ongoing astrobiological research and theoretical, as well as empirical, puzzles the advances in both astrobiology and numerical simulations are posing.

The different re-runnings of the tape are indistinguishable from an imagined ensemble of TwinEarths; of course, for each TwinEarth we could imagine much higher physical and ecological diversity. For example, TwinEarth could have a much smaller or much larger landmass, significantly affecting the degree of biogeographic isolation, and henceforth speciation and biotic competition. However, both proponents and opponents of Gould's view on contingency will hardly invoke that kind of difference as a crucial argument. Instead, the discussion has historically focused on (i) the sensitivity to initial conditions, (ii) mutational ordering, which can be construed as the history of external perturbations, and (iii) macroevolutionary punctuations, namely radiations similar to the Cambrian explosion and mass extinctions like the end-Permian or end-Cretaceous catastrophes. The sensitivity to initial conditions (i) is a generic characteristic of complex dynamical systems; often going under the name of 'chaos', it is hard to see how one could avoid it, except for the definitional fiat that the thought experiment must run from exactly the same initial conditions. Of course, the price to be paid for this is the decreased domain of applicability of the *gedanken* – and the decrease can go to zero applicability. But it is still interesting enough to provoke a lively discussion, even with this unrealistic constraint. The discussion centred on (ii), which is a form of perturbation constantly acting on any living system from the external world. While mutations can be considered random in the subjective (epistemical) and, perhaps, objective sense as well – since at least some of them are caused by objectively stochastic quantum-level events – the underlying source of contingency is the complexity of the genome on which the mutations act. In Beatty's example for mutational ordering, the key assumption necessary for the pairs of genes to act in conjunction is that the overall number of genes having this 'pairing potential' is such that the effect happens often enough within a tolerably low overall mutation rate. It seems that, once we fix the overall mutation rate to be such as not to jeopardize the survival of the species, the relative frequency with which (ii) happens is a function of the size and informational structure of the genome, not external circumstances. Thus, we might reasonably expect (ii) to be as important on the TwinEarth as it is on Earth – or as unimportant.

Of course, (iii) has been the most controversial for both Gouldian topics of contingency and punctuated equilibrium. It is an explicitly macroevolutionary effect, which at least in some cases could be initiated by *extrabiospheric* (in our particular case that would be extraterrestrial) agents. While there is no reason to believe that the ecological effects of a 10-km asteroidal impact on TwinEarth would be in any substantial detail different from one on Earth, indeed, similar to the one which created the Chixhulub impact crater 65 million years ago (and this should be true independently of our incomplete knowledge about the ecological consequences), the frequency of such impactors and relevant properties of their parent population may vary wildly. Of course, we may again take recourse in the definitional fiat and prohibit variations of the astrophysical surrounding as well, but this would surely defeat the purpose of the thought experiment, since cutting off so large a chunk of the parameter space would preclude many interesting insights. Besides, it would mute the point I am trying to make about the terrestrial biosphere being an open, not a 'closed-box' system, which is a mainstay of modern Earth-system science. Therefore, we should consider an astronomically plausible counter-*gedanken*:

Tau Ceti Centre: Close Sun-like star τ Ceti (accidentally, together with ε Eridani, the first target of a serious SETI search) possesses an Earth-like planet in the habitable zone. It also has a ten times as massive belt of cometary material analogue to the Solar System's Kuiper Belt. Since Kuiper Belt bodies and their analogues are the main source – and perhaps the ultimate source – of impactor populations in the inner planetary system in both Solar and the τ Ceti system, it is to be expected that the influx rate of such bodies will be by about an order of magnitude larger in the τ Ceti system than in the Solar System. This would imply an impact stress on the habitable planet an order of magnitude higher than the one suggested by geological record of the Earth, *ceteris paribus*.[20] Consequently, mass extinctions caused by large impacts are more frequent, and so, presumably, are changes of macroevolutionary regimes. It is even possible that the 'impact frustration' leads to very delayed biogenesis, or a family of parallel, simple and perhaps highly specialized 'shadow biospheres'.

It is worth emphasizing that while we have not found habitable planets in the τ Ceti system yet, the existence of an overly massive inner cometary ring is an empirical fact.[21] The puzzle why a Sun-like star would have such a massive ring of cometary material is compounded by the fact that τ Ceti is *older* than the Sun; commonsense and orthodox astronomy both suggest that the amount of cometary material in a planetary system should decrease with time, due to impacts on planets, evaporation of ices, loss of dust particles and occasional slingshot expulsions from the system.[22] Therefore, we would expect τ Ceti to have a *thinner* cometary disc in comparison to the one in the Solar System, and not an order of magnitude thicker. Although we do not have the relevant observational data yet, it is clear that the correct explanation needs to fall into one of two broad classes: either (A) the initial cosmogonic conditions were different, or (B) consequent dynamical evolution of the cometary population has been dramatically different (or perhaps some mixture of the two). Isn't that situation familiar? In 'the hardest of hard' part of physical science, celestial mechanics and dynamical astronomy, we need at least to entertain the possibility that evolutionary contingency gives the best explanation of the puzzling observational data. Obviously, the lesson of Proclus' Athenian eclipse is still with us, and we do not really know at present whether it is just a strange exceptional case or a common occurrence in the Milky Way. In the case of Regis III, it could be argued that it is a fluke – interesting both for its dramatic potential and as a Darwinian thought experiment – but there is no comparative intuition for the Tau Ceti Centre scenario. If anything, astronomical intuition would suggest otherwise, since τ Ceti is one of the closest Sun-like stars and the very fact that it was a target of the Ozma project testified that the community regarded it as a plausible candidate for an abode of life. If it was a rarity – say if the prior probability of an old star possessing so massive a cometary disc was 10^{-9} – it would have been difficult to explain why the 19th nearest star is such a rare fluke.[23] While the question as to what might be the correct explanation for the τ Ceti cometary disc mass is an immensely interesting astronomical puzzle, we should keep in mind that the scenario above is still just a thought experiment, and will remain so until we are able to decide empirically whether the τ Ceti system possesses a habitable planet or not. Its purpose is to illustrate that the macroevolutionary part of evolutionary contingency certainly is a part of the wider contingent behaviour of the astrophysical environment. In other words, what is often assumed as a given without further inquiry into the context of biological evolution could be a consequence of an interesting wider situation; again, we can see the necessity of abandoning old closed-box assumptions when habitable planet systems (per analogy with the Earth system) are concerned.

Is convergence useful?

Occasionally, scepticism about extraterrestrial life and thus, barring time travel, the only way of having anything even remotely empirically similar to replaying the tape of life, seems to be a source of the attitude toward contingency. Such is the case of Simon Conway Morris, who in *The Crucible of Creation* and *Life's Solution* argues both that evolution is predominantly convergent (and, consequently, that replaying the tape would yield roughly the same results) *and* that there is no extraterrestrial life, or at least complex extraterrestrial life.[24] In other words, while biogenesis (including the accompanying necessary stages of prebiotic evolution) is contingent and inherently improbable everywhere, the subsequent course of evolution is convergent and, indeed, likely to be similar to the terrestrial case.[25] 'Rerun the tape of life as often as you like, and the end result will be much the same.'[26] I shall return to the thrust of Conway Morris' work in the next chapter, since it belongs under the general banner of 'Rare Earth' hypotheses, but for the moment I wish to argue that, on an astrobiological level, this view seems implausible.

Consider, for instance, Conway Morris' own thought experiment: convergent life on a distant planet.

Threga IX: An Earth-like planet known to be inhabited, but with no details prior to arrival of our intrepid researchers. They soon find out that the life on Threga IX closely resembles life on Earth, for the space of biological possibility is tightly constrained. The array of biochemical, morphological, physiological, behavioural, and ecological trajectories describing various aspects of the Darwinian evolution passes through a much smaller part of the parameter space than hitherto assumed, and since the allowed region is much smaller, *eo ipso* the trajectories of two independent biospheres must be close. Thus, our researchers will not find animal locomotion based on wheels or sensory apparatus aimed at X-ray detection (more on X-ray vision later). Instead, they will find plants with nitrogen-fixing systems, photosynthesis, camera-like eyes sensitive in the optical part of the electromagnetic spectrum, and – crucially – bipeds with something approaching intelligence and even emotional and spiritual needs. "Whether the smiling biped on Threga IX offers us a tulip is not really relevant; remember, we are in pursuit of general biological properties."[27]

The underlying thesis here is quite strong: it goes further than the usual lack of sensitivity to initial conditions ('robustness of history'), which occurs in many places in science. Due to the velocity dependence of viscous friction force – often called 'Stokes' force' – all raindrops fall to the ground with roughly the same velocity, regardless of the varying heights of the originating cloud. Conventional opponents of contingency, such as Dawkins, would argue that natural selection acts similarly to Stokes' force, dampening trends inherited from the initial conditions. Some particular evolutionary trajectory is robust, according to this view, since subsequent motions towards peaks of the fitness landscape erase the direction and magnitude of the initial motion, if any. However, Threga IX requires more – a sort of 'galactic convergence' of habitable planets – since it is preposterous to assume that abiotic selection pressure is the same at any two different planets; this is clearly even truer for non-selective evolutionary mechanisms, such as genetic drift.[28] Of course, things would be further complicated by various feedbacks between abiotic and biotic environments of any particular species; if physics is of any guidance here, the trajectories of complex physical systems in phase space tend in general to diverge under the influence of strong feedbacks. Overall, the 'galactic convergence' is an extreme extension of the usual domain of convergence – almost straight into acausal mysticism.[29]

Now, compare the Regis III *gedanken* with the Threga IX. Based on Conway Morris' account, Regis III would hardly pass as a habitable planet at all. On the other hand, it is hard to pinpoint what specific criticism, on the evolutionary level, he could direct at Lem's scenario. The escape clause for Conway Morris now must be the different initial conditions. Regis III evolution (in its postbiological phase) has started from a unique and, indeed, contrived state. This pushes the key issue into the 'origins' domain, i.e., biogenesis. However, it is highly unlikely that whether the evolution of life on a planet in subsequent billions of years will be convergent, or not, depends on whether it originated *in situ* or was brought there via panspermia. There are at least some indications that panspermia needs to occur very early in terms of complexity of prebiotic evolution.[30] If that is indeed so, there is a 'slippery slope' kind of argument here: exactly how far back do we need to go to encounter the 'true' initial conditions (and the contingency contained in them)? Are such 'true' initial conditions contained in prebiotic biochemistry, such as Miller–Urey compounds, or at first enzymes' level, as argued by Jacques Monod and most of the 'lucky accident' school of biogenesis researchers, or in the RNA-world, or at the last universal common ancestor, or even, as some of the 'Rare Earth' proponents would prefer, at the proto-metazoan ('metazoa 1.0') level? Each particular answer poses some further problems, and neither looks obvious enough, nor provokes strong philosophical intuitions. In any case, it is worth re-emphasizing that the onus of proof here lies with those who would like to undermine Copernicanism by pushing back the realm of contingency to the unobservable 'origins' epoch. I shall argue this point further from the epistemological point of view in Chapter 6, related to the continuity thesis. For the moment, I suggest that we are not entirely condemned to the sort of agnosticism nicely formulated by Briggs and Fortey in a recent review:[31]

> The repeated evolution of sabertooth morphology among predatory mammals is doubtless an extraordinary phenomenon that might suggest some sort of tendency to evolve in a particular direction. However, the fossil record has also revealed designs that have never been

replicated. Graptolites, for example, colonial planktonic organisms with an extraordinary array of spiral, s-shaped, or branching rhabdosomes, have never had close homeomorphs in the 400 million years since their extinction. Is humankind a "sabertooth" or a "graptolite"? Unfortunately the probability that consciousness would have evolved even if the vertebrates had been eliminated by some late Cambrian extinction is unlikely to be susceptible to scientific inquiry.

I want to propose that astrobiology, in conjunction with massively parallel computer simulations, could actually give us an answer on a reasonable near-future timescale.

Suppose that during the next century, using advanced modes of astrobiological observations, human scientists discover not one, but a sufficiently large number, say, a hundred TwinEarths.[32] While this may look a tall order at first glance, we could think of this set of TwinEarths as a rather sparse subset of, say, a hundred million rocky, Earth-like planets, which are quite likely to exist in the Milky Way. This embedding set could be expanded, since this is after all a thought experiment, to include other nearby spiral galaxies if we feel the numbers are too constraining. As specified in the TwinEarth definition, they are all of roughly the same age as the Earth and are inhabited by complex metazoans. Suppose our astrobiologists know all that, but nothing about the specific morphologies of the inhabitants. According to Conway Morris, they would all be inhabited, if not by humans, then by humanoids or closely related 'bipeds': if the Threga IX conclusion holds, it must a fortiori hold for any TwinEarth, since the latter is any case more similar to Earth. Such a state of affairs would lead us to an empirical 'law' stating that the evolution of any life based on terrestrial biochemical building blocks progresses to the emergence of humanoids. While that seems quite implausible a priori, it seems unavoidable on the face of Conway Morris' position. Now, in contrast to the original 'replaying the tape' thought experiment, where one could always argue that the abiotic component of evolutionary competition has stayed the same throughout the history of our planet, in the version with TwinEarths we can continuously change the physical parameters of the environment. For instance, we could imagine much smaller or much larger landmasses, with different degrees of fragmentation, which will obviously affect biological diversity by boosting or suppressing allopatric speciation. Since there is no conceivable reason to regard these geological features of Earth as anything but contingencies of a planetary history with no general significance (determined, in turn, by the unknowable astrophysical and astrochemical conditions of a protoplanetary nebula out of which our planet formed, together with subsequent heavy bombardment), we do expect them to be quite variable on other planets.

But wait, we do not really need to sit down until the new generation of super-duper instruments come online, capable of observing geographical features of extrasolar terrestrial planets.[33] We can build detailed models of such planets in the digital medium of today's supercomputers. Such numerical simulations go a significant step beyond thought experiments – while, in some cases, extending and deepening the conclusions of thought experiments, e.g., in the famous 'Daisyworld' *gedanken* – often bringing entirely new and unexpected perspectives.[34] Some branches of science have been truly 'born again' by introducing large-scale simulations (cosmological structure formation, epidemiology, climate sciences); others have experienced significant refreshment and expansion of their domains (dynamical astronomy, pharmacology, geophysics). Evolutionary theory is, if not in the first, most certainly in the second category. Evolutionary simulations nowadays are in standard use to determine the probabilities of particular sequences of evolutionary events. Particularly relevant examples of the use of Monte Carlo and related numerical simulations in evolution are the tests devised by paleobiologist Daniel McShea to probe the relative relevance of passive and driven trends in macroevolution.[35] Although the subject is too complex (no pun intended) to enter into detail here, such simulated worlds offer us bright prospects for understanding the intricacies of large-scale evolutionary trends; such as those often cited and controversial increases in complexity, size, fitness, or independence from the environment, as simulations of cosmological structure formation have already given to cosmology, and discussed in Chapter 2. They probe the 'deep time' of geological and biological evolution, but have not yet been used in the complementary sense of probing the 'deep space' of the astrobiological landscape. In my view, it is a high time for astrobiology to consider that road much more seriously and, in particular, that astrobiological sector dealing with contingent or convergent evolution of life in the general cosmic context.

Geerat Vermeij has recently offered a particularly interesting argument supporting convergence – and, therefore, optimism in respect to independently evolved life in the Galaxy – based on the logical asymmetry between singular and multiple events in the incomplete terrestrial fossil record. If a character X demonstrably evolved multiple times, this clearly is an argument for convergent or parallel evolution; if X evolved only once, to the best of our knowledge, it is still *not* an argument for uniqueness, since the incompleteness of our record may hide repeated instances of X's independent evolution. In another affirmation of the key importance of observation selection, Vermeij shows that alleged singular innovations tend to be either more ancient or appear in small clades. Small clades would be undetectable if located in the more distant past, thus making the observed uniqueness only the *latest* instantiation of what is in fact multiple evolution of an advantageous character. 'Data and arguments . . . indicate that few, if any, innovations are truly unique. Indeed, the principles of physics and economics imply that many derived functional states are achieved many times in many clades because they impart substantial, widely applicable advantages to their bearers.'[36] Vermeij's argument gives powerful support for continuity of evolution, to which I shall return in the next chapter.

Testing convergence

There is another sense in which the question of contingency vs. convergence is astrobiologically relevant, which can be regarded as 'zooming out' beyond our geocentric and anthropocentric prejudices. Consider again the stock example of convergence in evolution: vision. It has been argued countless times among the proponents of strong convergence, that the fact that there are only a couple of types of eyes (camera-like eyes in vertebrates, compound eyes in arthropods) testify that evolutionary solutions to the problem of vision are indeed tightly constrained. Conway Morris goes one step further and argues that the very fact of visual perception evolving in so many distinct branches of the universal tree of life testifies to strong convergence. This claim, however, was persuasively criticized by the distinguished philosopher of biology Kim Sterelny in his review of Conway Morris' book, pointing out that it is only proper to consider these as examples of *parallel*, not convergent evolution.[37] Leaving aside this very pertinent problem whether most of the usually cited examples of convergence are indeed just examples of parallelism, let us 'zoom out' and ask *why has vision evolved in the first place?* Of course, countless bibliographic items teach us that this remarkable evolutionary invention has occurred since the amount of visual information potentially available to any living system is so vast, that its fitness is tremendously enhanced by being able to process at least a small chunk of that sea of information, using chemical mediators like the near-universal rhodopsin.[38]

Why is that amount of visual information so large? Nobody will deny it is a legitimate scientific question; and let us avoid hiding behind a disciplinary fence by claiming that it is not specifically a *biological* question. A straightforward answer clearly involves properties of visible light – and some less anthropocentric sources may include parts of infrared or ultraviolet light as well, known to be visible to some animals. Overall, this is still a minuscule part of the entire electromagnetic spectrum; and, needless to emphasize, in the hugely predominant part of it neither rhodopsin, nor other terrestrial light-recepting molecules work. 'X-ray vision' of Superman and other superheroes remains the stuff of comics; and well enough, since even in the wild diversity of the Milky Way and its Galactic Habitable Zone, there cannot be many places where X-ray sensory input could be of much use. The Sun and other main-sequence stars emit negligible amounts of X-rays; exotic places where there are significant amounts of X-rays – relativistic close binary systems, the vicinity of the central supermassive black hole of our Galaxy, etc. – are not likely to be habitable, anyway, at least not for the kind of life based on known chemistry.[39] So, the emergence of vision as we know it in the living world is at least partially a consequence of the astrophysical fact that main-sequence stars are balanced in such a way that their surface temperatures correspond to maximal emissivities in the visible and small parts of the IR and UV spectral regions. This, in turn, is ultimately – via the standard theory of stellar structure – a consequence of the properties of microphysical processes and fundamental physics; the very same that enables the universe to be habitable in the first place. (Of course, this does not preclude the existence of many living beings *without* vision on any habitable planet; after all, it certainly is cheaper in both the information and energy sense not to possess a complex function than to possess it – if *ceteris paribus* can be maintained. The latter could be a problem, as we shall see in Chapter 6.) On the other hand, the facts of astrophysics do make evolving an 'X-ray vision' sensory apparatus quite improbable, to say the least, even without entertaining the question whether constraints of form and development can make it outright impossible. For example, while many organic materials make excellent lenses for gathering optical photons, human engineers need heavy metals like gold, tungsten or iridium, entirely absent from living beings, to construct even the simplest 'grazing optics' for our X-ray telescopes and microscopes. Should, then, proponents of strong convergence rejoice if a living being capable of seeing, for example, red light at 6500 Å, is discovered on TwinEarth? Certainly not, if they wish the very concept of evolutionary convergence to retain its well-defined meaning. If we do not regard the empirical fact that most of the photons emitted by several times 10^{11} main-sequence stars in the Milky Way are optical (/near IR/near UV) as a confirmation that 'stellar evolution is convergent', it is unclear why would we would call the increase in fitness accompanying the capacity for processing information carried by these photons 'convergent'. Conversely, while it is normal in astrophysics to consider well-understood stellar evolution 'parallel' (up to the differences stemming from their different masses and original chemical composition), it seems that there is indeed a strong reason to consider the astrobiological consequences – in the loose sense of causality explained above – parallel as well.

The other side of the same coin – and a demonstration that it is not (at least not just) a semantic issue – would be the study of independent habitable sites where there is no such fitness increase associated with vision, or indeed any common and familiar organismic-level capacity. One example of such ecosystems would be Gold's 'deep hot biosphere', composed of lithospheric microorganisms.[40] Another would be a hypothetical 'shadow biosphere', the remnant of an independent terrestrial biogenesis.[41] A step further would be ecosystems potentially comprising metazoans that became isolated from daylight for a geologically significant time, such as the famous Lake Vostok in Antarctica. The disadvantage of these examples is, obviously, that either (i) they are very simple, or (ii) they share common ancestry with the rest of the terrestrial biosphere, so we cannot ever be sure that it was not similarity of initial conditions that is responsible for a particular observed convergence. Some extremely interesting tests of this sort will become possible with an 'independent Lake Vostok', namely the subglacial ocean on Jupiter's satellite Europa. A study of this sort has been proposed by Chela-Flores, giving a list of some possible tests of the generality of convergence if and when empirical observations from the Europan ocean become available.[42] Such an expansion of horizons on Earth and beyond would be very welcome in the evolutionary debate; as biopsychologist Lori Marino observes in one of the most detailed and articulate contributions to Chris Impey's anthology *Talking about Life*, 'The contingency versus convergence argument is endless', and she proceeds to show how we could extract new useful information from carefully analysing nature's experiments.[43]

The last chapter of Gould's *Wonderful Life* contains a beautiful discussion of 'seven possible worlds' in which terrestrial evolution was – anthropocentrically speaking – derailed from its historical trajectory.[44] In the first of these, explicitly discussed by Chela-Flores as a possibility in the Europan context, life on an alternative Earth never did evolve the eukaryotic cell and has remained prokaryotic 'forever' . . . or for how long, exactly? The answer to this question, Gould is fully aware, is not offered by evolution, or indeed, the entire realm of biology. Although his description of the events leaves much to be desired, and there have been some big strides made since the publication of Gould's book, it is clearly the astrophysics of our star, the Sun, which forces terrestrial ecology into closing whatever temporal window for evolving particular biological entities is relevant. The most important such temporal window is the one for habitability – again, at least from the selfish point of view of non-lithospheric non-extremophile life forms, like plants, animals or humans.

As vividly described by Ward and Brownlee, the old age of our planet will be characterized by the loss of a basic precondition for life – water:[45]

> The loss of oceans will be the single most dramatic step in the annihilation of life on planet Earth. We essentially carry seawater in out veins, as blood, and although long removed we humans, like all animals, are reminded of our marine ancestry through myriad biological ways. There is perhaps no vision more apocalyptic: the blue-green planet Earth turned brown, its surface water gone, most of its life extinguished . . . We can also propose that plate tectonics might end relatively soon in Earth's history, and when it does, there is a fair likelihood that all life on Earth will end soon after. The loss of oceans might be why plate tectonics ends on planet Earth. And with the cessation of plate tectonics, the last existing life on the planet is doomed once again.

This understanding follows the breakthrough results of Ken Caldeira and James Kasting, who first pointed out that it is the moist greenhouse effect which effectively limits the life span of the biosphere: global temperature rises slowly until the evaporation of ocean causes a runaway greenhouse effect based on water vapour.[46] These models show clearly that the remaining period of habitability – except perhaps for lithospheric extremophiles – is much shorter than the remaining main-sequence lifetime of our star, perhaps as short as 900 Myr from now.[47] Such fate, inescapable by any habitable planet in the Galaxy, provides the strictest closure to the temporal window for evolution. Of course, the window could be closed earlier by any natural or man-made disaster severe enough; the fact that we do not perceive such an event in our past is another example of observation selection effects – if such an event had occurred, nobody would have been around to discuss it.[48] Such events are more probable on Tau Ceti Centre, making evolutionary 'resets' more frequent, and the windows for completion of each macroevolutionary step shorter and more sharply defined.

Now, to return to Gould's possible world without eukaryotes, it has been a reality for so long a time (at least 1.1 Gyr, and possibly as much as 2.6 Gyr), that it is hard to see how one can find it surprising. After all, Copernicanism (and quite superficial and rough understanding of the ergodic theorem) would suggest that the 'average value' of Earth's astrobiological complexity is the world of prokaryotes, or at least something very similar to it. On some likely wide distribution of ages of hypothetical technological civilizations in the Galaxy, any visitor to our planet would have most probably encountered a prokaryote-only world. It has even become a sort of standard view in astrobiology itself with the rise of the 'Rare Earth hypothesis', which suggests that while simple life is ubiquitous in the Milky Way and, by extension, the universe, complex metazoans are rare or non-existent beyond our planet. While I shall soon present arguments against this hypothesis, it is worth keeping in mind that contingency could become empirically testable soon – for instance, in the abovementioned example of the Europan ocean.

One thing is clear: with the help of astrobiology, we could test empirically many hypotheses about evolution. As Derek Turner writes about Gould's 'rewinding the tape', the claim in this thought experiment 'remains an intriguing and largely untested live hypothesis about evolution'.[49] In the concluding section of that paper, it is also argued that a new class of statistical tests discriminating between biased and unbiased species sorting is necessary in order to understand how much contingency there is in evolutionary history. Taking a hint from the suggestion that there are partly 'sociological' reasons why progress has been absent from this issue (no pun intended!), I suggest that additional discouragement originates in a conservative reluctance to embrace the enormous expansion of the scope of evolutionary theory that astrobiology is bringing about on an almost day-by-day basis.[50]

A promising partnership

There are other classic puzzles in the philosophy of biology where ongoing astrobiological research could be of help. For instance, the problem of the level of selection assumes the existence of a biological hierarchy, but this working assumption is misleading, since it is exactly the origin of this hierarchy – on Earth and, presumably, other inhabited places in the universe – which needs to be explained through evolutionary theory. A philosopher of biology recently wrote:[51]

> Ideally, we would like an evolutionary theory which explains how lower-level entities became aggregated into higher-level entities, e.g. how independent genes joined up to form chromosomes, how organelles came to be incorporated into prokaryotic cells, how single-celled organisms gave rise to multi-cellular ones . . . we want to know how the biological hierarchy got there in the first place, rather than just treating it as a given.

In this, again, the help of astrobiology seems to be invaluable – analogously, as with Popper's problem of the origin of diversity and the contingency vs. convergence problem. Okasha proceeds to contrast 'synchronic' and 'diachronic' views of the levels of selection, but since the majority of the grand evolutionary transitions (with the exception, perhaps, of those involving intelligent observers, i.e., us) occurred in the distant geological past, the 'diachronic' view is at obvious disadvantage, barring either the invention of a time machine or an exploration of other biospheres in the Galaxy. I submit that the latter is both conceptually and practically much likelier.

It is safe to conjecture that at this stage of development, almost any breakthrough in astrobiology would have significant consequences for evolutionary theory. Such a breakthrough could be the discovery of living or fossil life forms on Mars or in subglacial oceans, the discovery of a 'shadow biosphere' on Earth, discovery of an oxygen-rich atmosphere on an extrasolar planet, deciphering a radio message of an extraterrestrial civilization, detecting an extraterrestrial artefact like a Dyson shell, or any other similar occurrence; a detailed taxonomy has been recently presented by Margaret Race.[52] Besides making headlines, any such discovery will become an important argument in any number of the debates about evolutionary theory. While most evolutionists are ready to admit that much when asked in principle, the ramifications of the discovery of extraterrestrial life still have not generated remotely as much theoretical and speculative activity as one might have expected. But the confluence of numerical models, evolutionary theory and a rapid stream of astrobiological discoveries will likely bring about some exciting sailing.

This relevance of astrobiology has only recently surfaced, if at all. For instance, Jean-Pierre Rospars in a very recent paper argues that the existence of universal biological laws would have a crucial impact on astrobiology in general, and SETI in particular.[53] After reviewing some of the arguments in favour of laws – for example, the efficiency of the universal genetic code in comparison with a large number of random variations – and some in favour of convergence, Rospars gives an interesting proposal that this immediately implies the necessity of *adding* another metric to the usual spatio-temporal distance between hypothetical intelligent species in the Galaxy, namely the cognitive distance. While this is not new in itself – indeed, Lem suggested something very similar in *Summa Technologiae* almost 40 years ago – the novelty lies in the very chain of reasoning, which starts entirely locally with the philosophical issue of biological laws and a new perspective on convergence.

Closely related is another aspect that makes the application of our evolutionary understanding of astrobiology exciting and a potentially fruitful research programme. It concerns the impact of the evolution of intelligent observers on their physical surroundings, and the possibility that through SETI studies we obtain a glimpse of such an impact, as mediated through artefacts of their material culture. While seemingly far-fetched, this option has gathered momentum, at least since Dyson's 1960 paper as mentioned above. Darwinian evolution of the terrestrial biosphere offers a basis for the results of human sociobiology and related disciplines, which try to play the integrative role along the biology ↔ culture axis. In the same sense, extended to the Galactic scale, Darwinian evolution could, at least in principle, offer a basis for studying the effects of culture in the most general sense, culture relieved of the parochial, confined attributes of 'human' or 'terrestrial'. While it is a very tall order, there is no obvious theoretical or empirical reason why this extension should not be contemplated.

On the contrary, the eventual inclusion of such potential manifestations of the most complex products of evolution (and naturalistic worldview) detectable from interstellar distances is symbolic for the process which, in the view proposed in this book, is an unavoidable consequence of progress in astrobiology: the expansion of the domain of biological and, possibly, postbiological evolution to encompass the entire Galaxy (and, by analogy, other habitable galaxies in the universe). It would apply, interdependently, on all levels of biological complexity, starting with prebiotic compounds slowly moving throughout individual planetary systems and even between them, via bacterial-level complexity and metazoans, to those characterizing intelligent observers and only vaguely discernible shapes of postbiological evolution. The end of the scale is, as briefly considered in Chapter 2, hard to elucidate, and only a synergy of SETI research and future studies may bring us closer to it. I shall return to this topic in Chapter 8. Such overwhelming expansion is a – perhaps, triumphant – consequence of the Darwinian universalism of those dignified concluding words of *On the Origin of Species*. After all, the basic impetus for Darwin's great theory came from travelling on board HMS *Beagle* (and Wallace's analogous stays in South America and the Mollucas) – which are travels *in space*, and in many ways from the point of view of a Victorian gentleman analogous to space travel today.

To summarize, astrobiology can be regarded as the ultimate testing ground of Darwinism. Only by finding and investigating biospheres, other than our own, will we be able to put an end to Popperian dilemmas about the historical and contingent nature of biological evolution, as well as many unanswered, theoretically significant questions, in particular those dealing with macroevolution. Success in this quest would utterly annihilate sceptical complaints, occasionally heard within the biological and philosophical community itself,[54] that we still do not have a *theory* of evolution, just the natural *history* of evolution, 'one damned thing after another'. Such Popperian echoes will fade out completely after the necessary help from astrobiology is not only acknowledged, but indeed forthcoming. This will put the study of living systems on exactly the same footing as the study of any other complex physical systems; of course, then it becomes truly unimportant whether we call that huge family of systems 'physical' or 'biological' or 'just complex' systems. As in Borges' famous 'Library of Babel' or Dennett's analogous 'Library of Mendel', parochial distinctions dissolve in the face of deeper symmetry. Thus, astrobiology not only promises further insight into the evolution of life, but also a self-reflective insight into the nature of biological thinking.

Of course, all these Darwinian considerations have applied to an imagined ensemble of habitable sites in the Galaxy. However, is there such an ensemble in the 'real world' of Milky Way astronomy, physics and planetary sciences? It is to this substantive question that I will now turn.

6

Rare Earths and the continuity thesis

> Have you entered into the springs of the sea, or walked in the recesses of the deep?...
> Have you comprehended the expanse of the earth?
> Declare, if you know all this.
>
> *Job 38:16–18*

A fascinating recent volume by Chris Impey entitled *Talking About Life* offers a rather unique perspective; in interviews with 37 of the world's leading experts in various subfields of astrobiology, masterfully conducted by the author, a reader gets both the flavour of science in action and the individual, often emotional, perspectives usually classified as the 'human interest' cliché.[1] Out of all the interviewees, the question about the 'Rare Earth hypothesis' (henceforth REH) of Peter Ward and Donald Brownlee was posed, directly or indirectly, to 22 of them. This is already a remarkable thing, when the extreme diversity of profiles the reader of this remarkable book encounters is taken into account; no other single *scientific* issue is as widespread in the interviews (in contrast to extrascientific issues, such as the personal origin of interest in astrobiology). Out of that number, a predominantly positive assessment of it was given by seven interviewees, a predominantly negative by eleven, while four were neutral/ambiguous. Of course, my subjective grading of replies is necessarily simplified here, and the individual accounts are wildly varying in length, scope and slant of their discussions. However, it is one of quite rare 'common grounds' in the vast spectrum of astrobiological activities; even if negatively assessed, it is obviously outstandingly important.[2]

While this is not, of course, a scientific poll in any conceivable sense, it is still indicative. Appearing at the very *fin de siècle* in 2000, with the astrobiological revolution in full swing, REH has become a prototype of a 'large-scale' astrobiological theory, representing the best-developed instance of what I have dubbed the astrobiological landscape. REH has many virtues, the most important perhaps being the courage to tackle big, Galactic-context issues. As such, it attracts widespread attention and both researchers and interested laypersons are challenged to think critically and give opinions thereof. Although obviously not a majority view – probably the same holds for the wider astrobiology community as well, outside of Impey's chosen interviewees – it still manages to be a sort of 'default' view, since no coherent alternative theory has been put forward, at least not with such eloquence and forcefulness as Ward and Brownlee have mustered. Such a kind of provocation is extremely welcome, especially in young fields of research.

Rare Earth – a key piece of the landscape?

So, what is all the fuss about? In a nutshell, REH is a probabilistic argument suggesting that, while simple microbial life is probably ubiquitous throughout the Galaxy, complex biospheres, like the terrestrial one, are very rare due to the exceptional combination of many distinct requirements. These ingredients of the REH are well known to even a casual student of astrobiology:

- *Galactic Habitable Zone*: A habitable planetary system needs to be in a narrow annular ring in the Milky Way disc, where chemical abundances and stability conditions for the emergence and evolution of life are satisfied.
- *Circumstellar habitable zone*: A habitable planet needs to be in the very narrow interval of distances from the parent star.
- *'Rare Moon'*: Having a large moon to stabilize the planetary axis is crucial for long-term climate stability.
- *'Rare Jupiter'*: Having a giant planet ('Jupiter') at the right distance to deflect much of the incoming cometary and asteroidal material enables a sufficiently low level of impact catastrophes.
- *'Rare nuclides'*: Radioactive *r*-elements (especially ^{238}U and ^{232}Th) need to be present in the planetary interior in sufficient amounts to enable plate tectonics and the functioning of the carbon–silicate cycle.
- *'Rare Cambrian-explosion analogues'*: The evolution of complex metazoans requires exceptional physical, chemical and geological conditions for episodes of sudden diversification and expansion of life.

There are other items on the list as well – in this sense, REH is an open theoretical system, since everyone is free to add items pertaining to a particular area of relevance and expertise. However, the general reasoning is that all these requirements are mostly independent and a priori unlikely, so that their combination is bound to be incredibly rare and probably unique in the Milky Way:

> The continued marginalization of Earth and its place in the Universe perhaps should be reassessed. We are not the center of the Universe, and we never will be. But we are not so ordinary as Western science has made us to be for two millennia. Our global inferiority complex may be unwarranted. What if Earth is extremely rare because of its animals (or, to put it another way, because of its animal habitability)?[3]

Ward and Brownlee even construct a symbolic equation they dub the 'Rare Earth equation' – by analogy with the Drake equation often (mis)used in SETI studies – which contains the requirements as fractions of the total set of Galactic planets.[4] The product of many small probabilities is a much smaller probability, so there are few places, if any, satisfying all the REH criteria. The fact that we observe such a place around us should not be surprising, since observation selection tells us that we could not observe anything else; there are no observers in places where the requirements are not met.

REH entails a particular type of astrobiological landscape: one in which almost all weight is assigned to simple, microscopic life forms everywhere. Ward and Brownlee make it very clear that they do not regard the transition between non-living and living matter as particularly difficult. In this regard, they follow closely the continuity thesis of Iris Fry, which I shall discuss later in this chapter: there is no big gap between non-living and living, and biogenesis will happen rather quickly where physical, chemical, geological, etc. conditions are satisfied. These vast lowlands in the astrobiological landscape are a reminder of Gould's counterfactual Earth without eukaryotes discussed above. It is complex life on Earth following the invention of multicellularity and, especially, the dramatic diversification in the Cambrian explosion, which is cosmologically rare. Therefore, it appears that REH is very easy to falsify by either finding an absence of simple life forms or finding traces of complex life forms (including intelligent ones; more on the operational definition of intelligence later). In particular, a clear prediction of REH is that we neither should be surprised to find living or extinct microbial life forms on Mars, nor biomarkers such as free atmospheric oxygen on Earth-like extrasolar planets; another prediction is that SETI projects, embodying searches for the most complex known life forms, will remain unsuccessful.

In addition, Ward and Brownlee break new ground by pointing out the importance of hitherto downplayed factors, such as plate tectonics, inertial interchange events, or 'Snowball Earth' episodes of global glaciation, for the development of complex life. This wide scope was perhaps the key factor why REH left such a strong impression in the decade following publication of their book, and the reason why it quickly became somewhat of a default position in many astrobiological circles. Thus, its challenge to Copernicanism has been partially accepted as sound in mainstream astrobiology – although, as argued by opponents, and as I will argue below, there are lower prices to be paid in the market of ideas. Even very strong criticisms of REH just underline its importance in exemplifying astrobiological theory and highlighting specific predictions. We have seen that something very similar occurred during the time of the 'Great Controversy' in cosmology: the steady state theory had been – although wrong – a crucial trigger of the great progress in understanding the large-scale structure of spacetime. Can REH play such a role, mutatis mutandis, in the astrobiological realm?

As far as the reasoning of Ward and Brownlee goes, one can hardly call it into question. As seen in Impey's book, not all practising astrobiologists openly subscribe to REH and, in fact, it is probably just a minority. How is the scepticism of the majority justified? There are essentially two lines of attack: (i) to deny the independence of the various particular requirements (terms in the 'Rare Earth equation'), and (ii) to deny that the particular requirements are indeed unlikely. In addition to these two, there is a vaguer sort of argument to the effect that the terrestrial biosphere might not be representative of the entire class of 'complex biospheres', so that the requirements for the terrestrial subtype might be more restrictive than in the general case. This last argument might be equivalent to calling upon what has been in the old-fashioned philosophical tradition often called the 'Principle of Plenitude'. I shall not discuss it much, since it belongs to the circle of concepts – together with, perhaps, narrow circumstellar habitable zones, or the Drake equation – which are being superseded by more serious astrobiological theory right now, while you are reading this book.

A very important general argument, mostly downplayed in astrobiology thus far, is hidden within type (i) criticisms of REH denying independence of the requirements. It deserves a separate name and a fuller discussion.

Unphysical *ceteris paribus*

Consider this, intentionally ludicrous, scenario:

Lockyer's conspiracy: Suppose you forgot primary school chemistry, but were still rather impressed by the fact that the universe is 75% hydrogen, 24% helium and just 1% of the remaining Mendeleev's Table stuff. Wise Joe, an amateur astrobiologist, thinks how it would be wonderful to have beings based on CHONPHe instead of conventional CHONPS; that is, having ubiquitous helium instead of rare and precious sulphur. Wise Joe has also heard some cosmologists proclaiming a spatially infinite flat universe and he reasons that *somewhere* among zillions upon zillions of galaxies in the infinite universe, there must be some small corner where such life forms originated to make use of the huge stores of helium or evolved to the same effect... But wait! Helium is a noble gas, unwilling to make any chemical compounds under the known laws of physics.[5] Certainly even seemingly all-powerful biological evolution cannot get around *that*?

This fallacy is a rather extreme version of what I shall call the unphysical *ceteris paribus* problem, quite pervasive in Rare-Earth thinking. (And pervasive in general, wherever the interface between historical and experimental disciplines appears.) Before we pass on to more palatable examples, it is important to clarify what exactly is ludicrous in Wise Joe's thinking. A quick-shot reply that it is his naive hope of circumventing the laws of chemistry and physics cannot be the whole story. Although we can regard the 'nobility' of noble gasses as an effective law of nature, it is clearly based upon deeper regularities, such as Fermi–Dirac statistics and other quantum considerations, which are not at all obvious, even to physicists. Rather, the Monty-Pythonesque element is introduced by his Panglossian belief that everything is for the good and his 'anything goes' approach to coherence of a hypothetical situation. Wise Joe's thinking contains a huge 'coherence gap' – in the words of the contemporary philosopher Ivan Havel:[6]

> In conceivable worlds of thought experiments, some states-of-affairs are, by design, the *same* as they are in our world, while other states-of-affairs are deliberately *different*... The crucial but often neglected feature of these worlds is that we seldom know what is the extent of the domain of "the same" and what is the extent of the domain of "the different", besides what is explicitly mentioned or used in the construction. Moreover, besides these two domains there is an inexhaustible realm of states-of-affairs that are *omitted* because they are believed to be irrelevant or because they are forgotten, obscured or entirely beyond the reach of human knowledge.

The omitted realm of states of affairs constitutes the *coherence gap*, and the question of its possible impact on our reasoning, the *coherence gap problem*.[7] We need to be highly cautious in evaluating any thought experiment not because it may entail empirical difficulties, but because the conceived world of the particular thought experiment may need additional assumptions or constraints in order for the desired outcome to occur. An iconic example of the coherence gap is the twin 'paradox' of special relativity: non-uniform motion of one twin in respect to another at the trip turnaround is the typical omitted state of affairs.

Further examples of the same tendency, slightly less amusing, are provided by *bad* thought experiments in 'virtual history' (or bad historical counterfactuals). In an example suggested by Niall Ferguson, 'no sensible person wishes to know what would have happened in 1848 if the entire population of Paris had suddenly sprouted wings, as this is not a *plausible* scenario.'[8] But the plausibility is, at least in part, in the eye of the beholder; while the implausibility of Lockyer's conspiracy and winged revolutionary Parisiennes seem obvious even to the layperson, there are certainly many cases of equal implausibility in 'exotic' fields of research – like astrobiology – which are not that obvious. Sometimes, they are extremely well hidden. Thus, the most prominent *proponent* of the use of counterfactuals to explain historical causation warns:[9]

> There are, in other words, two distinct kinds of counterfactual which have been used by historians: those which are essentially the products of imagination but (generally) lack an empirical basis; and those designed to test hypotheses by (supposedly) empirical means, which eschew imagination in favor of computation. In the case of the former, it is the tendency to rely for inspiration on hindsight, or to posit reductive explanations, which leads to implausibility. In the case of the latter, it is the tendency to make anachronistic assumptions.

There is a huge literature on this topic, mainly belonging to philosophy of history and historiography.[10] It warns a careful thinker that the use of counterfactuals is a delicate matter and should be approached with extreme care, in spite of their popular appeal. It is exactly these 'anachronistic assumptions' which, as we shall see, constitute a major part of some of the REH requirements.

In these examples, the coherence gap undermines reasoning by hiding additional assumptions of a contingent or empirical nature (as contrasted with logical). In the context of historical sciences dealing with the material world – such as palaeontology, cosmology or cosmogony of the Solar System – I shall call this problem the unphysical *ceteris paribus*. It is a particular case of the wider fallacy of neglecting the context. If we are certain that A and B entail X, but for some reason we can observe or take into account only A – because of observational limitations, theoretical incompleteness or computational intractability – then it is often the case that observation of some other $A' \neq A$ would lead us to believe that $\neg X$. But the reasoning is incorrect, since it might be the case that some other B' holds, and A' and B' also entail X. In this simplified model, B is the *ceteris paribus*, which we wish to retain in order for our counterfactual thought experiment to be meaningful. However, it might be that the coherence gap makes the state of affairs (A', B) incoherent.

In the context of REH, unphysical *ceteris paribus* is visible in many places, since many of the individual Rare-Earth requirements are formulated in terms of Earth-related counterfactuals: *if such and such were different, the Earth would not be hospitable to complex life forms*. Thus, since such and such is prima facie improbable, this decreases the overall probability of complex life forms evolving anywhere. I submit that most of these arguments are unsound, since the implicit *ceteris paribus* cannot be maintained, which is visible in the following well-known instance.

The issue at hand is the famous argument about Jupiter being the optimal 'shield' of Earth from cometary/asteroidal bombardment. A smaller part of the problem is that the claim might be empirically wrong. The common-sense conclusion about the role of Jupiter – employed by REH – has been brought into question by the recent work of Horner and Jones, who use massive numerical simulations to show that the conjecture that Jupiter acts as a shield against bombardment of the inner Solar System is untrue in a large part of parameter space.[11] Moreover, they conclude, 'that such planets often actually increase the impact flux greatly over that which would be expected were a giant planet not present.'[12] In other words, Jupiter might have a detrimental effect on the habitability of Earth! If the results of Horner and Jones withstand the test of time and further research, it is a serious blow to REH and the related way of thinking. However, it would still be a major score for astrobiology as a field, since it will demonstrate maturity of the discipline in which intuitively solid prejudices could be rejected based on precise, quantitative work.

A still bigger part of the problem with the 'rare Jupiter' requirement for REH is that the *ceteris paribus* state of a 'Jupiter-less' Solar System is unphysical. REH theorists argue that:

1. Both Earth and Jupiter exist in the Solar System.
2. Jupiter deflects a fraction of potential impactors from collision trajectories.
3. With more impactors on collision trajectories, Earth would suffer a higher frequency of catastrophic impacts.
4. A smaller frequency of catastrophic impacts increases the habitability of any planet.

Hence,

5. Earth's habitability is increased by the presence of Jupiter.

Horner and Jones investigate (and seem to refute) premise (2); but even if we retain it, the argument is incorrect, since the meanings of 'Earth' in (1) and (3) are different. In fact, premise (3) is self-contradictory if we specify Earth as the planet we live on today and know reasonably well. There is no 'Earth without Jupiter', since by definition Earth is a planet formed and evolved through a complex historical process in which Jupiter played an important role. *No Jupiter = no history of the Solar System as we know it = no Earth.* The two possible historical trajectories – the history of the Solar System with Jupiter and without it – are incommensurable. The contrary assumption – that the Earth must be one and the same in (1) and (3) – is analogous to the 'anachronistic assumptions' warned about in the historical context by Ferguson.[13]

A desperate proponent of REH could still claim that Earth is unimportant as a specific entity, and that we should anyway use it as a placeholder for something like 'a rocky planet in a habitable zone of its parent star'. However, this is self-defeating, since we lose the option of using observation selection to account for the a-priori minuscule probability of finding ourselves on an 'Earth', once we find out that 'Earths' are incredibly rare. In other words, we need to account for the alleged increased habitability due to the presence of 'Jupiter' of *this particular planet, i.e., our Earth*, and not just any Earth-like planet.

Unphysical *ceteris paribus* is usually a consequence of neglecting the causal context or oversimplifying the complexity of an observed situation. In supposing how the state of affairs could be different, Rare-Earth theorists assume simple, linear change, not taking into account the self-organizing nature of the relevant physical system, where a very small change at time t could cause dramatic divergence at some later time, $t + \Delta t$. Asking about the fate of Earth in the absence of Jupiter is self-contradictory; Earth is a part of the complex system that includes Jupiter as a major component, so there are no guarantees that Earth would have existed at all if Jupiter were not present. Even if a similar planet existed, we would have to account for many other differences between that counterfactual situation and the actual one, so the question as to what degree it is justified to call such a body 'Earth' would be very pertinent.[14] There is, of course, no way of magically removing Jupiter at some point in the history of the Solar System, but the conclusion of unphysical *ceteris paribus* would have stood even if there were such a supernatural occurrence. After all, the mechanism for enhancing Earth's habitability suggested by Ward and Brownlee is a protracted affair; even a magical vanishing of Jupiter would not really influence our planet until a geologically relevant period of time has elapsed. Since most processes in the Solar System (and, indeed, anywhere in nature) are such that small uncertainties or perturbations in the initial conditions lead to exponentially divergent subsequent histories, the amount of changes caused by Jupiter vanishing after some relevant time would be huge; it is by no means clear whether the habitability of Earth would increase or decrease.

The same reasoning applies for some other REH requirements, notably the one dealing with the 'rare Moon'. As pointed out by Christopher Chyba and Kevin Hand in their fine review, the claim of chaotic obliquity fluctuations being supressed by the size of the Moon, leading to the habitable Earth of long rotational period we inhabit today, is something completely different from the counterfactual claim that the Earth would not be rotationally stable if the Moon never formed.[15] Although it is difficult to estimate Earth's primordial rotation in the absence of a hypothetical Moon-forming impact, it is quite possible that a very small rotational period of the early Moon-less Earth would have ensured stable obliquity in any case.[16] Again, the issue as to the extent that such a body, having a drastically different history (no large impact, no Moon, no subsequent tidal slowing down of fast primordial rotation), deserves to be called 'our' Earth is open to debate. Ditto for the consequent increase or decrease of habitability.

These are examples of a wider intellectual malady of our age: the tendency for over-atomizing complex things, searching for shortcuts through the complex causal webs of history, and imagining individual pieces as entirely changeable while magically preserving *ceteris paribus*. In a biological context, the same tendency has led to excesses of zealous ultra-adaptationism, and especially their misapplication in pop sociobiology and evolutionary psychology, which have been criticized by many.[17] It is the idea that advances in science are similar to putting together a child's puzzle – starting from one corner and advancing until just a few of the pieces are missing from the overall picture. However, this cute metaphor is a rather misleading account of the actual scientific process, for the shape of the puzzle picture, as well as the form of the remaining 'hole', changes continuously and as a function of the piece we are currently holding in our hand. Just as our theories determine even the basic concepts we use in formulating specific research problems, nature in turn shifts the ground beneath our theoretical constructions in a complex and unpredictable manner. We may *imagine* keeping everything else fixed and changing just a single parameter of our model – but in conducting such a thought experiment (and REH theorists tend to forget it is a thought experiment) we may wish to be careful not to fall into the coherence gap. The grand puzzle cannot be solved by changing the shape of any individual piece while keeping all the rest the same, since the shape of the remaining empty space changes with various approaches.

Further arguments against REH

Since REH has provoked so many discussions and comments, it is not surprising that many critical reviews have appeared in the literature. They range from charges of its extrascientific agenda to rather complex technical arguments, like the already mentioned studies of Horner and Jones. I shall not dwell much on the first kind, although connections between REH and various forms of hidden creationism or 'Intelligent Design' are many. They might not be so ominous, as suggested by David Darling in his popular book on astrobiology,[18] but the curious fact is that some REH proponents do not make a particular secret out of them. Conway Morris, for instance, whose work on evolutionary convergence was discussed in the previous chapter, gives many REH-style arguments for the uniqueness of Earth (and approvingly quotes Ward and Brownlee), while at the same time concluding that it is high time for God's return to science from where He was allegedly exiled by vulgar materialists and mechanicists.[19] Guillermo Gonzalez, a co-author, with Ward and Brownlee, of the seminal paper on GHZ, went a step further and wrote a popular science book arguing for 'Intelligent' Design, among other things, on the basis of REH.[20] While this might be regarded as obvious misuse, it is important to keep in mind that such things are bound to happen as long as we do not take both naturalism *and* Copernicanism seriously enough.

On the technical side, a recent study casts doubt on the rarity of plate tectonics in the set of extrasolar planets.[21] According to Diana Valencia and her collaborators, models of recently discovered 'super Earths' (planets with up to ten terrestrial masses) indicate an early development of convective processes in such bodies. In turn, such convective processes will inevitably lead to plate tectonics, which thus could be ubiquitous in the Galaxy, obviously contradicting REH. It is possible that we shall soon see further such instances undermining REH claims. The Potsdam group of astrobiologists have built numerical models of various REH requirements, with the typical outcome that complex life habitats are on the average 100 times rarer than the simple life ones. While two orders of magnitude less is 'rare' by human everyday standards, the Potsdam group model still leaves us with millions of planets with complex biospheres in GHZ.[22] Even if the slant of forthcoming research is more REH-friendly, it is clear that new tests involving extrasolar planets – such as the one on the 'rare Moon' requirement suggested by the study of Dave Waltham[23] – will bring us closer to understanding how a particular dynamical variable (in this case, planetary obliquity/rotation rate) contributes to the overall shape of the astrobiological landscape. Here it is hard to deny the large and positive role played by REH in provoking much of the ongoing research.[24]

There is a perceived connection between REH and the work on GHZ, which was also originated by Ward and Brownlee (with Gonzalez), but the connection is neither logically necessary nor does it play the role critics ascribe to it. For instance, Cohen and Stewart – in what is perhaps the best example of 'radical' criticism of astrobiology as 'an unimaginative approach to a subject that absolutely cries out for imagination'[25] – repeatedly argue that habitable zones, both circumstellar and Galactic, are anthropocentrically defined and largely useless. Counter-examples to the classic understanding of a habitable zone, such as subglacial oceans on Jovian moons, are used, in such a view, to argue against REH in general. However, similarly to most of the other claims of Cohen and Stewart, although this argument contains a grain of truth, it is too extreme and overdrawn. Habitable zones are part of the overall quantification of habitability. What is the alternative? Cohen and Stewart do not offer much in this regard, except for nice SF-inspired stories and a lot of rhetoric about the need for a wider and more liberal view of possible exotic life forms. The truth is, as usual, somewhere in between: neither reification of habitable zones, nor vague preaching about the ubiquity of strange life. We need quantitative models, out of which habitable zones would emerge as just rough, rule-of-thumb approximations. This is exactly what the astrobiological landscape approach, sketched in Chapter 3, is all about. The landscape encompasses all the different variables of relevance for habitability – one of them is certainly the temperature range for finding liquid water, but this should not be reified and canonized (and here Cohen and Stewart are on the right track). The landscape view offers a sorely needed opportunity for quantifying astrobiology and, as a flexible framework, it can accommodate both REH and its more Copernican alternatives.

Some authors, usually those hostile to anthropic reasoning, have criticized REH because of the exceedingly low a-posteriori probability of almost any historically contingent event. If you ask the question, What is the probability of you being born on the date of <YOUR BIRTHDAY>?, you may end up with a very low probability when you reckon all the chance events required by that contingent fact: your parents conceiving you at a small interval of time compatible with your subsequent day of birth, their having met before at particular time interval, etc. This is a reflection of the obvious mathematical fact that the probability of any sufficiently long chain of probabilistic events is low. In the case of evolutionary processes taking place on very long temporal scales, containing a number of key milestones (or 'critical steps', or 'hard steps', or 'major evolutionary transitions', or whatever various authors used, either as well-defined operational concepts or as mere metaphors), it is easy to 'reconstruct' even smaller outcome probability by inserting another contingent event somewhere in the middle of the chain. However, the very broadness of this criticism shows how little one could learn from it in any particular case. It is indisputable that some of the REH requirements (terms in the 'Rare-Earth equation') are overlapping and in any case are unclear and vague enough that any assigned probability is highly speculative. On the other hand, if post-positivist epistemology teaches us anything it is that we need to start somewhere and to have at least some hypothesis giving potentially testable predictions. In this case, the supposed weaknesses of REH are more like opportunities for further testing – the story with the steady state theory all over again.

Ward and Brownlee are honest enough to admit that the main threat to their view comes from lack of imagination as far as the scope of complex life is concerned.[26] Astrobiological quest is bound to be the quest for strangeness; what else could it be? Heraclites expressed it 25 centuries ago in the form of his famous aphorism on expecting the unexpected (I shall return to it in the next chapter). One of the great visionaries of the twentieth century, J. B. S. Haldane, emphasized the same point in stating '[my own suspicion is] that the Universe is not only queerer than we suppose, but queerer than we *can* suppose'[27] – which is sometimes jocularly dubbed Haldane's law. Thus, while lack of imagination is certainly a failure, it is a type of failure very hard to quantify by any reasonable metric.

Finally, I wish to mention another politicized instance of the suggestive use of language. Proponents of REH share with opponents of astrobiology (and 'pluralism' with regard to life in the universe in general) an unfortunate tendency to pose as 'sceptics', in contrast to their opponents, who are usually labelled as 'believers'. The implication – or insinuation – here is that it is *extraterrestrial life* (in the REH case, complex/intelligent life) that constitutes the scales upon which the terminology is measured. Even a person in the street knows that scepticism is desirable in science (and belief is appropriate to other spheres of human activities, like religion, politics, or filling in tax forms). Let me offer a 'modest proposal' to reconsider things, and make the existence of life on Earth the reference point. Now, proponents of REH could be reasonably labelled *believers in the uniqueness* of the Earth and its complex biosphere, and opponents are *sceptics* as to the uniqueness claim. It is much easier to disprove empirically the belief in the uniqueness of X than to disprove the belief in non-uniqueness. The latter would be impossible in an unconstrained situation. The saving clause for REH proponents could be the fact that they usually constrain the field by arguing for 'X is unique in the <PARTICULAR EMBEDDING SET>'; specifically, Earth's complex biosphere is unique in the Milky Way. While the discovery of an Earth-like planet with a complex biosphere, or a SETI signal, or an astroengineering project, will be a strong rational reason for abandoning belief (if, admittedly, not the full-scale falsification), it is at least awkward to argue that any empirical finding gives rational reasons for abandoning scepticism.

The continuity thesis: 'neither chance nor design'

A great advance in the nascent philosophy of astrobiology has been introduced by the work of Iris Fry, who in several papers and an excellent book, *The Emergence of Life on Earth*, elaborated a key principle for the scientific study of biogenesis.[28] This principle she calls the *continuity thesis*:[29]

> This common element, which I will coin "the continuity thesis", is the assumption that there is no unbridgeable gap between inorganic matter and living systems, and that *under suitable physical conditions the emergence of life is highly probable*. It is the adoption of the "continuity thesis", concerning the philosophical dimension of life–matter relationship, which has turned the origin of life into a legitimate scientific question, and which constitutes a necessary condition for any scientific research in this domain.

In essence, the continuity thesis encompasses almost all modern approaches to the origin of life (including those postulating the origin of life elsewhere and its transport to Earth via some form of panspermia). Its substance is evocatively summed up in the title of Chapter 13 of her book: 'Neither by chance nor by design'. Fry persuasively argues that the continuity thesis is the only meaningful way to proceed if one wishes to remain in the scientific domain; both the 'lucky accident' naturalistic account of biogenesis and supernaturalistic design lead us into the same blind alley.

The continuity thesis is mainly an epistemological necessary condition, but it can be applied as heuristics as well, both in a positive (encouraging particular research programmes) and negative (eliminating other particular research programmes) sense. Clearly, as Fry emphasizes, the ultimate test can come by either repetition of biogenesis in the laboratory or astrobiological research in the origin of life *elsewhere*. However, it is important to grasp the central epistemological point on which everything else hinges. As the saying (sometimes attributed to Richard Feynman) goes, 'we don't really understand something if we can't make it'. When we apply this maxim to biogenesis, it is clear that the 'lucky accident' school of thought, if correct, blocks any chances of repeating biogenesis in the laboratory; hence, it obviates all prospects for really understanding life. In this respect, which is vital from the point of view of scientific methodology, Monod and other subscribers of the 'lucky accident' position do not differ too much from creationists. This embarrassing similarity in style rather than in substance has been repeatedly and very forcefully elaborated by Fry; indeed, creationists often use the arguments of the 'lucky accident' school in their various writings. If the probability of biogenesis – even under favourable physical and chemical preconditions – is astronomically small, say 10^{-100}, any laboratory attempt at re-creating life is bound to fail. Thus, the continuity thesis is a necessary prerequisite for most of the empirical work on the origin of life.

Even smaller probabilities have been occasionally cited in the literature, both by naturalists and proponents of creationism. Thus, Manfred Eigen, a Nobel laureate chemist, cites the probability of random assembly of a polymer with a thousand nucleotides corresponding to a single gene as 1 part in 10^{602}.[30] Hoyle and Wickramasinghe calculated that the probability of random assembly of the complete set of enzymes in even the simplest known bacteria is of the order of $10^{-500\,000}$. These sorts of 'hyperastronomical' numbers have led Hoyle and Wickramasinghe to invoke either an eternal universe – in contradiction with standard cosmology – or a creative supernatural agency. The (in)famous metaphor of a tornado assembling a Boeing 747 out of a junkyard, often cited by Sir Fred Hoyle, nicely expresses this sort of desperation, which has, luckily enough, been overcome in modern theories of biogenesis.[31]

Superficially, there are similarities between the continuity thesis and the strong anthropic principle (SAP) of Barrow and Tipler, which states, 'the universe must have those properties which allow life to develop within it at some stage in its history'.[32] But they are indeed very different. The continuity thesis is a description of an evolutionary process, not a teleological *prescription* for it, like SAP. While building on the environment described by current cosmological theories, the continuity thesis does not favour one or another cosmological model. In fact, it is disteleological in the sense that it is perfectly compatible with an empty universe or a universe in which all mass resides in black holes. Since there is no physical condition for evolution to start, there is no life or observers in such universes, of course, but the continuity thesis reflects that case as well, not denying them, as SAP does (at least in a single-universe or 'sparse multiverse' version).

If we accept the continuity thesis, there is no reason to consider other habitable places in the universe as anything other than an extension of the habitable space on Earth, in terms of all particular properties of biogenesis. Does it apply to further 'major evolutionary transitions' as invoked by Maynard-Smith and Szathmáry? In particular, we are entitled to ask whether one of the mainstays of REH – the alleged rarity of 'Cambrian explosions' and accompanying surges in biospheric complexity – is a 'lucky accident' of the Monod or Hoyle–Wickramasinghe sort. Here one should keep in mind that it is immaterial what *kind* of improbability underwrites our judgment that something is 'rare' or 'unique'. Considering the fact that molecular clocks give consistently very high ages for the first metazoans (significantly more than 550–570 Myr before present) and that more recent research suggests that the period of complexification was a protracted affair, it seems that on the empirical side we are also approaching a negative consensus. In the domain of psychology and the philosophy of the mind, we are also witnessing a long process of *downplaying* the role of mental phenomena traditionally associated exclusively with humans (like consciousness). Thus, there is no reason to avoid extending the powerful argumentation of Fry to those cases.

Here, I suggest that the continuity thesis should be extended to encompass a wider range of major evolutionary transitions, or 'crucial steps' in overall cosmic evolution. In particular, I suggest an *extended continuity thesis*, which brings the reasoning of Fry to its logical conclusion, namely that there are no unbridgeable gaps between simple life and a complex one, and between complex life and an intelligent one (possibly also between biological intelligent life and a postbiological intelligent one). Whenever and wherever physical, chemical, ecological, etc. conditions are suitable, the emergence of complex life is highly probable – and ditto for intelligent life.

There are at least three reasons why we should accept the extended continuity thesis *as a working hypothesis*: two major and one auxiliary. The first reason is that, as long as we stick to the naturalistic account of all major evolutionary transitions – and do not assume, for example, as Wallace did, that the origin of mind requires a supernatural boost – Fry's reasoning continues to work inexorably. If we wish to truly understand metazoan complexity, we need to be able to, at least in principle, replicate the Cambrian explosion in the laboratory (or the emergence of Vendobionta, or whatever was the relevant episode in the history of life). If we wish truly to understand intelligence, we need to be able to construct it in a (computing) laboratory. If either of those is a 10^{-100} 'lucky accident', then the story of the scientific understanding of these crucial steps is over. Since we are far from being forced into such a quandary, the alternative – accepting the extended continuity thesis – should be provisionally accepted. This is, of course, crucial for the field of artificial intelligence; any degree of optimism regarding

AI cannot be justified without at least tacit acceptance of the extended continuity thesis.

The second reason is, of course, Copernicanism. Increases of our empirical knowledge which have traditionally supported the Copernican assumption should not be confused with moves (such as the one performed by the 'lucky accident' school) which block our empirical work by fiat. Occasionally, this is presented as referring to 'mediocrity', but it is, in fact, just the Copernican principle, applied in a partly new astrobiological context. Until we understand the structure of complex life and intelligence better, it is only prudent to follow the prescription that has worked since the dawn of modern science.

Finally, we have some practical reasons for holding on to the extended continuity thesis. As I have argued in the previous chapter, if evolution is truly to be Daniel Dennett's 'universal solvent', then it needs to be universal indeed – meaning that many specific predictions of evolutionary hypotheses can only be given in the astrobiological context. This is a clear and specific methodological situation in which the continuity thesis enables us to connect fruitfully to, say, the research on possible Martian biogenesis. But that should not be the end. Since most of the *interesting* evolutionary hypotheses deal with metazoans (or entities specific to intelligent observers and cultural evolution), using the extended continuity thesis as a working hypothesis – and this can never be emphasized enough – seems fully warranted. A similar case applies to Chaisson's concept of 'cosmic evolution'; although that author professes agnosticism with respect to the issue of extraterrestrial life,[33] it is hard to see how to avoid the loss of explanatory power if the alternatives ('lucky accident' or supernaturalism) are accepted. As Tsokolov points out, difficulties in defining life originate in the continuity of transitional evolution; since the continuity is an objective property of nature, by acknowledging it we are bound to make progress in the formal definitional project.[34]

Haldane's ladder and noogenesis

By *noogenesis*, I denote the origin of intelligence, understood operationally as whatever mental capacity defines intelligent observers: self-awareness, self-reflection, tool-making capacity, or even consciousness (all those and many others which have been proposed in literature should be taken very loosely). The term has a long and interesting history, being apparently first used by Pierre Teilhard de Chardin in his posthumously published treatise, *The Phenomenon of Man*.[35] But the term itself has no necessary links with Teilhard's wider eschatological concerns or his Omega-point theory. Although Teilhard was interested in the origin of modern humans (to an understanding of which he substantially contributed as a palaeontologist), the term has been, in my view, very happily chosen for wider use. After all, its very usage in parallel with the term 'biogenesis' is one step further towards relinquishing anthropocentrism; of course, just as in the case of biogenesis, one should avoid the reification trap and fixate on it as a particular real event, something which either happened or not at any given point in time in any given place. Instead, it should be thought of as a process, or a family of processes, with poorly understood internal stages. As usual in science, our theories determine the contents of the concepts we are using. In particular, the acceptance of the extended continuity thesis implies that noogenesis will reach different particular internal stages at various regions of the astrobiological landscape, at any particular 'snapshot', because local astrobiological clocks tick at different rates (see also the discussion of Carter's argument in Chapter 7 below).[36]

It was J. B. S. Haldane who vehemently insisted (in many of his writings, and especially in his futurist thinking in *Daedalus* and *The Last Judgement*) that cosmological, astrophysical, chemical, biological, psychological and cultural evolutions are just man-made, selected and labelled parts of a single continuum.[37] I call this modern form of the old Empedoclean picture, *Haldane's ladder*. This simple and yet so powerful concept not only directly inspired his pioneering work (with Oparin) on biogenesis which defined the entire field, but encouraged him to speculate on the evolution of humanity in the next no less than 40000000 years. This Janus-like quality, boldly facing both distant past and distant future armed with well-founded philosophical guidelines is exactly what contemporary astrobiology seeks to capture.

In accordance with the basic plan of 'zooming in', I turn now to the narrowest filter, created by noogenesis – intelligent observers and their technological civilizations. As usual, I shall regard the existence of terrestrial intelligent observers and our present technological civilization as a fully naturalistic consequence of the evolution of matter in the cosmic context. Here it is necessary to part company with many illustrious thinkers encountered thus far; for example, we recall how Wallace has insisted on the necessity of the 'disembodied mind' in order to explain the evolution of human intelligence and consciousness. This amounts to cutting through Haldane's ladder at some particular point, either before the emergence of complex metazoans, as the 'Rare Earth' theorists prefer, or at some point before noogenesis or even later, as argued by a diverse group of critics whose views I shall explore in the next chapter. However, Haldane's ladder extends still further – first to the realm of what we have called intelligent observers, and probably even further, to the realm of postbiological evolution. How much further crucially depends on being reasonably clear about what we mean by intelligent beings. This is far from being a simple problem, nor could we hope to bypass it by using the recipe of Judge Stewart on obscenity. In the rest of this book I shall sidestep the tough definitional issue and will use the syntagma 'intelligence' (and 'intelligent beings/observers') to denote whatever character it is which enables logical reasoning, intentional behaviour and the development of material culture. In this manner, it will serve as a placeholder for a whole bunch of different terms occasionally used in literature to denote the same mental phenomena, like 'self-awareness', 'tool-making capacity', 'intentionality', even 'consciousness' (although it is clear that many of them are used only loosely). For example, we shall justify the existence of 'intelligence' in the SETI label by recalling that what we have been seeking thus far are entities possessing material culture (enabling the creation of sources of radio or optical signals) and intentional behavior (enabling modulation of such signals to reflect some meaning, either for internal or external use).

Observe that the working philosophy of both proponents and opponents of REH can be formulated in terms of the extended continuity thesis. While proponents obviously deny the extension between biogenesis and subsequent complexification giving rise to all metazoa, there can be various alternative views, supported by various opponents of REH, who would prefer cutting the ladder somewhere higher (or not cutting it at all!). Thus we could have an astrobiological landscape dominated by metazoans, but with exceedingly rare instances of noogenesis; or a landscape with many 'noogenetic peaks', corresponding perhaps to the 'Galactic Club' of intelligent species; or, maybe, a landscape which is so dominated by old peaks/plateaus, corresponding to postbiological stages of evolution, so different from us that there is only a very slight chance we would ever recognize them as such. I shall return to this topic in Chapter 8.

A critic might argue there is a sleight of hand in the discussion of the grand cosmological scene thus far, where the Archipelago of Habitability is confused with what should be called the 'Archipelago of Observership'.[38] In other words, there should be an Archipelago *within the Archipelago*, containing a subset of the universes capable of supporting not just any kind of life, but intelligent observers as well. The reason why I do not consider this cheating – and in fact not particularly problematic – is that I do not find it plausible to cut Haldane's ladder too soon and lose the powerful epistemological support which the extended continuity thesis offers us. As usual with the astrobiological landscape, the mathematical apparatus of the future theory of astrobiological dynamics will hopefully provide us with the relative frequencies of, for example, 'simple life', 'complex life' and 'intelligent life' habitats. At the same time, such envisioned future theory will enable us to understand better the very concepts we are using. Since this is still in the future, we cannot be certain one way or another, but for practical purposes (to which I shall pass in the next two chapters) whether the habitats for intelligent life are an order or a couple of orders of magnitude sparser than, say, the habitats of 'just' complex life forms, makes no substantial difference.

If the continuity thesis extends to noogenesis – and, if we follow Haldane's recipe still further, to the origination of material culture – it has important consequences in both theoretical and practical domains. One controversial practical consequence relates to SETI projects: in contrast to the 'lucky accident' view, the continuity thesis, coupled with our understanding of the age-distribution of habitable planets, suggests that the number of viable SETI targets in the Milky Way is substantial. Of course, saying that is not very precise; and it stands in sharp contrast to the clear prediction of REH that the number of such targets will be zero (or very close to zero). The asymmetry is due to the same fact that there is no general theory of complex parts of the astrobiological landscape – no theoretical framework for what I have called 'astrobiological dynamics' creating a reference space for the astrobiological landscape (see p. 79 above). Jorge Luis Borges in the famous short story *Tlön, Uqbar, Orbis Tertius* describes a world in which no book that does not contain its own anti-book is considered finished.[39] So far, there has been no anti-book, in a serious sense, to *Rare Earth* – and this enormously complex task remains a challenge for astrobiologists of the future.

If, ironically, SETI has the potential to quickly falsify REH and reaffirm Copernicanism, we need to take it very seriously indeed. Why does, then, even a cursory look at literature and curricula – not to mention the science press corps – show that it is patently not the case? The subject which is, by any reasonable metric, still in the embryonic stages of its development is, instead, either considered outdated or it stays on the level of freakish curiosity, which is best relegated to recreational contexts. And, curiouser and curiouser, the attitude is present within contemporary *astrobiological* circles as well. Can we make a rational case for SETI that will include neither naivety nor anthropocentrism?

7

SETI and its discontents

It is what opposes that helps.

Heraclitus of Ephesus

The next crucial step or major evolutionary transition after noogenesis is the formation of material culture and technology. Is there a common ground on which such a degree of astrobiological complexity can be meaningfully subject to scientific discourse, no matter where and when it appeared in the Milky Way astrobiological landscape? A tradition that has recently celebrated its half-centennial gives a strongly affirmative answer: SETI was set up by its pioneers, Morrison, Cocconi, Drake and others to search for particular manifestations of other technological civilizations in the Galaxy. In other words, SETI probes that particular high-complexity region of the astrobiological landscape. It is still our best hope to answer the third of the Canonical Three questions of astrobiology: *What is the future of life and intelligence on Earth and in space?*

This is not prejudicating whether SETI has performed its task poorly or admirably thus far. As I will amply demonstrate, there is a great deal of room for improvements and even a radical re-orientation. However, these practical issues are less important than the ground truth that, in the same way as noogenesis is a continuous part of universal evolution, SETI is a continuous part of the overall astrobiological enterprise. Contrary to widespread prejudice, it does not matter whether intelligent beings are intrinsically frequent or rare. Mammals are rare, while bacteria are frequent. Galaxies are few, while grains of interstellar dust are numerous. Does that make microbiology or the study of interstellar dust more important and worthy of funding than mammalian zoology or Galactic astrophysics by any sensible metric? Obviously not. There is no such 'numerical parsing' of nature. The search for microbial life on Mars or Europa, representing a stand-in for low-complexity Galactic life, and the search for manifestations of high-complexity life, such as radio-signals and artefacts, belong to the same continuum of investigations of the astrobiological landscape. As I have argued in relation to REH, to reject this is to reject the continuity thesis and abandon hope of rationally understanding and simulating either biogenesis or noogenesis, or other such crucial steps. This is a steep price to pay – and for nothing except exclusivist and anthropocentric illusions.

SETI studies (defined operationally here as all research on the high-complexity end of the astrobiological spectrum, not necessarily just those studies that have been actually undertaken) do have some specific and apparently peculiar problems and issues. Part of the reasons reflects the specifics of the targets of search – technological civilizations – but part is a historical artefact of the development of science. The questions motivating SETI are very old, going back to ancient philosophical discourses, but the search itself emerged as a scientific discipline in 1959–1960, which, in turn, is much older than the ongoing astrobiological revolution. This particular chronology is at least partially responsible for some of the tenets of SETI and the 'new' astrobiology being out of synch, the state of affairs I shall discuss at length below. We are now in a position to overcome this burden of the past. Before that, though, it is important to discuss the other side of the story: the *inherent* problems facing the search.

The 'Big Three': classic anti-SETI arguments

First, I will consider three classic arguments for scepticism regarding the existence of SETI targets. These are, in the conventional chronological order:

1. Fermi's paradox;
2. The argument from biological contingency;
3. Carter's 'anthropic' argument.[1]

Fermi's paradox (originally formulated by Enrico Fermi during a lunch at Los Alamos in 1950)[2] consists in the absence of extraterrestrial intelligent beings and their manifestations or traces in our past light cone, coupled with our knowledge of the spatio-temporal scales of the Galaxy, and some additional philosophical assumptions (naturalism, some kind of realism, etc.). I shall not explicitly discuss Fermi's paradox here, since it is a vast area, deserving of much deeper scrutiny, especially since progress made in astrobiology has actually made it *more* disturbing in recent years, and we are nowhere near resolution. Some comments about modern forms of Fermi's paradox will be given in surveying philosophical criticisms of contemporary SETI projects, but since this topic is excessively complex, to try to do more within the scope of the present book would do injustice to the problem. *Everything should be as simple as possible, but not simpler* (Einstein).

The second and third arguments are less widely known, but they have played – and are still playing – an immensely important role in the SETI-related discourse. The argument from biological contingency, also known as Simpson's argument, was present, as we have seen, in a vague form in Wallace's 1903 book, but was posed in its specific form by the great palaeontologist George Gaylord Simpson in 1964 in a paper indicatively titled *On the Nonprevalence of Humanoids*. It argues that the probability of anything even remotely similar to humans evolving on any other planet is minuscule, since biological evolution is dominated by contingency, and even if conditions anywhere else were at early times similar to those on Earth, the outcome of evolution would be entirely different. As I have discussed in Chapter 5, this is equivalent to Gould's 'replaying the tape' many times throughout the Galaxy.

Carter's argument, the latest of the 'Big Three', is based on our ignorance about the characteristic temporal scales of biological evolution leading to the emergence of intelligent observers. On the other hand, the temporal 'window of opportunity' for the emergence of such observers on any particular planet is defined by rather well understood astrophysical and ecological processes. How come, then, that the two relevant timescales are very close in the case of Earth? By understanding this as another 'anthropic coincidence', to be explained as an observation-selection effect, Carter proceeds to show how the existence of other intelligent observers in the Galaxy *at the present time* is improbable.

Before we go into more details about these fascinating arguments, it is worth re-emphasizing that such working examples clearly show why philosophy is relevant to astrobiology. None of those arguments can be resolved by any conceptually simple empirical observation or experiment. Of course, we do hope to be able to resolve the problems derived from the 'Big Three' in an empirical manner – but there is no consensus about what kind of empirical research will lead to resolution, what the desiderata and modes of setting up are for such a research programme, or which outcome will count as a legitimate solution. Clearly, these are philosophical issues. It is no coincidence that we have encountered such deep conceptual issues in a young and immature field. This parallels the Kraghian discussion of the relative importance of philosophical arguments at the time of immaturity of cosmology, as encountered in Chapter 2.

Another caveat is necessary in order to be able to assess these sceptical arguments without much background noise. While the sceptical arguments are entirely legitimate scientific arguments, it is hard to avoid the impression that at least a part of their use in debates and controversies about astrobiology and SETI has been motivated by the idea that official sanctioning of scientific work on extraterrestrial life – no matter that it is only a part of the whole of astrobiological endeavour – indirectly supports various sorts of pseudoscience and charlatanism: UFOs, 'ancient astronauts', etc. Of course, one does not need sophisticated philosophical analysis to see that this 'argument' is illogical. Some analogies, for instance, are to hold hydrodynamics responsible for speculations about 'memory of water', solid-state physics for the 'healing power of crystals', or quantum mechanics for various items of quantum mysticism. Any topic should be approached with extreme caution, but in any case, *quacks should not be given enough power to exert influence upon the course of scientific research.*

The argument from biological contingency

In a 1964 paper, Simpson argued that ubiquitous biological contingency prevents the emergence of any species capable of meaningful communication with us; thus, contingency makes SETI projects unlikely to succeed.[3] While he did not argue that this would make such projects unjustified in principle (after all, they could be construed as *testing* the importance of biological contingency), subsequent authors who used this argument definitely argued for such a stronger conclusion. Let us try to formulate the argument in syllogistic form (while keeping in mind that Simpson did not formulate it in this manner), using the conventional concept of evolutionary morphological space:

1. That part of the morphological space representing intelligent, communication-capable species is congruent with the part of the morphological space representing humanoids.
2. The part of the morphological space representing humanoids is of negligibly small measure within the entire morphological space.
3. The rate of independent evolution of intelligent, communication-capable species on habitable planets within a sufficiently large ensemble of such planets is proportional to the size of the part of the morphological space representing such species.
4. *Homo sapiens* are an intelligent, communication-capable species currently existent in the Galaxy.

Hence,

5. The probability of *Homo sapiens* detecting another intelligent, communication-capable species within the ensemble of habitable planets in the Galaxy is negligible.

Much is uncontroversial here. Premise (4) is obviously true.[4] On the other hand, premise (3) is a reasonable mathematical consequence of our conventional understanding of stochastic events and our concept of the biological morphospace. It is similar to the phase-space interpretation of the famous Boltzmann formula

$S = k \ln W$ relating entropy and statistical weight ('probability' in a loose sense). The 'size' of any part of the morphological space could be understood as its n-dimensional volume. Note also that for conclusion (5) to follow it is not necessary that proportionality is a linear relation; we can use weaker versions of (3) in which the rate of evolution of intelligent species is simply any non-decreasing function of the size of the corresponding chunk of morphospace. Premise (2) is nothing less than Copernicanism applied to biological evolution. The late epoch of evolution of hominids clearly testifies to its truth; Simpson and other evolutionary biologists sceptical toward life and intelligence on other worlds list many additional arguments for this premise.

Thus we come to premise (1), which is the prime suspect, and not only because its formulation is by necessity a complex one. Gould, in his criticism of Simpson's argument and its use by Tipler and other SETI-pessimists, invoked exactly the distinction between the general and particular claim as the underlying logical fallacy.[5] The particular claim is the supposed low relative frequency (= rarity) of humanoids; the general claim is the supposed low relative frequency of any kind of intelligence. Gould and many proponents of SETI deny the general claim. This would amount to the alternative hypothesis that the part of the morphospace corresponding to intelligent observers is significantly larger than the part corresponding to humanoids. These alternatives are shown in Figure 7.1 as options (A) and (B). Simpson's argument corresponds to case (A), in which the darkened part of the morphospace representing intelligent beings is extremely small, and Gould's rejoinder corresponds to case (B), in which the same part is of moderate size. Of course, since morphospace is many-dimensional, with perhaps a very large number of dimensions, it is impossible to represent graphically other than to use symbols. We might be sure that by far the largest part of it corresponds to simple life forms, i.e., microorganisms. Now, the question on which Simpson's argument hinges is whether the reality corresponds to case (A) or case (B). In the final analysis, it depends on how high the acquisition of intelligence moves the given life form towards the peak of fitness landscape. If intelligence is very advantageous, then we could be right in speculating that there are many different evolutionary trajectories leading to such an important trait. If, on the other hand, intelligence is only weakly advantageous or neutral, or – in the extreme pessimist view which should not be discarded – maladaptive, it will be ephemeral (on an evolutionary timescale) on Earth and its chunk of the general morphospace is negligible. We can think of many ways intelligence could be advantageous, but any such discussion is extremely prone to anthropocentrism; not surprisingly, such views are many. On the other hand, there is one particularly disturbing family of scenarios for the *negative* adaptive value of intelligence – those of anthropogenic existential risks, where humans destroy themselves or permanently curb their creative potential. Obviously,

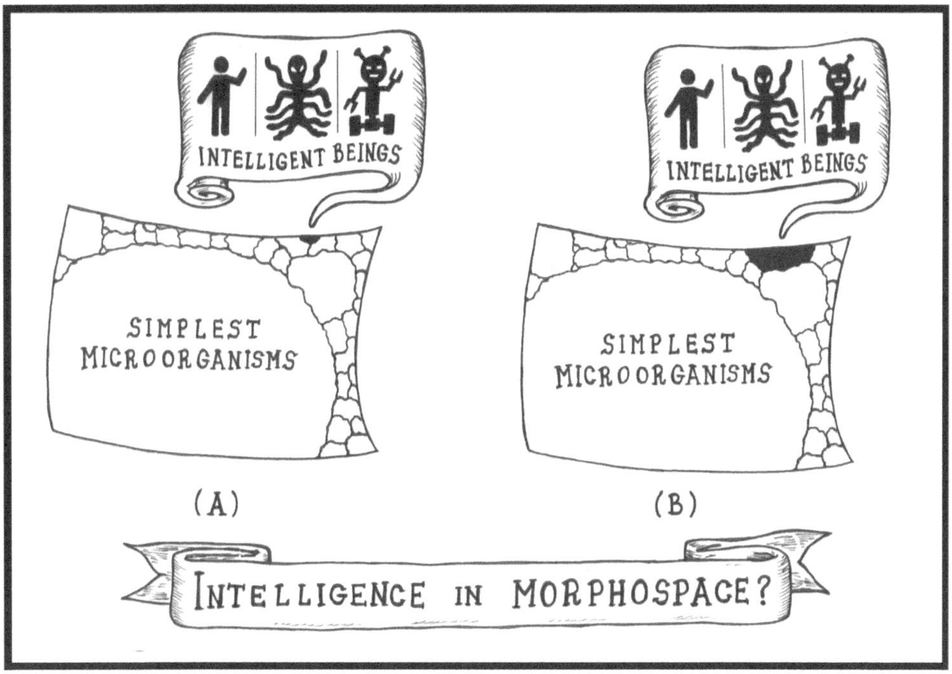

Figure 7.1. A schematic representation of alternative ways of looking at the evolution of intelligence. In case (A), intelligent beings, humanoid or non-humanoid, occupy a negligible part of the evolutionary morphospace, making the emergence of intelligence elsewhere highly unlikely (even the size of that part is here greatly exaggerated for graphic purposes – it is best to think about it in terms of something that is 10^{-100}, or some similar astronomically small fraction of the total). In case (B), intelligence of various kinds is represented by a sizeable chunk of the morphospace, making its parallel evolution in various habitable locales likely. (Courtesy of S. Popović.)

if any of the adverse outcomes are realized, it will be due to our intelligence, one way or another, so an impartial observer would be forced to assign a large negative value at least to our kind of intelligence. Here, it is easy to notice how tightly astrobiological thinking is tied to future studies – roughly, optimism towards the future of intelligence on Earth tallies well with case (B), which in turn would imply high relative frequency of emergence of intelligent observers and justification for SETI. Conversely, pessimism with respect to intelligence on Earth could be associated with the validity of premise (1) in Simpson's argument and negligible chances for SETI success. It is clear that much further research about human evolution (both biological and cultural) and the nature of intelligence is necessary before we could empirically discriminate between cases (A) and (B). In the meantime, I think it is very difficult, to say the least, to use the argument from biological contingency as an argument for cessation of practical SETI research.

Let me sketch one of the specific ways in which premise (1) might be undermined. The encephalization quotient (EQ: the ratio of cerebral mass to the body mass of an organism) is often used as a metric, which at least gives some pointers to a physical description of intelligence(/self-awareness/tool-making capacity). EQ has generally increased since the Cambrian explosion, which might be regarded as an instance of the general increase of complexity in that period. However, its rate of increase was faster until 251 Myr BP, i.e., until the famous end-Permian mass extinction episode, the harshest mass extinction in the Phanerozoic eon. Afterwards, it continued to increase, to reach the historical maximum in *Homo sapiens*, but at a slower pace.

Nowadays, no serious palaeontologist would argue that the end-Permian catastrophe, 'when life nearly died', to quote a title of Michael Benton's popular book, was *necessary* in the historical sense.[6] (The qualification 'in the historical sense' is required here in order to avoid fruitless debates about metaphysical determinism.) Therefore, it is possible to argue that the rate of EQ increase during the Palaeozoic could have brought about the emergence of intelligent observers much earlier if the end-Permian cataclysm did not happen. This statement ought to be distinguished from the 'naive' convergentism criticized in Chapter 5; there is certainly no reason to believe that intelligent observers which would have evolved in such a manner (without the end-Permian extinction) would have been mammals or remotely similar to humanoids. Thus, if confirmed by further palaeontological research, the existence of a 'knee' in EQ history could be construed as indirect evidence that noogenesis can be approached from different evolutionary directions, undermining the argument from biological contingency.

Another problem with the usual formulation of the contingency argument is that it is 'timeless' – it does not take into account the fact that intelligent beings also evolve, so the corresponding chunk of the morphospace may not stay the same at all times. This is, admittedly, of minor importance when it comes to practical SETI activities, since evolutionary timescales are so much longer than those associated with practical searches, but it is still rather undesirable in a general argument.

There is another version of Simpson's argument, which is not based on the rate of evolution of intelligent species per se, but on *our capacity* for detecting radically different forms of intelligence and *their capacity* for communication – and the duration of both. I shall call it the Fiasco argument, since in the last novel of Stanislaw Lem, *Fiasco*, a version of this argument has been advanced.[7] It can be schematically represented in a similar manner as the above:

1′. That part of the morphological space representing intelligent species detectable by humans (the 'cabin of contact') is not much larger in comparison to the part of the morphological space representing humans.

2′. The part of the morphological space representing humans is of negligibly small measure within the entire morphological space.
3′. The rate of passage of intelligent species through a part of the morphological space corresponding to intelligent species detectable by humans is proportional to the size of that part of the morphological space.
4′. *Homo sapiens* are an intelligent, communication-capable species currently existent in the Galaxy.

Hence,

5′. The probability of *Homo sapiens* detecting another intelligent, communication-capable species in the Galaxy is negligible.

This is still closely related to Simpson's argument (we note that premises (2) and (4) are identical to (2′) and (4′) above), but has some specific advantages. First, the Fiasco argument explicitly takes into account the fact that intelligent species themselves evolve – without prejudicating which sort of evolution guides further changes (socially mediated epigenetic rules? Lamarckian cultural evolution? post-biological evolution? etc.). The 'cabin of contact' is defined in an operational way, without prejudicating upon the deep definitional issues of 'what counts as an observer?' type. We can imagine the evolution of the high-complexity sector of the astrobiological landscape as intelligent species emerging, evolving at different rates, entering the cabin of contact, and subsequently at some point exiting it. We can speculate what happens next, but not really know it without empirical research *in situ*, which obviously is on a completely different timescale to SETI projects. After exiting the cabin of contact, intelligent communities may become extinct, or devolve to non-intelligence again, or achieve some sort of 'transcendence', either technological (as in the popular Singularity-type scenarios) or mystical, or completely unimaginable to beings with limited cognitive abilities such as ourselves. In any case, they move beyond the domain of the operational definition – beyond being viable SETI targets.

Therefore, we may prefer the Fiasco argument to the original version of Simpson, since the former does not constrain astrobiological complexity per se. Similarly, it cannot be easily linked to either pessimism or optimism towards the future of intelligence on Earth. It is easy to see why this is so: if civilizations can leave the cabin of contact without being destroyed (by either external or internal agencies), there is no obvious reason why they could not continue to increase in complexity. Many 'transcendence' scenarios are exactly of this type.[8] Such a fate for humanity could be appealing to many cultural optimists, as well as to the critics of SETI.

Can we escape the Fiasco argument and retain meaningful and practical SETI discourse? The price paid by the sceptic here is that the cabin of contact is hard to define precisely. Intuitively, it seems a more flexible entity than those set by purely biological considerations – there is no obvious reason for doubting that it will expand with diversification of the scope of practical SETI techniques. Although there may be an absolute limit to the cabin of contact, it is hard to tell a priori how far the limit is in comparison to the clearly narrow scope of historical SETI activities so far. It seems that only diligent and persistent work on finding new and applying already known methods of search can give us the correct answer. Overall, while both Simpson's and the Fiasco argument are incredibly important and thought provoking, they are hardly as decisive as some of the critics have presented them. Rather, they are wake-up calls for more serious work on both the theoretical and practical sides of SETI.

Carter's argument

The well-known argument against the existence of extraterrestrial intelligence, proposed by the astrophysicist Brandon Carter,[9] can be characterized as follows:

CA: If astrophysical (t_*) and biological (t_b) timescales are truly uncorrelated, life in general and intelligent life in particular form at random epochs with respect to the characteristic timescale of the astrophysical environment (notably, the Main-Sequence lifetime of the considered star). In the Solar System, $t_* \approx t_b$, within a factor of two. However, in general, it should be either $t_b \gg t_*$ or $t_b \sim t_*$ or $t_* \gg t_b$. The second case is much less probable a priori in light of the independent nature of these quantities. Carter also dismisses the third option, since in that case it is difficult to understand why the only planetary system known to be inhabited (that is, the Solar System) exhibits $t_* \approx t_b$ behavior. On the contrary, one would then expect that life (and intelligence) arose on Earth, and would predict that it arose at other places in the Solar System, much earlier than they in fact did. This provides a weak probabilistic reason to believe that $t_b \gg t_*$ (in which case the observation selection effect explains very well why we do perceive the $t_* \approx t_b$ case in the Solar System). Thus, extraterrestrial life and intelligence have to be very rare, which is the reason why neither has been observed to date, in spite of the conjecture that favorable conditions for them exist at many places throughout the Galaxy.[10]

In other words, the discrepancy between the two timescales makes our appearance on the cosmic scene exceptionally early (which should not surprise us, when anthropic self-selection is taken into account) in comparison to the bulk of the distribution of ages of intelligent observers in a sufficiently large volume of space-time. The right question to ask is not *Are we alone?*, but rather (just like the very first sentence of Steven Spielberg's famous 1977 movie, *Close Encounters of the Third Kind*), *Are we the first?* Carter's argument has been given a great deal of attention in astrobiological and SETI-related discourses. Many diverse authors hold CA in high regard.[11] SETI sceptics, such as Frank Tipler, routinely use it to support the claim that we are alone in the Galaxy.[12]

The crucial assumption of $\mathbb{CA}$ is that there is no a-priori reason for correlation between t_* and t_b. Astrophysicist Mario Livio pointed out in 1999 that this is the main weakness of this argument; processes that *induce* correlations between the two timescales, like the oxygenation of the atmosphere on terrestrial planets, undermine the argument.[13] Notably, if stellar UV radiation prevents the appearance of land life due to high absorption by nucleic acids and proteins, then it is critical that a sufficient ozone layer occurs before land life could appear. This, in turn, might induce a correlation between the astrophysical and biological timescales, since various stellar masses (and thus various stellar lifetimes) will generate different amounts of UV radiation and dictate various rates of oxygenization of the atmospheres of hypothetical planets in their respective habitable zones. With the elimination of this essential assumption of the independence of the two timescales, $\mathbb{CA}$ is significantly undermined.

Some coincidences are not really related to any observation-selection effects, but instead are consequences of our poor knowledge of the range of the relevant parameters. When Kepler noticed that the number of planets in the Solar System is equal to the number of regular Platonic polyhedra, this coincidence impressed him so much that he constructed a whole mystico-cosmological system around it. We know nowadays that there are more than five planets – and even that the definition of a planet is not God-given or given by nature, as the debate over the status of Pluto has shown in recent years – but the number of Platonic solids remained the same, so Kepler's coincidence simply evaporated. I submit that Carter's coincidence is more of the Keplerian kind than, for instance, the Dirac–Dicke coincidence considered in Chapter 3.[14]

$\mathbb{CA}$ relies on the assumption that there are fixed (or at least well defined) and roughly known timescales for astrophysical processes. The alternatives encompass (i) external physical forcings acting on local biospheres throughout the Galaxy, (ii) elements of complex evolution, namely quasi-periodicity, stochasticity, change of macro-evolutionary regimes and secular evolution with cosmological time, and (iii) biotic feedback, in the sense that timescales may change 'on the move', before noogenesis, due to interaction of the biosphere with its physical surroundings.

Let us first redefine the astrophysical timescale t_* as the timescale of *continuous habitability* of a terrestrial planet in the Milky Way galaxy. The difference may sound pedantic, but is in fact crucial when it is recognized that (astro)physical processes other than the evolution of the parent star can influence the habitability of a planet. For instance, the 'Snowball Earth' episodes that occurred at least twice in the geological past represent global catastrophes that may have annihilated all life, except for the small habitats around marine volcanoes and hydrothermal vents.[15] It is entirely plausible that similar episodes of severe global glaciation could have annihilated all life at Earth-analogues elsewhere, so that the 'astrobiological clock' gets a complete reset; possibly even without any external causative agent, but due to an unfortunate combination of plate movements and Milanković cycles.[16] Any such large-scale trends make $\mathbb{CA}$ a posteriori less appealing, since they induce further correlations and have their own quasi-deterministic timescales, which thus undermine the independence assumption. In particular, the need to abandon the 'closed box' astrobiological picture of Earth (and terrestrial planets in general) is emphasized in a number of recent studies from different points of view. For instance, habitable zones are *defined* as spatio-temporal regions where conditions for life arise due to *correlated* processes. As far as prospects for SETI are concerned, the relevant region is GHZ, which occurs as a consequence of roughly understood processes of the chemo-dynamical evolution of the Milky Way and its stellar populations. Even more telling is the concept of the Cosmic Habitable Age as the period in astrophysical history in which habitable planets occur.[17] Insofar as habitable zones are an unavoidable part of the modern astrobiological discourse, any argument based on the independent development of biospheres automatically looses force.[18]

Biological timescales, obviously, are still very poorly understood in the most general cosmic context; a large part of the astrobiological revolution deals with the attempts to elucidate them. Palaeontologist Dale Russell has claimed that the appearance of intelligent beings occurs on the average about 3 Gyr after the initial stages of planetary formation.[19] Much shorter timescales have been proposed: astrobiologist Christopher McKay argued that plate tectonics actually *delayed* the appearance of complex life forms on Earth by keeping the level of oxygen low for a long time. According to that idea, the duration of environmental conditions such as those that were experienced during the Precambrian could be as low as 10^8 yrs on planets without plate tectonics, such as Mars.[20] This, in turn, would significantly accelerate the emergence of sufficient complexity as a precondition for intelligence. It is in this astrobiological key that we can reiterate part of Wilson's philosophical criticism of $\mathbb{CA}$ contained in the following passage:[21]

> At first glance, the claim that t_e [the duration of evolution on Earth up to noogenesis] should not differ from a given value of $\bar{t}$ [the likeliest biological timescale for noogenesis] seems to be equivalent to the claim that $\bar{t}$ should not differ from a given value of t_e. But these claims are fundamentally different. The reason the latter one is invalid is that $\bar{t}$, insofar as it represents the time that evolution is intrinsically most likely to require, is a probabilistic or statistical quantity. Our knowledge of the value of such quantity cannot be significantly enhanced by the evidence of a single case, especially the nonrandomly chosen one of our own evolution. Only if we were to become aware of a large number of actual cases of extraterrestrial evolution and their corresponding timescales, or if we were to advance our knowledge of the timescales governing various evolutionary mechanisms, could we provide a reasonable estimate of $\bar{t}$, and perhaps eliminate values of $\bar{t}$ much less than τ_0

[Sun's lifetime]. But given only $t_e \sim \tau_0$, we cannot on the basis of this single evidential sample conclude much at all about $\bar{t}$. We certainly cannot eliminate, as Carter thinks we can, the possibility that $\bar{t} \ll \tau_0$.

To see further weaknesses of $\mathbb{CA}$ we need to abandon one of the most cherished prejudices of the nineteenth and most of the twentieth century, which is uniformitarianism of rate (or gradualism). It is not only that the habitability of an astrobiological site is not constant in time, but the frequency of important events (like the mass extinction episodes) is changing with the evolution of the Galactic system. As already mentioned, there are a host of recent indications that the Solar System is in fact an open system, strongly interacting with its Galactic environment.[22] Interactions induce correlations; correlations ruin arguments based on independence assumptions and coincidences.

When we reject hidden uniformitarianism, even the $t_b \ll t_*$ case of Carter's dilemma is not to be rejected so lightly. An obvious counter-example in this respect is the much-debated 'impact frustration' of early life forms.[23] It is conceivable that early terrestrial life appeared independently several times, only to be destroyed by catastrophic impacts during the epoch of the so-called late heavy bombardment. Only when the frequency of impacts decreased sufficiently were early life forms capable of spreading, diversifying and evolving in such a way as to produce a rich and complex terrestrial biosphere. If this was so, $t_b' \simeq 10^7 - 10^8$ yr $\ll t_*$, where t_b' denotes the timescale for biogenesis. While we cannot still infer anything about the timescale for *noogenesis* (t_b), if for any reason there were an upper limit to t_b' (related perhaps to the chemical evolution of Earth's atmosphere or surface and the increase of the incident solar radiation flux), it could be perfectly conceivable that life could entirely miss the temporal window of opportunity due to the impact interruptions. This scenario is instructive, since it shows both (a) the strong coupling between life forms and physical environment, and (b) the timescale forcing through which a physical timescale – the interval between major impacts – actually becomes the relevant quantity.

If we accept that $\mathbb{CA}$ is at least severely limited by the non-uniformitarian history of life, it is natural to ask for more details about the major non-uniformities that impede its 'monotonic' progress.[24] As mentioned in the discussion of Olum's problem, modern astrophysical research offers much in this respect. An important paper of astrophysicist James Annis opened a new vista by introducing (though not quite explicitly) the notion of a *global regulation mechanism*; that is, a dynamical process that prevents or impedes uniform emergence and development of life all over the Galaxy.[25] In Annis' model, called the phase-transition model for reasons soon to be obvious, the role of such global Galactic regulation is played by gamma-ray bursts (henceforth GRBs); colossal explosions caused either by the terminal collapse of supermassive objects ('hypernovae'), or mergers of binary neutron stars.

Astrobiological and ecological consequences of GRBs and related phenomena have been investigated recently in several studies.[26] To give a flavour of some of the results, Arnon Dar calculated that the terminal collapse of the famous supermassive object Eta Carinae could deposit, in the upper atmosphere of Earth, energy equivalent to the simultaneous explosion of 1-kiloton nuclear bombs per km^2 over the surface of the hemisphere facing the hypernova.[27] According to some calculations, a Galactic GRB can be lethal for eukaryotes up to a distance of 14 kpc.[28] Thus, this 'zone of lethality' for advanced life forms is bound to comprise the entire GHZ whenever a GRB occurs within the inner 10 kpc of the Galaxy. Since the regulation mechanism exhibits secular evolution, with the rate of catastrophic events decreasing with time, at some point the astrobiological evolution of the Galaxy will experience a change of regime. When the rate of catastrophic events is high, there is a sort of quasi-equilibrium state between the natural tendency of life to spread, diversify and complexify, and the rate of destruction and extinctions. When the rate becomes lower than some threshold value, intelligent and space-faring species could arise in the interval between the two extinctions and make themselves immune (presumably through space-faring technology and astroengineering; see Chapter 8) to further extinctions.[29]

GRB regulation has an important correlation property: the rhythm of biological extinctions should be synchronized (up to the transport timescales of $\sim 10^4$ yrs for both γ-rays and high-energy cosmic rays) in at least part of the histories of all potentially habitable planets. A bold hypothesis has been put forward recently by Adrian Melott and collaborators that a known terrestrial mass extinction episode, one of the 'Big Five' (the late-Ordovician extinction, *c.*440 Myr ago), corresponds to a Galactic GRB event.[30]

Such an extinction would undermine Carter's argument. With a set of modest additional assumptions, it is possible to show it quantitatively for the simplest cases. For instance, in Figure 3.3, we have seen the results of a toy numerical experiment designed as a simple realization of Annis' astrobiological regulation model. The system exhibits a shift of behaviour as we move from small values of Q (gradualism) to large ones (catastrophism). At large Q, we have a step-like succession of astrobiological regimes, governed by external timescale forcing. In each regime, it is obvious that the ages of inhabited planets are not independent; on the contrary, they are temporally correlated. In other words, (neo)catastrophism removes, ironically enough, the basic tacit assumption of $\mathbb{CA}$. If the agents of extinction are correlated over the spatial scale of the GHZ, timescale forcing undermines Carter's reasoning in a natural way. (It is important to emphasize that in the simulation above, all sources of correlations between the life-bearing sites barring GRB regulation are neglected. Some processes – like panspermia, either natural or directed – certainly deserve to be taken into account; these will only strengthen the correlations, thus decreasing our confidence in $\mathbb{CA}$.)

It seems too early to draw sceptical conclusions about the abundance of extraterrestrial life and intelligence from our single data point via Carter's argument. In addition to other deficiencies of the argument pointed out in the literature, we emphasize that a picture in which regulation mechanisms reset local astrobiological clocks (which, consequently, tick rather unevenly) offers a way to reconcile our astrophysical knowledge with the idea of multiple habitats of life and intelligence in the Galaxy. In other words, Earth may be rare in time, not in space. The unsupported assumption of gradualism is identified as the main source of confusion and unwarranted scepticism. The astrobiological picture presented here can be understood by means of loose analogy with the much-discussed theory of punctuated equilibrium in evolutionary biology. Seeking to explain the evident stop–start nature of the fossil record, Eldredge and Gould proposed the theory of punctuated equilibria.[31] According to this theory, species tend to remain stable for long periods of time ('stasis'). The equilibrium is punctuated by abrupt changes in which existing species are suddenly replaced (on a geological timescale). The astrobiological analogy of palaeontological stasis can be found in the large Q parts of Figure 3.3, where we see long periods (of $\sim$1 Gyr duration) with the same number of inhabited planets before a sudden change.

This undermining of Carter's argument accords with Copernicanism, showing simultaneously how a Copernican tendency comes as a *consequence* of the physical discourse, not as a sacred dogma. A natural generalization of the history of the terrestrial biosphere to the case of the Milky Way from the astrobiological point of view entails the acceptance of a sort of Galactic punctuated equilibrium: episodes of multiple noogenesis separated by a period of stasis. In this picture, in contrast to $\mathbb{CA}$, there is no fixed, unique, reified timescale distribution. The tremendous progress in astrobiology clearly demonstrates that the oversimplifications inherent in $\mathbb{CA}$ are no longer tenable; if it is largely, as Carter himself admitted, an argument from ignorance, then any decrease in our ignorance ought to prompt its reassessment and re-evaluation.

Beyond epistemology: the context of the campaign

Now we dive into murkier waters. There are three types of 'foundational' criticism of SETI, not to be confused with the 'Big Three' anti-SETI arguments I have discussed above. These pertain to the field as a whole, and are encountered in the literature over the course of the last half century:

(1) Criticism motivated by open or thinly veiled extrascientific beliefs and goals. Some traditional critics of pluralism with respect to life and intelligence in the universe invoked religious reasons, like the uniqueness of Christ's life and mission, to argue against deity creating multiple planets with intelligent observers.[32] Vestiges of such a form of thinking can be recognized in some modern criticisms ostensibly based on naturalism, but still retaining religious anthropocentrism and endowing humanity with a cosmically significant mission. We have seen the example of Wallace's protoastrobiological theory as one such statement, though the same basic style of thought is present much closer to our time in, for instance, Tipler's *Physics of Immortality*.[33] To this category also belong criticisms based on the magnitude of the US federal debt (!), or other accusations of wastefulness.[34] Of course, irrespective of its scientific status, SETI is a social activity that cannot be entirely separated from its cultural environment, but this is an old-hat truism.
(2) 'Philosophical' criticism denying the existence of SETI targets, our justification for engaging in the search, and/or the value, either cognitive or practical, of SETI projects on general grounds. The common denominator of this category is immanent criticism and the downplaying of empirical evidence. As we shall see, the basis for this is usually a curious mixture of scepticism, historicism, misunderstanding of evolutionary contingence and semantic confusion; in addition, it is often accompanied by heavily biased, or even hysterical, rhetorics against the denounced 'believers'. The campaign which labels SETI as pseudoscience on the basis of an alleged consensual view of what science must be belongs to this category.
(3) Criticism of the actual practices and methods of SETI, which accepts its general justification, but aims at a more appropriate or efficient practical approach. Propositions suggesting that, for example, optical or infrared surveys are more appropriate than conventional radio searches, belong to this category, as does the seminal 2003 study of Steven J. Dick on the necessity of re-evaluation methods and priorities in light of our increased awareness of the possibility of postbiological evolution.

There is no sharp boundary between these three types and we usually encounter some mixture of all three.[35] I wish to avoid discussing type (1) criticisms, but discuss in some detail type (2), and cautiously endorse type (3). Type (1) criticisms are hard to discuss, even when entirely sincere, since extrascientific elements are either impossible or at best very difficult to judge rationally; whether the Creator exclusively chose Earth and us for salvation, whether savings on SETI or, indeed, any kind of science savings (against a background of skyrocketing military expenditures) could play any role in balancing any budget, or whether we should relish or shudder at the prospect of (post)humans colonizing the empty Milky Way might be important issues, but are beyond the domain of present-day science and naturalistic philosophy. Even brute intuitions of various people will diverge wildly on all of these. On the other hand, type (3) criticisms are desirable in any field; that half a century of SETI practice has not been crowned by success makes this type of criticism to some extent unavoidable, even when the underlying SETI philosophy is provisionally accepted. I shall show how some tenets of the type (3) criticism may help alleviate the worst excesses of philosophical scepticism, and will develop its twin theme of detectability/recognition of the artificial in the next chapter.

Insufficiency of philosophical criticism

If you do not expect the unexpected, you will not find it; for it is hard to be sought out and difficult.[36] Contemporary philosophers critical of SETI seem to have forgotten this famous, almost 26 centuries old, dictum of their Ephesian colleague. As prototypical for this genre of criticism, I shall consider three highly visible publications, the 1993 paper by Roy Mash published in the prestigious *Philosophy of Science*, the 2006 book by the historian and philosopher of science George Basalla, entitled *Civilized Life in the Universe*, and a section dealing with SETI in the recent popular philosophy book, *Nonsense on Stilts*, by a distinguished philosopher of biology, Massimo Pigliucci.[37] Of course, this is a subjective choice; other representative choices could be made, since a small cottage industry of criticisms of SETI has emerged recently, somewhat similar to the cottage industry of anthropic reasoning critics discussed above.[38]

First a disturbing note. The tone of SETI criticism is often mocking or even dismissive, in contrast to the professed aims and style of philosophical discourse. The language is telling: proponents of the existence of extraterrestrial intelligent observers are commonly called 'enthusiasts', or 'believers', or 'SETI honchos'. Basalla's book excesses in this type of rhetoric, in particular, many irritating alternations between locutions 'SETI researchers' and 'SETI believers' applied, as clear from the context, to the same group of people. Such rhetorical usage is, essentially, aimed at stigmatizing philosophical opponents and implying that belief – which is, of course, an integral part of scientific endeavour – is somehow more central to SETI than to quantum field theory, or geophysics, or botany. Such language would be universally considered inappropriate in other fields of study, even in those fields that can hardly be considered less speculative by any reasonable metric. It is rare to encounter talk about 'superstring believers' or 'directed mutation enthusiasts', not to mention 'artificial intelligence honchos'; even when such qualifications are used, they are marked by ellipses or similar cautionary signs and are not repeated throughout the discourse. This is not only a measure of the distribution of power within the scientific community and those sectors of society related to academia, but it also undermines the basic rules of rational discourse in an atmosphere of intellectual honesty.

The paper of Roy Mash ostensibly deals only with the arguments using big numbers coming from astronomy as a rationale for SETI. Although there have, indeed, been many such arguments in published literature, it is important to clarify that this sort of numerical impression is not *necessary* for SETI to be a meaningful activity. A single supercivilization of Kardashev's Type 3, or Type 2.x, could be a much better target then a thousand civilizations similar to ours; to be more precise, it might contain a much larger number of potentially detectable targets than any large number of young technological civilizations taken together. Again, the context of time and (astrobiological) history is at least as important as the sheer number of habitable or inhabited sites in the Galaxy or in the observable universe. There is no logical connection between the belief in ETI being common or numerous and the belief in SETI as a justified scientific enterprise.

Of course, it may not sound that way to fundraisers and politicos. Indeed, we get a sniff of extrascientific motivation straight away: Mash does not fail to mention, in the very first paragraph of the paper that '[a]s of 1992, Congress began funding a new and ambitious ten-year search project at a cost of one hundred million dollars.'[39] He proceeds to analyse two versions of the 'big numbers' argument, the first being one he ascribes to Lucretius, which is similar to the famous 'infinite monkeys writing *Hamlet*' cliché: in an infinite universe everything that is allowed by the laws of physics will happen, so there will be an infinite number of extraterrestrial civilizations, including Earth's exact copies (and many inexact ones). Except for an inexplicable omission of the obviously very relevant issue of cosmological horizons, there is not much to add to Mash's treatment of this Lucretian argument. As Mash himself notices, 'as far as the current ETI debate is concerned, [it] is irrelevant'.[40]

Much more promising is the discussion of inductive arguments, first investigating the bad old Drake equation, concluding that 'we simply lack the theory to attach warranted values to most of the terms'. Then he passes on to an even vaguer terrain of weak induction based on the Principle of Mediocrity – a feeble version of Copernicanism – where large numbers of potentially habitable sites are invoked to justify various generalizations from single cases, namely those related to Earth and its environment and history. Here, Mash offers two related strategies: (1) to deny the validity of the Principle of Mediocrity in its 'pure' form, and (2) to try to show that, even in its 'diluted' form, the Principle entails unpalatable consequences. I shall try to show here that, while the first strategy is at best irrelevant for assessing SETI, the second is more a rhetorical trick than a sound argument.[41] This will offer me a clear shot at clarifying some general issues in the relationship between SETI and the rest of astrobiological research.

In the course of his attack on the 'pure' Principle of Mediocrity, Mash describes a thought experiment (in the form of an amusing dialogue between Aggie and Maggie) in which a huge number of boxes of various sizes are stored in a warehouse. An observer opens a box and finds a black marble – does she really have a good reason to believe that some of the other boxes, too, have black marbles inside? At least one among the huge number of all kinds of boxes? Aggie says yes, Maggie says no. Mash admits that his heart is with Maggie, who argues that there is no consequently definable probability shift occurring due to the observation of a single black marble and that, therefore, no inductive inference from the single case could be made; of course, nothing is wrong with that. What is wrong, though, is to create the impression that this abstract discussion is relevant for practical issues which are, after all, directed by theory and observation. And while Mash a few pages back rightly complains that the Drake equation is lacking theory (such perspective being more justifiable in 1993 than now), in Aggie vs. Maggie, suddenly, the theory is excised with all the 'background knowledge' and only logically pure argument considered.

When the double standard is rejected, it is not difficult to see why this thought experiment is inapplicable to anything even remotely similar to the SETI enterprise: there is no logical or physical connection between the warehouse, boxes in the warehouse, and the black marble. The object found – the black marble – is ingenuously chosen to be quite unrelated to boxes and warehouses, and what we expect to be stored in warehouses. As explicitly stipulated by a protagonist, 'I know nothing at all about the warehouse, what's it used for, who owns it. I know nothing about the boxes, whether they've come from the same place, how they got there.' Such professed ignorance, obviously, cannot be analogous to our astronomical database. On the other hand, Mash performs a balancing act, since not only later in the dialogue a bit of empirical information (or 'background knowledge') is, in fact, admitted in the form of marbles usually possessing gouges and nicks, but there is a hidden assumption Mash's protagonists do not discuss at all: since boxes are 'of various sizes', we can immediately reject the subset of boxes in which a marble does not fit – for instance all boxes smaller than 1 cm. Since this is a thought experiment, there is no reason to reject the possibility that most boxes are very small – in fact, the 'big numbers' premise would be excellently served by the existence of many zillions of boxes the size of a nanometre. After all, Aggie and Maggie do not wish to discuss empirical issues at all – so they have no reason to deny the existence of nanometre-sized boxes in a warehouse! However, such small boxes, regardless of their numbers, *are* excluded from the hunt for black marbles from the start.[42]

Mash justifies this by appealing to an 'undiluted' Principle of Mediocrity; but very soon afterwards, he is forced to admit that such an 'undiluted' form appears very rarely in the SETI-related discourse. True enough, and with a very good reason. Attempting to expel empirical knowledge from a field that has been, for some time prior to 1993 and to a still larger degree now, empirically rooted, is bound to be a dry and fruitless enterprise. As an attempt to encompass perfect abstraction, it fails, since even his spokesperson Maggie asks about black marbles with gouges as a subset of black marbles – and not, for example, black marbles with small hooks made of the metastable nuclear isomer technetium ^{99m}Tc. We expect, on the basis of our empirical knowledge, that marbles tend to have gouges and not small hooks made of extremely rare radioactive isotopes. Of course, this also prompts the question why should we not go some steps further and use the Principle of Mediocrity (in something I am tempted to call an *undiluted*$^{+}$ form) to argue that human beings could live in intergalactic space, since our spatial location obviously is not typical, as explained in Chapter 3 in the discussion of observation selection effects. The thesis becomes preposterous *only* on admission of empirical information: that human beings require things like air, food, water and a reasonable range of temperatures, all being duly absent from intergalactic space – and all contrary to the author's stipulation that 'considerations of theory and background knowledge are ruled out from the start'.[43] According to Mash, this would be a 'dilution' of the purity of the Principle; clearly, such a 'pure' principle is worthless in any case.

How could one make the warehouse thought experiment more similar to what really is going on in the SETI discourse? Suppose that we observe the environment a little bit more carefully and find out that the building next to the warehouse is a marble factory – would we still conclude that our intuitions that at least one other box contains a marble are unsupported by the previous observation of a marble-containing box? Marbles are expected outcomes of the processes occurring in the neighbourhood. Would Mash's heart still be with Maggie?

Alternatively, suppose that instead of a marble we find a telephone cable in a randomly chosen box. There are no other buildings around the warehouse, but further observations reveal a plaque attached at the warehouse entrance which reads *AT&T*. Again, should we expect that at least in some boxes we might find another telephone cable? Now, this version of the thought experiment is an improvement upon the previous one, for it follows more closely the disteleology of Darwinian evolution. The purpose (or goal) of the marble factory is the production of marbles; however, the purpose of *AT&T* is not producing or storing telephone cables. Telephone cables are just pieces of equipment used in the activities of that company for the proximate goal of establishing telephone connections, offering services of such connections and, perhaps, the ultimate goal of making profits. But there is both a logical and a physical connection between the discovered telephone cable and the properties of the place where the discovery was made.

Even if we still think the probability of finding a similar telephone cable in another box is too low, what about the probability of finding *any* piece of telephone equipment in another box? One could argue that it is not just larger by virtue of the $A \subset B \Rightarrow P(x \in A) \leq P(x \in B)$ probability rule (which Mash, surprisingly enough, seems to consider an empirical matter), but larger over and above it due to (i) the genetic relationship between the members of the 'telephone equipment' overset, and (ii) the logical connection between this overset and the warehouse plaque. In terms of astrobiology/SETI research, we do not need to expect – as argued by Gould against Simpson – to find *Homo sapiens* on other planets, or even some generalized version going under the title of 'humanoids'; we expect to find agents possessing general properties like observership and intentionality. Entities we call 'a telephone cable', 'a telephone switch', or 'an RJ11 jack' stand in internal relationship analogous to entities we call '*Homo sapiens*', 'intelligent observers (2)', or 'intelligent observers (3)' – they share some minimal set of common properties. In both cases, the elements are part of a wider process (a telephone system operation in the former case, physical/chemical/biological/cultural evolution in the latter).

The main omission, then, in this part of Mash's discourse is an artificial severance of the link between the find and its context. Mash's thought experiment fails to adequately reproduce the situation encountered in SETI studies, since black marbles are not a naturalistic outcome of the evolution of boxes and warehouses. The word 'outcome' is *numerically neutral* here – it does not mean either frequent or rare, in the same manner as both arthropods (frequent) and mammals (rare) are outcomes of the same Darwinian evolution on Earth. In contrast to black marbles, both individual observers and their civilizations are a naturalistic outcome of the evolution of planets, stars and galaxies. We can dispute the likelihood of such evolution taking place in a particular region of the overall parameter space in a given interval of time – this exactly being the subject matter of emerging *astrobiological dynamics* – but we should not deny the existence of such a logical relation.

Mash's second strategy is to ask to what degree specification of SETI targets – technological civilizations in the 'window of communication' – determines the strength of the inductive inference from large numbers. Since our planet contains such specific items like the English language, *Brandenburg Concertos*, or velcro-clad, skateboarding teenagers, would the inductive inference from a large number of Earth-like planets that the same items exist at some other planet in the universe be justified? Intuitively, the answer is negative – 'monkeys' *Hamlet*' arguments aside – and Mash attempts to capitalize on such a firm intuition of ours. This strategy for undermining inductive inference from the Principle of Mediocrity can be adequately called 'overspecification'.

Overspecification is a clever strategy, but it fails on two counts, one general and one particular. The general issue is similar to the one we encountered in discussing observation selection effects: it is always possible to decrease formally the prior probability of any historical occurrence. To use Mash's own example, Earth contains English speakers, but two randomly chosen English speakers Alice and Bob are, in fact, very different when their molecular composition, their development from the embryonic stage, the environments in which they live and work, the details of their personal and ancestral histories, etc. are taken into account. Their difference is such that one could naively conclude that so heterogeneous beings are very unlikely to have anything in common. Yet, lo and behold! Both Alice and Bob speak English and are able to communicate very well. Of course, it is clear *to us* that some properties – like the detailed spatial distribution of various molecules on Alice's and Bob's foreheads – are unimportant, while some other properties, for example, the fact that both Alice and Bob were born in a particular geographic area of Earth's surface, are important to their capacity to communicate using the English language. Unfortunately, for Mash's strategy, the insight into which properties are important and which are unimportant is far from being trivial, and it is impossible to reach an insight from logic and probability theory alone. In a completely foreign context – which is exactly the 'working context' of astrobiology and SETI studies – our commonsensical intuitions, so dear to philosophers, are bound to fail. So, in a sense, Mash's strategy is simply too strong and leads to a *reductio*: if the presence of English speakers somewhere else in the Galaxy is intuitively improbable, how exactly is their presence *on Earth* not improbable as well when we discard the parochialism of our intuition? (And, by the way, if you meet some velcro-clad *and* English-speaking teenagers, ought you not be flabbergasted by the product of two astronomical improbabilities resulting in an even more astronomically improbable observation? Pascal's wager, anyone? Remember that in Mash's account there are no chains of evolutionary or otherwise physically connected states of affairs. Thus, cultural causality which, according to our common sense, explains our observation of such teenagers, is inadmissible as well.)

The particular side of Mash's overspecification strategy fails more easily. He claims that there is no justification for not applying the Principle of Mediocrity to velcro-clad teenagers and English speakers without some additional constraining principle. Moreover, he makes a great deal of the alleged absence of such a principle. He is wrong – such a principle exists, and it is the very same extension of Fry's continuity thesis discussed above: there is no unbridgeable gap between inorganic matter systems and living systems and intelligent systems and cultural systems. These are steps in the evolutionary chain of truly 'cosmic evolution'.[44] In addition, it is exactly the continuity thesis that gives us the 'missing principle' or 'additional condition' whose alleged lack is deplored by Mash, Kukla and the rest. It has been around for a long time, at least since Haldane. As Kukla writes:[45]

> Since its unconstrained use leads to absurdity, there must be some additional condition for the applicability of the mediocrity principle. But there's no hint of what this additional condition might be in the ETI literature . . .

When he says this, it is hard to avoid citing (perhaps with a shrug) John Earman's dictum that 'the philosophy of science quickly becomes sterile when it loses contact with what is going on in science'.[46] Once again, the problem is not taking our theoretical thinking too seriously – the problem is not taking it seriously enough.

Any sensible understanding of evolution, including a strongly convergentist view such as that of Conway Morris, does not deny many contingencies of the overall evolution of any sufficiently complex system, including biological systems found in nature. Neither Conway Morris, nor anyone in SETI circles (where people, naturally, incline towards convergentist views), would insist that there is a way of deriving an English-speaking property from the properties of the human genome, or of terrestrial biochemistry, or of the laws of physics characterizing the low-energy world we are observing. Nobody argues that English speaking is a convergent property of human evolution. But neither is there a sense in which these items are strictly independent: the world with different laws of physics may lie outside of the Archipelago and thus be, of course, devoid of English speakers. Even if we fix such 'low level' requirements for life and observers, in the world of evolutionary contingency it only makes sense to talk in terms of relative frequencies, which can be, in turn, translated into talk about relative volumes of compartments of the morphological space. A compartment of the evolutionary morphospace containing English speakers is much, much smaller than the compartment containing all intelligent observers. This will be true irrespectively of the absolute size of the latter, i.e., whether the true place of intelligence in evolution is better sketched as (A) or (B) in Figure 7.1. Moreover, it will be true not only on logical grounds, but also on a number of physical, biochemical, biological, sociological and linguistic grounds as well.

This sort of failure is all pervasive when it comes to the philosophical assessment of astrobiology and SETI studies. So Mash writes:[47]

> The most important criticism to be brought against the Drake equation is that we simply lack the theory to attach warranted values to most of the terms. The equation thus tends to promote a false sense of rational hope insofar as it suggests that scientific methodology is at last being brought to bear on the problem, and that a well-founded answer, if not within our grasp, is close.

Here he repeats an old adage, which is a major part of the critics' repertoire: SETI allegedly lacks theory – its only theoretical constraint is the Drake equation. But the Drake equation is itself old hat, and is – contrary to endlessly repeated assertions of the critics – hardly used as anything else but as a simplistic rule of thumb. It is not necessary to use it at all in discussing SETI; and, even if it is used, we today understand the origin and theory behind many or most of its terms, and the theoretical modelling of the rest of them is currently under way.

Mash repeats the same charge towards the end of his paper, in a slightly different form, but again as a deplorable lack of theoretical background:[48]

> In the absence, however, of a principle that will govern unconstrained application of the assumption of mediocrity – a principle which will relate big numbers to real possibilities in more than just an intuitive fashion – there is little hope of rationally negotiating the slope between the attitudes of the believer and the sceptic. To date no such principle has been offered, nor, I submit, is any in the offing.

The last statement is wrong, at least in AD 2012. The principle is the thesis of evolutionary continuity, which pervades the whole field of astrobiology and related sciences and puts various exclusivism to rest (with respect to the Earth, terrestrial biogenesis, terrestrial noogenesis, etc.). To put it simply, there is a major evolutionary transition (or a 'critical step') in noogenesis and the appearance of material culture of intelligent observers, potentially detectable from afar; there is no major evolutionary transition in the appearance of English speakers or velcro. Hence, the justification for SETI, which is steadily approached by more quantitative and precise analyses.[49] Even the quibbling about the Principle of Mediocrity is much less important here. If, as suggested repeatedly in this book, the spatio-temporal astrobiological landscape of the Milky Way is such that an adequate measure of astrobiological complexity is, on the average, increasing throughout the Galaxy, it is essentially unimportant whether this trend started from a few locations (or even a single location), or is better represented by many evolving and possibly coalescing 'local bubbles' of high-complexity life and intelligence. The Principle of Mediocrity pertains only to the latter question, not the substantive issue.[50]

To be fair towards Mash, it was more forgivable to neglect the (astrobiological) context in 1993, before the great strides made in addressing the Canonical Three, than it is now. The same cannot be said for the book by the historian of science and technology, George Basalla, published in 2006. Although the book is entitled *Civilized Life in the Universe*, and although the author explicitly states in the introduction that '[b]ecause I concentrate on civilized beings, I have little to say about the origin of primitive life on Earth or the possible existence of microscopic life elsewhere in the universe', the contents do not entirely reflect this. On the contrary, the book almost exclusively deals with the *history of our ideas* about extraterrestrial life and extraterrestrial intelligence, without distinguishing the two most of the time.[51] As a piece of historical scholarship, Basalla's book is an interesting contribution to understanding the belief in many inhabited planets – what is usually called 'pluralism' in the historical context. But it is not a book about extraterrestrial intelligence; this is only a part of the overall problem with Basalla's approach, which is far more than just grumbling about the title. Since, in contrast to postmodern relativists, Basalla is representative of the philosophical tradition that can be vaguely labelled as 'realist', he certainly does not confuse our opinions about X with X itself. There are many obvious examples of this throughout *Civilized Life in the Universe* – for instance, it is clear that he does not endorse Herschel's idea of living beings inhabiting the Sun and other stars, nor does he think that Shklovsky's hypothesis about the artificial origin of Phobos is just a form of 'discourse'. Therefore, the caveat about dealing with intelligent beings and not other aspects of astrobiology should be observed together with the title of the book – and its table of contents. Remember the sketch of the history of physical cosmology given in the Chapter 2; how preposterous would a claim sound that these same milestones are major events in the history of the entire material universe! Yet such a ludicrous approach is the norm with Basalla and it is hard to see how the immaturity of astrobiology (in comparison to physical cosmology) could justify that.

There is more, though. By making the caveat, Basalla implies that he – in contrast to the criticized SETI scientists – knows (i) why terrestrial biogenesis is different from biogenesis in any other cosmic context, and (ii) where that sharp distinction between intelligent and non-intelligent forms of life is located. Such claims, against the background of a naturalistic and evolutionary account of the origin of life and intelligence, as well as the results of contemporary astrobiological research, seem to me no less bold than the claims of Drake, Sagan, and other SETI pioneers and enthusiasts, whom Basalla is eager to criticize. In general, *Civilized Life in the Universe* clearly shows why it is dangerous to take individual aspects of astrobiological research and analyse them in isolation.

After a fairly detailed history of 'pluralist' views in the Western intellectual and scientific tradition, Basalla moves on to the main dish: criticism of contemporary/recent past SETI activities. Here, all means are allowed, including *ad hominem* attacks – the favourite target being Carl Sagan, whose automobile licence plates, both desired and real, have been found suspicious as betraying his unscientific and irrational nature.[52] Basalla also deploys a particularly bizarre form of guilt by association where, for example, a report of a workshop has its scientific value decreased if it cites, as a coda and very clearly separated from its main text, a science-fiction novel.[53] All this is gathered as fuel for Basalla's main hypothesis, which essentially states that SETI is the continuation of a mythological, mystical and religious tradition of populating heavens with superior beings. Since the thrust of this hypothesis clearly has little to do with either astrobiology or the philosophy of science in the strict sense, I shall refrain from analysing it further, and in addition to the arguments listed thus far, I will only mention two minor points. First, the quotation of David Hull about naturalism from p. 23 above is applicable here as well. It is Basalla, the critic of SETI and not its practitioners, who violates the spirit of Hull's dictum, for instance, when he writes that 'extraterrestrials discussed by scientists are as imaginary as the spirits and gods of religion or myth'.[54] Second, the approach to this sociology of science criticism is obviously marred by Basalla's insistence on personal quirks and idiosyncrasies as the main motivation of scientific activity, an attitude that is not only demeaning to the many scientists mentioned, but is also outdated and prone to deep misunderstandings. One example of such a misunderstanding is Basalla's utter incapacity to distinguish between 'scientists engaged in SETI' and 'SETI believers'; there is no logical connection between the two, apart from the historical fact that some of the people with a scientific interest in SETI were not, in fact, 'believers' or 'optimists' about the likelihood of finding extraterrestrial intelligence any time soon. The best-known example is Freeman Dyson, but there are others, including, at least in their later years, such pioneers of SETI as Iosif Shklovsky and Sebastian von Hoerner. Their attitude could have been described at best as *agnosticism* towards the issue of likelihood of SETI success, especially if the qualificative 'soon' is added.

Much more disturbing than the rather weak thesis about SETI as quasi-religious mysticism is a dose of hypocrisy in Basalla's account when it comes to (mis)using the history of ideas, a propensity he shares with other critics, both those discussed here and writers like Nicholas Rescher and Ernst McMullin. When, for instance, Basalla writes that '[i]ntelligent alien beings remain as elusive, and controversial, in the twenty-first century as they were in ancient times when Greek thinkers first wondered about their existence',[55] he is, of course, factographically correct if we equivocate on the notion of 'intelligent alien beings' or allow its loose usage. However, taken in the context of the history of ideas *sensu stricto*, this rhetorical manoeuvre is at best sophistry and at worst outright intellectual hypocrisy. Imagine Basalla's reaction to the claim that, for instance, 'unified field theory remains as elusive, and controversial, as it was in ancient times when Greek thinkers first wondered about its explanatory power'. I suspect that he would, together with most philosophers and historians, be quick to cry foul and point out that our modern concept of unified field theory is something entirely different from the musings and teaching of pre-Socratics about *arche*, and even Empedocles'theory of elements and forces. It would be ahistoric, anachronistic, 'Whiggish', naive, etc., on the speaker's part to claim a conceptual continuity between Hellenic cosmological thought and modern theoretical physics. An avalanche of patronizing lectures on the correct meaning of a term and its changing and evolving over the course of history, undoubtedly, would ensue. And fairly enough, since in a serious scholarly approach to the history of ideas, there is a world of difference between ancient materialist perceptions of the unity of the world and modern quantum field theories and their talk about 'unifying the forces of nature'. Nevertheless, how can that same ahistoric and 'Whiggish' mistake be lauded when it comes to extraterrestrial intelligence? It is completely obvious that the concept of extraterrestrial intelligence of Lucian or Plutarch or even Bruno was something entirely different from the similar-sounding concepts used by Carl Sagan or Nikolai Kardashev or Freeman Dyson. This is valid at least to the same extent as the debate about fundamental interactions in the works of Steven Weinberg, Richard Feynman, or even Roger Boscovich is incommensurable with the debate about forces in fragments of Empedocles or Anaxagoras or the Ionian atomists.[56] In fact, we should expect such a 'Whiggish' reading of history to be even less comparable for complex entities like intelligent observers than is the case for relatively simple concepts like fundamental interactions, since complex entities require more historically rooted interpretations and are likelier to be misunderstood in some crucial aspect or detail. Since no serious post-Kuhnian historian of science will talk about ancient physics and its relation to modern physics in such a simplistic and ahistoric manner, it is indeed puzzling why such a double standard should pass muster where the search for extraterrestrial life and intelligence is concerned.

(To reinforce the parallel between astrobiology and physical cosmology discussed in Chapter 2, one may note that there was a series of similar episodes of philosophical backlash in the history of modern cosmology as well. In perhaps the last such egregious episode, the distinguished philosopher of science Stephen Toulmin has argued, in a book co-authored with June Goodfield, as late as 1964, less than a year before the discovery of the cosmic microwave background radiation (!), that the origin of the universe is as mysterious as it was for Immanuel Kant in the late eighteenth century.[57] At least Toulmin and Goodfield did not go back to ancient Greece, as SETI detractors regularly do. The great historian of cosmology Helge Kragh comments upon this episode with charitable understatement: 'It was not an unreasonable attitude in the mid 1950s, but to repeat it ten years later was a sign of ignorance about the progress which, after all, has taken place during the decade.'[58] This is a rather appropriate comment on some of the philosophical criticism of astrobiology and SETI. A decade is quite a long period in revolutionary times and, as the astrobiological revolution continues, in our epoch of rapid exchange of information one may find such ignorance even less excusable.)

Usually, the double standard is justifying by insisting, as Basalla repeatedly does, that SETI studies are 'burdened by the past' to a high or unusual degree.

How are we to understand that claim? One would expect the real situation to be exactly the opposite. SETI has been portrayed by the 'founding fathers' as a 'wave of the future' and the proverbial man in the street, when asked about SETI will, if anything, connect it with relatively modern and sophisticated technologies like large radio telescopes, or with undoubtedly forward-looking and optimistic personalities represented, for instance, by Carl Sagan. As a matter of general principle, I find it entirely natural that SETI studies belong to the small bunch of disciplines that try to answer the last of the 'Canonical Three' defining questions of astrobiology: *What is the future of life and intelligence in their cosmic context?* Now, one is fully entitled to be very critical about the ways 'real', i.e., historical SETI projects and studies so far (in the specified 1960–2012 period), have justified – or not – such an important mandate. For the same reason, one may criticize practical SETI methodology and point to the alternative and potentially more fruitful ways of gathering information about the distribution of intelligent life in the universe, one could and should criticize the lack of future-oriented strategic thinking in planning and executing SETI programmes. This is nothing new. Such an approach was taken by the great historian of science Steven J. Dick, who in 2003 published a very important paper arguing that some of the premises behind the orthodox SETI and Drake equation are based on prejudicating intelligent observers as biological entities, neglecting the very real possibility (and prospect, as far as the future of humanity is concerned) of *postbiological* evolution. Interestingly enough, in spite of the explicitly professed reverence for the historical work of his colleague, and several of his publications cited, most notably *The Biological Universe*, Basalla does not cite Dick's 2003 paper.

Dick's criticism – in sharp contrast to 'philosophical criticism', which I labelled type (2) above – is not at all hostile to astrobiology and the SETI endeavour, quite the contrary. It firmly belongs in type (3): criticisms aiming to improve the efficiency of the practical side of SETI work. It shows that some aspects of the working philosophy of searches so far – and those envisioned in the near future – are tacitly based on unwarranted assumptions of biological nature. This is appropriate only insofar as we neglect the possibility of a radically different evolutionary trajectory of intelligent species – the one Dick calls postbiological evolution, which is a much-discussed prospect in future studies when applied to the future of humanity. There is more on this in the next chapter, but at the moment it is worth noting that Dick's insistence on *rethinking* SETI in terms of a vastly expanding parameter space of postbiological evolution tallies well with the previously mentioned calls for practical pluralism in empirical SETI activities.

In contrast to Dick, Basalla's massage of history is occasionally almost comic. He goes out of his way to construct an elaborate hypothesis to explain the fact that most SETI researchers assume the main SETI targets to be significantly older than human civilization. Part of the explanandum is an unremarkable selection effect: since we are a very young technological civilization, it is to be expected that any possible contactee is older than us. There are physical reasons as well for such a rational belief: astronomers have known for quite some time that the age distribution of Sun-like stars in the Milky Way indicates that most of these stars are older than the Sun (and it is uncontroversial to assume that their habitable planets are also correspondingly older!).[59] Since Basalla has attempted to reduce SETI to a quasi-religious conspiracy, such astronomical arguments are, sadly, of no use to him.

In addition, although he does mention an antireligious remark about the exploration of the cosmos made by the Soviet leader Nikita Khrushchev, Basalla tries to compensate for it in the very same sentence by quoting Carl Sagan about the spiritual aspects of the space travelling experience. However, the major point proving detrimental for Basalla's hypothesis about SETI as a quasi-religion is the fact that Soviet scientists not only engaged in practical SETI projects, but were among the first to give SETI full academic legitimacy. Those scientists not only lived in the shadow of an atheist ideology, but were, for most of the cases known in sufficient detail, sincere atheists. And all of this occurred in an otherwise conservative Soviet academic establishment. Of course, a rather closed-minded hypothesis of Basalla would be able to wriggle out of this serious objection; perhaps by postulating that, while tirelessly working on SETI-related issues, Shklovsky, Kardashev and the rest were, in fact, clandestinely subverting the official ideological position!

Finally, I shall consider the most recent account of SETI in philosophical literature, and although it is given only briefly and in passing, it is instructive of the rather unhealthy state of type (2) criticism. Pigliucci's book deals with the issue of pseudoscience and most of it is devoted to such usual suspects as creationism (in various guises), climate-change denialism, and postmodern social constructivism. It is extremely well written and most of its points are salient and quite timely. However, in the second chapter, tellingly entitled 'Almost science', the author (a distinguished philosopher, mainly involved in the philosophy of biology) devotes several subsections to the fields which are, in his opinion, neither pseudosciences, nor fully legitimate members of the scientific family. Here he puts SETI studies in the illustrious company of string theory, interpretations of quantum mechanics, evolutionary psychology and history (of the 'synthetic' kind done recently by Jared Diamond). While the club is certainly fun to be in – and only a staunch conservative does not expect great breakthroughs to come right out of one or more of these domains in the next few decades – the justification offered by Pigliucci in the case of SETI is weak, outdated, and reflecting particular philosophical prejudices similar to the ones described above in Mash and Basalla.[60]

Similarly to the Toulmin–Goodfield case previously mentioned, philosophical criticism of SETI today fails, as a norm, to adequately follow and engage research literature (including relevant fields such as animal intelligence and communication, the astrophysics of stellar and planetary systems, and postbiological evolution); in such a situation, later sources are obviously guiltier than the earlier ones. Mash, writing before the full swing of the astrobiological revolution and at the very dawn of the WWW era, could argue that there is a notable lack of astrobiological theory, but the same claim is simply poor scholarship in the case of Pigliucci, writing in 2010. For some strange reason, the internet revolution did not bring enough close following of the literature as one would naively expect; it seems that virtues other than the accessibility of information are still more important in this respect.[61] Thus, Pigliucci again states that the theoretical basis behind SETI is (not 'was', since in contrast to Basalla, the author of *Nonsense on Stilts* does not take a historical approach) nothing more than the Drake equation. Not only is this irritatingly dated, but it also gives a misleading impression about the whole endeavour. Since such a simple message needs repeating, I shall repeat it here: Drake's equation is a rough and trivial rule of thumb, worthy of an acknowledgement for its historical role and, especially, in provoking its own constructive criticism.[62] However, the instrumentalization of it in the hands of philosophers for denigrating the entire SETI enterprise clearly shows that it should not be cited any more without deeper theoretical justification – and, indeed, in the important research literature published in the last decade or two it is only infrequently cited.[63]

The reality is very different from that painted by the critics: while we are still far from an extensive and precise SETI theory, significant steps have been made, which should not be ignored. Some have been discussed in more detail in previous chapters (with the primary literature cited there), but at least some important signposts need to be repeated here. They include understanding GHZ as a spatial and temporal distribution of habitable planets; understanding the probabilistic implications of early biogenesis on Earth for biogenesis elsewhere; an improved understanding of exotic habitats, like subglacial oceans on Jovian planets' satellites; the steps forward made in understanding humans' phylogenetic past and, thus, understanding noogenesis; the deployment of at least crude numerical models and simulations of habitable planets, both individually and in the vast set comprising the astrobiological landscape of the Milky Way; and many others.

The other important aspect of the failure of philosophical criticism is its oversimplifying approach to SETI *practice*. Almost all critics repeat that SETI is the search for intentional radio emissions. (Basalla does briefly discuss other approaches, like optical SETI or the search for artefacts, but presents them like oddities and aberrations almost laughable in themselves.) This is a significant distortion, and one in which, unfortunately, the majority of the SETI community must take the brunt of responsibility. It is true that SETI efforts are often *presented* by researchers in the community as being based on systematic listening for intentional messages from the vicinity of Sun-like stars.[64] But the duty of a critic, in particular a critic with philosophical pretensions, is to pierce the veil of presentations and conventional wisdom and obtain as wide as possible a cross-section of the criticized phenomena. In this particular case, the extension of searches beyond the radio waves and intentional messaging was present in various writings of the founding fathers like Sagan, Shklovsky and, in particular, Dyson.[65] Optical SETI has already become part of the mainstream approach, and surveys in other parts of the electromagnetic spectrum have at least been proposed in theoretical studies. This applies particularly to surveys in the infrared, since those are the likeliest to detect products of extraterrestrial macro-engineering, such as Dyson shells, artificial planetary rings, O'Neill habitats and Matrioshka brains. A specific form of γ-ray SETI search has been carried our by M. J. Harris, using data from the Compton Gamma-Ray Observatory to search for signposts of antimatter burning in the Galaxy.[66] This case is an illuminating example of how philosophically uncontroversial a test of a SETI-related hypothesis can be. The reasoning is fascinatingly simple: astrophysics clearly tells us there is no bulk antimatter (individual antiparticles created by cosmic rays do not count) to be found in the Galaxy. Therefore, if very penetrating γ-rays created by bulk antimatter burning – for instance, by using it as a fuel for spaceships – are found, they will likely be telltale signs of engineering by technologically advanced extraterrestrial intelligences. The negative result of Harris' study puts upper limits on the amount of (by default, artificial) antimatter burning in this part of the Galaxy; future γ-ray telescopes will enable still stricter constraints. The result has been obtained with a minimum of theoretical assumptions and at no expense of taxpayers' money, since it used the existing CGRO database. *What more could one desire, as far as scientific methodology is concerned?*

Elsewhere, I have argued in detail that exactly the same reasons that make orthodox radio SETI impractical, do in fact support alternative approaches.[67] Irrespective of their specific merits and demerits, and in particular irrespective of establishment views, these alternative approaches are part and parcel of the SETI endeavour, especially where criticism of the entire field is concerned. On the other hand, although the context in which it (re)appears may be new, the approach itself is old news – Dyson's celebrated 1960 paper being contemporary with the Ozma Project and the beginning of scientific SETI. Therefore, the excuse that this was unknown at the time of writing of Basalla's, Pigliucci's, or any similar criticism is hardly acceptable.

On Fermi's paradox, Pigliucci writes:[68]

There are two significant problems with Fermi's paradox, which again shouldn't require a rocket scientist to figure out . . . The most obvious problem with the paradox is that it simply assumes, without justification, all too familiar human psychological attributes (particularly, a penchant for exploration and/or conquest) and projects them on other life forms of which we know absolutely nothing. This is the by now familiar principle of mediocrity, but on steroids! Perhaps the simplest explanation for Fermi's paradox is that other civilizations just don't behave like Spanish conquistadores . . . The second problem with the Fermi paradox, it seems to me, is that it is not exactly logically tight. Let us assume for the sake of argument that there is, in fact, another technological civilization out there, at about the same stage of development as ours (by the way, whoever decided that civilizations have to go through stages similar in any way to our own history?). Now imagine that somewhere on the mother planet of that civilization, a well-known physicist named Henry Stop is having a nice lunch with colleagues, and they are casually discussing the possibility of the existence of other civilizations just like theirs. Henry listens to his optimistic colleagues, who think that the galaxy teems with intelligent life, and quips: "Then where is everybody?" Henry Stop has just "demonstrated" that we on Earth do not exist. Now, that is what I call a paradox.

I have quoted this *in extenso* just to show the breadth of prejudice permeating the philosophical type of criticism of SETI. Of course, it does not require a rocket scientist – a careful thinker acquainted with the contemporary literature is enough – to see that both 'problems' are at best semantic hair-splitting and at worst failures of logic. Both issues hinge on the extremely literalistic reading of Fermi's paradox, something that has been abandoned, at least with the Brin's important 'Great Silence' paper in 1983; of course, serious thinkers understood it much earlier – thus Stanislaw Lem wrote about *silentium Universi* in his massive overview of the future of intelligence and technology in the mid 1970s.[69] The locution 'Fermi's paradox' should not – if we wish to have a substantive criticism of ideas, and not just a scholastic discussion about words and names – be used literally for Fermi's lunchtime remarks, *whose exact content is anyway unknown*, but as synonymous with the more general and precise Great Silence paradox.[70] The Great Silence paradox has nothing in particular to do with the exploration of conquest, and even less with any form of human psychology or human history (not to mention using such a cliché as the 'Spanish conquistadores', predictably bound to cause adverse reactions in this politically correct epoch of ours – perhaps chosen exactly for that reason). The Great Silence paradox – which is, I wish to repeat and emphasize, synonymous with the reading of Fermi's paradox in the majority of serious studies on the subject – has to do with the general notion of *detectability* of other intelligent species in the universe. Detectability is a wide term, encompassing many options. Some are, obviously, exploration and conquest, but others are about sending intentional or unintentional messages in various parts of the electromagnetic spectrum, building Dyson spheres, Matrioshka brains, shell-worlds and other Lucretian 'eternal monuments', creating γ-ray signatures of antimatter burning, or sending inscribed-matter packages.[71] Or even simply putting a huge neon sign across the Galaxy, in a humorous version of the great postmodernist prosaist, Italo Calvino[72] – all that *and much more*, since the branching of detectability options over interstellar distances and across the array of detecting techniques remains to be fully classified. It is important *not* to close the list of possible manifestations and traces of extraterrestrial civilizations exactly in order to counter anthropomorphic biases against which philosophical critics ceaselessly preach – but they preach to the choir, since it was SETI researchers who pointed that out first. As mentioned above, Fermi's paradox in this general and serious sense would still be valid if, for some reason unknown to us, we are indeed *currently* the only technological species in the Galaxy. We would be fully entitled to ask about and search for 'interstellar archaeological' remains of previously existent extraterrestrial technological species, and to ask for an explanation for their absence. Detectability deals with the interaction of an intelligent species with its environment – there is no 'human psychology' in stating that at least a large subset of any intelligent species influences its environment (one is tempted to say all intelligent species, but the temptation should be resisted since our imagination is certainly limited). Their capacities for such influencing may change non-monotonically – this answers Pigliucci's rhetorical question about 'going through similar stages'[73] – but if they pass some particular threshold, the resulting change in the environment becomes detectable over large distances. It might not be detectable immediately, it might not be detectable by any particular method or approach, and it certainly is not detectable in the general case via radio-listening surveys, but there is a sense in which detectability can be defined without referring to any particular human historical or technological peculiarity. Moreover, I wish to go a step further and state that *detectability can indeed be defined without reference to any form of the Principle of Mediocrity whatsoever*, constrained or unconstrained, pure or diluted. Therefore, Fermi's/Great Silence paradox makes sense, even if we relax the Copernican assumption – it is enough not to substitute some anti-Copernican assumption for it. An example of the latter would be, for instance, the statement that all, except for the measure zero subset, intelligent observers dwell in deep ocean and, consequently, never develop astronomy or space travel.

What the thought experiment with Henry Stop shows is that critics do not appreciate the probabilistic nature of Fermi's paradox; this is coupled with unclear notions about the spatial and temporal scales of our Galaxy. We notice the unrealistic expectation that potential SETI targets are 'at about the same stage of development as ours', while astrophysics and astrochemistry tell us exactly the opposite should be the case: a large majority of Earth-like planets are much older than Earth, so Copernicanism would suggest older intelligent beings as well. But Henry Stop would not sound that persuasive if Pigliucci were to say that he's having lunch in a civilization 1.7 billion years older than humanity (according to the results of Lineweaver and collaborators discussed above). A thinking reader could justifiably worry that, in such a situation, dear old Henry would have been able to check empirically upon our existence or non-existence without any noticeable interruption in eating, or whatever it is that members of billion-years-old supercivilizations do at lunchtime.

Nevertheless, let us suppose that everything in that cute story is just as Pigliucci says. Why would Pigliucci comment on Henry's erroneous conclusion with 'Now, that is what I call a paradox'? Presumably, it is because the inference about the specific fact of our existence is clearly (and funnily enough) wrong. Henry Stop is thus alleged to have concluded that there is only one civilization in the Galaxy – instead of what exactly? Apparently, instead of *two* civilizations in the same Galaxy: Henry's and ours. Is the difference relevant?

I argue that it is not. In the case of a thought experiment involving only a single other extraterrestrial civilization, Henry's reasoning is rather unsurprising and non-paradoxical. The Milky Way with two civilizations – of about the same (and low) age, as stipulated in the setup – is essentially a dead place as far as SETI is concerned. It is a realization of Carter's gloomy argument discussed above. In this context, as in many astronomical and, indeed, scientific contexts, two is sufficiently close to one. Neither of us have had time to reach the other one, either by electromagnetic waves, inscribed-matter packages, modulated neutrinos, or spaceships, nor did any have time to engage in macro-engineering detectable over interstellar distances. Such a situation would be perhaps unexpected (not entirely close to the heart of either contact-optimists or complete pessimists), but it is entirely non-paradoxical. Henry made an error, and a small one (in spite of hurting all our anthropocentric pride) – he did not really conceive a paradox.

But that is, presumably, not what Fermi had in mind, nor is it what Pigliucci wanted to say. The strange situation arises because the question, allegedly, is not about having *another* civilization, but about having *many* civilizations. Yet all philosophers know that 'many' is vague (and is, indeed, the source of a class of paradoxes known as the *sorites*). A natural step would be a generalization: instead of a single Henry, we are asked to imagine a large number of Henries all over GHZ, say Henry_1, $\text{Henry}_2, \ldots \text{Henry}_N$. Now the thought experiment begins to fall apart, and for two reasons. The first one is that the assumption that all N civilizations containing Henries are of the same age is not only implausible, but in fact comes very close to the unphysical *ceteris paribus* error discussed in the previous chapter. In contrast to 1950 – the year of Fermi's Los Alamos lunch – today we are not in total darkness about the age and spatial distributions of Sun-like stars and Earth-like planets in the Galaxy. The elucidation of GHZ, and the complex theory of chemical and dynamical evolution of our stellar system behind it are great achievements of contemporary astrobiology and should not be discarded lightly, even in thought experiments. We have a rather specific age distribution for Earth-like planets and thus are able, in the conservative limit of searching for life most similar to the terrestrial one, to define our expectations under a broad Copernicanism. Accordingly, Henries are a priori unlikely to be in the same or an even remotely similar epistemic state. To neglect this and still assume the same age (or very similar, it does not change anything), or the state of cultural evolution including friendships and lunchtimes would be very close, awkward as it sounds, to using the Principle of Mediocrity on steroids! Therefore, Pigliucci has been swinging an unwieldy double-edged sword here.

The second reason for the *Gedankenexperiment* falling apart is that it neglects the statistical character of Fermi's paradox. In other words, we should talk about the astrobiological landscape, as discussed in Chapter 3 above – a complex set of astrobiological trajectories, which behaves in some regards like complex many-body system. Such systems cannot consist of just two subsystems; there is no gas made up of two (or five or ten or 562) molecules! The astrobiological landscape could be dense or sparse, depending on the particular values of yet unknown dynamical parameters, notably those describing biogenesis and noogenesis, but it is defined over a particular spatio-temporal region (for Fermi's paradox it is the Galaxy since the thin-disc formation) or it's not useful at all. The real question is, 'what is the ratio k/N, where k defines the subset of Henries such that neither Henry_1 nor Henry_2 nor... Henry_k observe any other civilization in their past light cone?' If k/N approaches one, we have a real paradox. But there is no real argument, logical or empirical, to drive k/N to unity – it seems preposterous to prejudice observations made by observers in other civilizations, especially those by definition not exchanging any information with us! Therefore, to the extent that the purpose of Pigliucci's thought experiment was to demonstrate a paradox inherent in Fermi's reasoning (and, even more to the point, the thinking of subsequent researchers of the paradox, such as Brin), it fails.[74]

Towards a coherent philosophy of noogenesis and SETI

The depressing failure of philosophy – as embodied by the described case studies – to adequately engage deep issues related to the SETI sector of astrobiology should not leave us in an entirely negative mood. While it seems obvious that a serious synthesis of views from epistemology, the philosophies of physics and biology, and some sectors of the philosophy of mind, necessary for the foundational thinking on astrobiology and SETI, has not occurred as yet, this should be regarded as a challenge, not an obstacle. There is a huge open space for work, and the prospect of opening quite new problem fields – such as Olum's problem discussed in Chapter 4 – is exciting, to say at least.

We can also think about the philosophy of cosmology (as usual, to be distinguished from philosophy-in-cosmology), such as it was in its early decades, from the 1920s to the 1950s, or even at the time of the Toulmin–Goodfield book. This philosophy (works by Herbert Dingle, Milton Munitz, or Michael Scriven may serve as case studies, together with occasional remarks by Sir Karl Popper or Hans Reichenbach) was characterized by an oversimplifying and usually negative approach towards 'speculation' and 'enthusiasts' in physical cosmology. It promoted scepticism, emphasized uncertainties, castigated the use of imagination in speculation, deplored the small empirical database, invoked comparisons with religious doctrines or mystic sentiments, and significantly lagged behind the research literature on the subject. Sounds familiar? Luckily for us, the philosophy of cosmology subsequently recovered, as testified, for instance, by the contributions of important thinkers like Adolf Grünbaum, John Earman, John Leslie, Quentin Smith and Nick Bostrom in the last three decades. Could we plausibly expect that the philosophy of astrobiology/SETI passes through such a 'growing pains' phase as well?

There is a degree of indisputable awkwardness in attempts to join SETI with the rest of astrobiological research, as openly testified, for example, by Peter Ward, one of the authors of REH and a prominent astrobiologist.[75] For proponents of REH it is, of course, a natural attitude, but it is often shared by people in the SETI community as well, which is much harder to explain. I submit that the awkwardness comes mainly from our explicit or tacit resistance to the continuity thesis. Since Haldane, there has been a great deal of resistance on many levels, including the anthropocentric biases discussed in previous chapters. The resistance is weakening at the level of metazoan complexification (as it has all but vanished at the level of biogenesis), but it is still quite strong at the level of noogenesis. In addition, the cards are more mixed in this game, since there are many sincere naturalists and future-oriented thinkers who share the reluctance of uniqueness proponents to accept continuity. In the end, I find that there is a big job – and indeed a duty – to cleanse the remnants of wishful thinking, exclusivism and anthropocentrism of any sort from our astrobiological discourse, which will automatically smooth the joining of SETI research with the mainstream of the ongoing revolution.

The current and unfortunate 'disjoint' nature of SETI studies makes them easier targets for superficially philosophical criticism, but it would be an oversimplification to conclude that this state of affairs is only due to hostile critics and science journalists/writers lacking understanding of the continuity thesis and oblivious to what is going on in science at the moment. Unfortunately, a large part of the orthodox SETI community contributed to this situation in many ways: by blindly following the dicta of the 'founding fathers'; by pompously quoting the Drake equation and other improvised rules of thumb as sources of eternal wisdom; by methodological conservativism (insisting on radio searches for intentional messages, neglecting optical/IR searches, and neglecting the development of theory); by incautious, irresponsible, easily misused public statements (e.g., the infamous 'immortals' paper of Frank Drake)[76]; and, most of all, by failing to actively and critically engage with anti-SETI arguments, notably the three classical ones, which can give us hugely important new insights in the field. The last item reduces to the failure on the part of the community to actively join and promote the astrobiological revolution of the last *c*.20 years. The defects are all curable – what is really needed is fresh energy, ideas, and the will for a cure. The last item is, as is often the case with illnesses, the scarcest to find.

Clearly, the noogenesis sector of astrobiological thinking is particularly fascinating for us, as intelligent observers. However, this should not eclipse the basic embedding of this story in the much wider and deeper story about life in a general cosmic context. SETI is just the study of a particular small piece of the astrobiological landscape – no more and no less. As I shall try to show next, there are intriguing questions touching upon deep issues in other fields of philosophy, including ontology and aesthetics, into which SETI could provide important insights.

8

Natural and artificial: the cosmic *Domain of Arnheim*

The world is not only queerer than anyone has imagined, but queerer than any one can imagine. This is a most disturbing thought, and one flees from it by stating the exact opposite.

J. B. S. Haldane

There is a 1930 painting by Max Ernst entitled *Anthropomorphic Figure with Shell Flower* (Figure 8.1);[1] like most surrealist paintings it does not really depict the content of the title . . . or does it? On the top of the flat, right-angled body, resting on a rostrum-like base, sits a small, childishly drawn chicken's head. There are elements – or quotations from other paintings – which could be interpreted as arms and legs of the bizarre creature; some are symmetric, some are not. So, is the 'anthropomorphic' in the title indeed a joke? Would such a creature pass as a 'humanoid' in the sense of Simpson's 1964 article? On the other hand, is it just an easel on wide feet? Can we entirely divest the contents of the work of art from the anthropomorphism of its creator?

The relevance for astrobiology is obvious: while 'some strangeness in proportion' is encountered in any science, in astrobiology – and especially its SETI sector – strangeness must be the norm. Moreover, strangeness that is produced as a part of an extraterrestrial culture is likely to be the most impenetrable at all. As we have seen in the previous chapter, there have been philosophical attempts to show the implausibility of the SETI enterprise on various grounds; while most of these are unsuccessful – and not very good philosophy either – there is one salient point common to most of them, and it is quite disturbing even to the fiercest proponents of the search. This point can be succinctly put in form of the question: *How shall We recognize Them?* Of course, this does not pertain just to biological form, but even more to the products of material culture (most generally speaking) of extraterrestrial intelligent observers, be it radio messages or Dyson spheres or any other artefacts. This is connected from one side to the question of how big is that part

Figure 8.1. Max Ernst (1930): *Loplop Presents a Flower, or Anthropomorphic Figure with Shell Flower*. Loplop was a kind of imagined friend or a familiar of Ernst's, appearing in many of his paintings, especially in the 1930–1949 period.

of the morphological space containing intelligent observers (*How weird can They be?*) which, as we have seen, plays the key role in Simpson's argument. On the other side, it connects to various controversial 'sociobiological' issues, again generalized to the large, astrobiological scales (*How diverse can Their culture be? How is it constrained by biological form?*). While this is a gold mine for philosophical exploration, surprisingly little has been actually done in this regard.[2] Partly, this is so because critics have rationally concluded that since SETI is unjustified on other grounds, elaborating on these issues would be a waste of time and energy; but also, supporters are often afraid that a bold inquiry into these highly controversial issues would further expose their discipline to charges of being too speculative or even pseudoscientific. Therefore, there is still much space for posing the questions – and we are, as I shall try to convey in this chapter, very far from the answers. (Besides, as already mentioned, both critics and supporters – insofar as the groups can be distinguished, since there is an overlap – have already manifested a lack of imagination, the failure certainly more damaging on the part of supporters.) Just as I have so far discussed arguments allegedly showing the non-existence of SETI targets, the goal here is the investigation of problems resulting from their possible *existence*; such Borgesian irony is what makes the whole enterprise appealing in the first place.

Unconscious intelligence?

If a lion could talk, man would not be able to understand her; or at least so argued Ludwig Wittgenstein in his posthumously published *Philosophical Investigations*.[3] Lions are apparently produced by the same Darwinian evolution that produced us, and are indeed located quite closely to us in the biological morphospace. Even if other biospheres are functioning with the same biochemistry and even if Darwinian evolution is truly universal, as argued in the Chapter 5, there is no reason to believe that the distance in morphospace between us and intelligent extraterrestrial observers will be smaller on any reasonable metric. Should we give up hopes about SETI then? Even if the answer is negative, as I argue here, we need to understand the problem in depth, along the lines of the discussion of the 'Big Three' classical sceptic arguments (pp. 149–161 above). There are two parts to this problem. One part is the nature of the communication apparatus (the counterfactual in Wittgenstein's aphorism); the other is the semiotic difficulty of recognizing and understanding this independently evolved and utterly strange intelligence. In this section, I discuss the former, while the rest of the chapter is devoted to the latter.[4]

In contrast to life, the concept of intelligence is much less intuitive, as testified by fierce debates – many of them ongoing – on the topics of animal or machine intelligence. Ambiguities inherent in the concept might limit our practical searches. Perhaps the most pertinent scientific account attempting to resolve these ambiguities is a brief paper given by David M. Raup, one of the leading palaeontologists of the twentieth century, at the 38th International Astronautical Congress at Brighton, UK, in 1987, under the indicative title 'Unconscious intelligence in the universe' and published five years later in *Acta Astronautica*.[5] For much of his illustrious career, Raup has been one of the major contributors to our understanding of the great episodes of mass extinctions of species in Earth's past, as well as a champion of the extraterrestrial causes for at least some of those, thus exemplifying the multidisciplinary character of the astrobiological enterprise. He has been a player in the transitional period in which gradualism relinquished its oppressive hold on thinking in Earth and planetary sciences. While indications that the outcome of the great nineteenth-century battle between gradualists and catastrophists could be overturned have appeared throughout the twentieth century, two developments have been crucial. First, the gradual (no pun intended!) acceptance of plate tectonics as the large-scale theory for explaining many complex phenomena in Earth sciences. And second, the evidence that the history of Earth and the terrestrial biosphere exhibits sporadic and dramatic convulsions known as mass extinctions (or *mass extinction episodes*, a case where *nomen est omen*, since it is their episodic nature which undermines gradualist dogma). The key year here is 1980, when Luis and Walter Alvarez and their collaborators published the discovery of the so-called iridium anomaly at the K/T boundary, demonstrating for the first time clear empirical evidence for a catastrophic mass extinction brought about by extraterrestrial agency.[6] Raup was one of the referees of that seminal paper, as he has candidly described in detail in his instructive book, *The Nemesis Affair*, subtitled *A Story of the Death of Dinosaurs and the Ways of Science.*[7]

In this particular SETI-related study, Raup speculates that metazoa on other planets may have evolved, by natural selection, the ability to communicate by radio waves. Biologists have not yet found creatures using radio waves for communication on Earth, though the use of electromagnetic phenomena in general by life forms widely separated in morphological space, such as magnetotactic bacteria and electric eels, is well documented. On other habitable planets in the Galaxy, evolving radio communication could not be excluded – and would not even be regarded as a particularly radical evolutionary inovation. Our own SETI projects are still mostly radio-based. Radio communication in such non-intelligent organisms as proposed by Raup would persist much longer than radio communication developed by intelligent beings, which would be ephemeral due to cultural changes. After all, the Earth has already become a weaker source in some electromagnetic bands since the expansion of cable TV, cable internet and cable telephony. Strategies for SETI should take into account the possibility of such biologically evolved radio – and, by analogy, optical, etc. – communication.

(A terminological note: Raup himself calls such cases instances of 'unconscious intelligence' – I adopt here the designation, with a cautious proviso that this does not presume upon the very controversial issue of the nature and the origin of consciousness. Better terms, perhaps, would be 'a simulated intelligence' or 'non-intentional intelligence', although they have their own cache of problems: simulation implies both a designing agency and a purpose, while intentionality is perhaps too anthropocentric a criterion.)

Raup's unconscious intelligences are an additional source of the noise we need to filter out in our SETI efforts. Persistence of this kind of noise over long (evolutionary/geophysical/astrophysical) timescales is the major confusing factor, since such noise is likely to last much longer than any or most intentional efforts. Thus, in a large enough sample of habitable target planets, biologically evolved emitters might dominate over the intentional ones. However, the problem does not seem insurmountable – once we are aware of the issue, we need to apply more stringent search criteria, designed to eliminate such 'unconscious' noise from our candidate SETI signals. Two decades after Raup's study, there is no reason to abandon his then cautious optimism. Of course, much further research is necessary, especially the modelling of those evolutionary trajectories likely to lead to the emergence of unconscious intelligence. For instance, the choice of emitting wavelength bands could be very different for low-intelligence biological transmitters from high-intelligence, intentional technological observers. It is similar to the other kinds of noise – as we learn more about it, we learn how to eliminate it more efficiently.

But suppose that we manage to build non-anthropocentric SETI projects capable of clearly separating manifestations of real intelligent observers like us from the noise originating from unconscious intelligences, be they radio, optical or something else, completely unexpected. We would still have overcome only the instrumental part of Wittgenstein's challenge. Unfortunately, there is another major stumbling block we need to face on the road to a fully meaningful SETI: how to recognize such manifestations in the general case, which includes the most exotic parts of the astrobiological landscape.

'. . . It was only by analogy that they called it colour at all'

This characterization was, as noted at the end of Chapter 1, ascribed to *scientists* studying the strange meteorite in *The Colour Out of Space*. At a glance, it is not much – after all the analogy is usually assumed to be imprecise, and in Lovecraft's narrative this is strengthened by the emphasis on 'only'. On the other hand, it is important enough to warrant the title of the entire story – because, at least to those present, *there was no other handle in our language!* I submit that the dichotomy is crucial for the second, (bio)semiotic part of the Wittgenstein problem as applied to SETI. What information can be extracted about the peaks of the astrobiological landscape might be interpreted – in at least a part of the strangeness spectrum – only in terms of analogy, which can be misleading by definition. In particular, the capacities of observers to influence their surroundings may mimic the 'apparent design' exhibited by natural selection (and other evolutionary forces).

This is pertinent for that old philosophical/theological chestnut, namely the distinction between 'natural' and 'artificial', be the latter caused by a supernatural Designer or lesser designers, i.e., intelligent observers. In the naturalistic context, the capacity for making this distinction – or the lack of such capacity – is bound to play a role in astrobiology. Since Darwin, we have learnt to recognize the *apparent design* in living nature as the logical consequence of the inexorable march of natural selection. However, intelligent observers – humans, thus far – also possess a specific capacity for *intentional design*. On Earth, and in most contexts familiar to us, we have no difficulty in distinguishing between the two; indeed,

Figure 8.2. René Magritte's *The Beyond* (1938). The juxtaposition of an artefact and an apparently inhospitable planetary surface poses the age-old problem of the distinction between 'natural' and 'artificial' in the general context (it is remarkably prescient of our contemporary photographic images of the Martian surface; compare with Figure 8.3).

this is usually done almost reflexively.[8] Such quick recognition presents more of a hindrance than an advantage in the strange contexts characterizing the search for life in the wider astrobiological narrative – especially if the context includes, or at least cannot exclude, the intentional action of other kinds of intelligent observers.[9]

As often happens in the history of ideas, artists have best expressed this dichotomy. Consider the painting of the great Belgian surrealist René Magritte, *The Beyond* (*L'au-delà*, 1938), shown in Figure 8.2. At first glance, its simplicity is completely at odds with the complexity of surrealist tradition, known to a layperson mainly from the oeuvre of Dali, Ernst, Miro, Ray and their pop-art followers: a single rectangular construction (a tomb with no inscription?) on a dry, reddish, rocky surface. The sky is clear, yet ominous; a small solar disc surrounded by a massive halo. There is no trace of any movement – an observer can hardly avoid being stunned by the untimely, eternal quality of this atmospheric painting, achieved

Figure 8.3. The surface of Mars as seen by the *Spirit* rover. (Courtesy NASA/JPL-Caltech.)

with a minimal number of elements. The contrast of the two major components can hardly be more pronounced: a grim desert landscape without any sign of life – and an equally still but completely unnatural artefact with pronounced sharp edges and rectangular surfaces, further emphasized by the contrast between the clear white marble or concrete and the dusty surface. Obviously, the artefact reminds *us* of a tomb or a sarcophagus, since it is a 'culturally filtered' form, common to most human civilizations and present in highly iconic examples all over the world. But what would it suggest to a non-anthropomorphic extraterrestrial intelligent observer?

(Curiously enough, the scenery uncannily resembles the surface of Mars – the reddish dust and rocky desert familiar from countless images of the post-*Viking* era – painted almost four decades before such photographs became available. Compare it with the surface features in Figure 8.3, imaged by NASA's *Spirit* rover and with the famous Martian sunset on the edge of Gusev crater (front cover). Even things like the closeness of the horizon and the impression of a small angular size of the Sun are associated by a present-day observer more with the specifics of a known location, Mars, than to a vagueness of some place 'beyond' experience.

Figure 8.4. René Magritte's *The Domain of Arnheim*, first version (1938).

This eerie coincidence – or artistic precognition – may remind us that 'alienness' is a finely graduated continuous property; if not for the tomb, there would have been no inscription on the desert surface in Magritte's painting, in the same manner as there is no inscription on the tomb itself.)

Nevertheless, Magritte was almost obsessed by the same motive; consider *The Domain of Arnheim* (Figure 8.4, and the back cover illustration), a painting with no less than nine versions, the first and the most famous one created in the very same 1938 and inspired by the eponymous story of Edgar Alan Poe.[10] Poe's story deals with landscape architecture, a discipline by definition bridging the chasm between the natural and the artificial. At one point, he writes:[11]

> No such combination of scenery exists in nature as the painter of genius may produce... And yet how unintelligible is this! In all other matters we are justly instructed to regard nature as supreme. With her details we shrink from competition. Who shall presume to imitate the colours of the tulip, or to improve the proportions of the lily of the valley? The criticism which says, of sculpture or portraiture, that here nature is to be exalted or idealized instead of imitated, is in error... In landscape alone is the principle of the critic true.

Consider the three key compositional elements of the painting: an eagle-shaped mountain ridge, two eggs, and the modernist facade of the building that contains the observer. Which of the three is natural and which is artificial? *We* recognize the head of an eagle in the shape of the mountain and intuitively expect that the probability of orogeny and weathering spontaneously creating such a shape is so small as to be practically nil. Therefore, we expect that an extraterrestrial intelligent visitor to Earth would conclude the same about the Mount Rushmore faces upon first encountering them and would set out to search for the intelligent designers of the monument.[12] But would we have the same intuition about the ridge in Magritte's painting if we had previously never set eyes on an eagle, or indeed any bird? Note that this is largely independent of the amount of background non-context-specific knowledge; *we* are still far from completely understanding all the details of orogenesis (mountain formation), let alone having a quantitative theory giving the (im)probability of forms such as the cliff in Magritte's painting or the Mount Rushmore faces appearing without a designer. So, how to estimate this improbability?

Instead, let us consider the inverse situation: suppose human astronauts land one day on a distant Earth-like planet, after previously detecting (in orbit?) indications of a long-vanished civilization of non-anthropomorphic intelligent beings. There are no preserved visual images of those ancient beings (we might imagine millions of years have passed since their extinction and erosion has destroyed all surface material artefacts) – but there are plenty of mountain cliffs around. Do some of them contain images of ancient political or cultural leaders carved in hard rock? Suppose that the answer is affirmative. Would such images strike us as manifestly artificial, as in the Mount Rushmore case? The answer is almost certainly no, if that single piece of visual information is all we have. This does not contradict what I have argued in the previous chapter concerning Roy Mash's use of overspecification. The probability here is so small and the connection between the mountain's morphology and the builders' morphology itself is empirically unknowable. Although thought experiments of this sort do possess an unsettling nuance, in contrast to the general SETI case to which overspecification allegedly applies, there are no practical consequences here.

On the other hand, SETI might have some similarity to the 'ornithology textbook' case: suppose that intelligent extraterrestrial observers (or newly constructed AI disconnected from a non-specific empirical database) first see *The Domain of Arnheim* and subsequently discover an ornithology textbook. Neglect for the moment the relationship of eggs and birds (and neglect the non-trivial issue to what extent a textbook itself could be intelligible out of context) and let us consider just the mountain morphology. With the addition of this morphological information on birds as a family of specific biological forms, coupled with the assumed dynamical knowledge of planetary geology, alien observers could rationally build a probabilistic argument that a very small probability of non-intentional ('natural') occurrence of the observed form suggests that it is in fact artificially constructed, i.e., that they see an artefact of macro-engineering. Even this may turn out to be a situation rather more optimistically clear-cut than one might expect, since it depends on the ability of observers firmly to establish causal independence of the processes involved. While for orogeny and avian evolution it may be easy, in the general case it might not be (think about landscape architecture using advanced biotechnology, for example).

Now, consider the relationship of the eagle (as a representative of birds in general) and the eggs – having a background knowledge of the lifecycle of this particular taxon of terrestrial animals helps us observe Magritte's aquiline cliff and the eggs as deeply connected visual symbols. Without specifying, or indeed knowing anything further about the painter, his style, the circumstances of the painting's creation, critical reception, etc., we may speculate that the connection is a metaphor of birth and death, maturation, the passage of time or perhaps its cyclic nature. All those could easily have a universal significance, or at least be significant to beings very different from us. There is nothing, however, in the painting itself that would enable an alien observer, completely ignorant of the terrestrial biosphere, to come up with such a metaphoric interpretation: to her/him/it the visual elements would remain separated and their conjunction meaningless. The connection between the eagle – or something in the shape of an eagle – and the eggs seems logical to us, human observers. It is again *we* who have insight into the deeper genetic (in the logical, not the biological sense of the word) dependence: every bird, including any eagle, is born from an egg.[13] Even if the alien observer somehow infers (or is told) that the shape of the cliff corresponds to an animal living in mountains, there still will be no connection with the eggs unless deeper information on that animal's biology is obtained from some other source (say an ornithology textbook).

The eggs lie on a massive concrete window or terrace sill with regular grooves, obviously part of a building where the viewer of the landscape is located. Again, it is clear to *us*, human observers, partaking in the same cultural and historical environment, that this visual element is obviously part of architecture and design, obviously artificial. Such an obvious impression is not at variance with the fact that there are many regular-looking objects in nature. Popular examples include the *Nautilus* shells (near-perfect logarithmic spirals), the shapes of many species of diatoms (see Figure 8.5 for a striking example), going all the way to biomorphic prototypes of tree-like columns, bone supports, mushroom domes, and rib-shaped rafters used by Antoni Gaudí in his fabulous buildings. The celebrated volume of D'Arcy Thompson is a classic compendium and discussion of many such examples, in spite of its now outdated formalist concept of evolution.[14] Is there a clear cognitive reason why we tend to separate the regularly shaped diatom from the concrete window sill and put them into distinct categories of 'nature' and 'artefact'? Probably not, since part of our process of sorting must be culturally conditioned; this is safe to say without, of course, giving any credit to social constructivism and

Figure 8.5. Natural or artificial? Many diatoms, like this oceanic *Thalassiosira* (a major participant in the biological circulation of carbon), in the construction of their exoskeletons manifest uncannily precise geometric regularities, usually assigned only to products of industrial design. (Courtesy of http://sciencephoto.co.uk.)

similar fashionable doctrines. Many times, we hear that an artist, or an architect, or an engineer, or a high-fashion designer was 'inspired by nature' or by some more specific natural objects (especially products of Darwinian evolution). In a similar way, we are sharing a part of that strange tension – since the outcome of the creative process is by definition an artefact.

Regularity here is exhibited in particular morphologies. This means that perceived morphological properties are universal geometric properties, like being spherical, being logarithmically spiral, etc. This is different, though not necessarily at first glance or by any particular observer, from the cases of culturally specific morphologies, which have been historically invoked to prove particular design arguments. A paradigmatical example was the great Jesuit, Athanasius Kircher, who in 1678 jubilantly published the discovery of a block of calcite, many millions of years old and certainly untouched by human hand, which contained inscriptions of all letters of the Latin alphabet, including a diphthong Æ.[15] The 'last Renaissance man', and the father of Egyptology, did not need any additional comment or interpretation of this empirical fact – is there a better proof of the benevolent Divine providence?

Nowadays, 'God's graffiti' are, of course, met with a condescending smile (with perhaps a whiff of nostalgia for the simpler times), but any such reaction should be framed by (1) a clear grasp of what exactly is funny about Kircher's conclusion, reached on the basis of firm – in both senses of the word! – empirical evidence, and (2) the rather obvious inference that with respect to naturalistic, but sufficiently strange non-anthropomorphic, designing agents, we could find ourselves the object of the same 'Whiggish' amusement. We should not mock the illustrious Jesuit for his identification of apparently random grooves and scratches and dents in calcite as intentional artefacts. What is really funny is the 'background' assumption that the apparently powerful designing agent – not necessarily all-powerful; it applies to 'lesser Designer' or 'basement universe' cases as well – uses a Latin alphabet of the kind used in learned theological tomes. It is, of course, prima facie improbable that a random set of scratches looks like the alphabet. Maybe one could expect cuneiform as more probable? This is not entirely jocular; in a possible world in which the Battle of Kadesh ended in a complete triumph for the Hittites (or, alternatively, in which the Neo-Assyrian ruling classes were more culturally conservative, rejecting the Phoenician alphabet as a dangerous innovation) would we really have less reason to accept the Designer explanation? As in the case of Mash's conversationalists Aggie and Maggie, it comes down to cases in which we cannot determine the absolute values of probabilities but can – or cannot, according to the sceptics – tell that there was a probability shift due to a particular observation.

Note the key difference between the design in Kircher's calcite and the apparent design of living species brought about by natural selection. Both are illusory *as design* since we are nowadays amply justified in our rational belief that neither supports the existence of a designer. Otherwise, they can hardly be more different. While the apparent design of Darwinian forms is inextricably linked to the functions of living organisms, Kircher's scratches are entirely random – unconnected to anything else relevant to science. (Such a form of gratuitous design would have been recognized if we were to find, for instance, that every planetary system in the Galaxy possesses a prime number of planets.[16]) The same conclusion would have been valid if Kircher's rock were found on Mars, or in the interior of a meteorite or, as a long-term prospect, on an extrasolar planet.

The relationship of this dichotomy of natural vs. artificial to practical SETI activities is not as remote as one would think. One natural aspect of the synthetic view of astrobiology is the convergence of SETI and future studies in searching for material artefacts of cultural evolution potentially detectable over interstellar distances. This is the subject of the thriving field of macro-engineering/astroengineering studies.[17] There are some pragmatic suggestions, too. In 2005, the French astronomer Luc Arnold published a fascinating study (in, by default, the somewhat conservative *Astrophysical Journal*, the premier research journal in astronomy and related fields) dealing with the possibility of indirect detection of large artefacts of advanced technological civilizations in the Galaxy.[18] The idea is beautifully simple: astronomers

Figure 8.6. Natural or artificial? This beautiful image of Pearl-Katar, an artificial island in the Persian Gulf, could look puzzling to a completely alien intelligent observer, mixing features of both spontaneity and artifice. (Courtesy of the Middle Eastern Dredging Company.)

have been searching for transits of planets across the discs of their parent stars for some years already, perfecting the techniques for detection of minuscule changes in stellar brightness accompanied by such occulting events. All theoretical models on which the interpretation of the observations is based are made for spherical planets – large non-spherical objects would cause different brightness curves and thus would, in principle, be detectable. Since we do not expect large non-spherical objects in nature, any such anomalous transit detection would be a strong indication of extraterrestrial intelligent beings capable of feats of astroengineering. Arnold argues that such detections of *artificial transits*, in cases of favourable geometry, are distinguishable by current space missions like *CoRoT* or *Kepler*. The key underlying assumption, that we do not reasonably expect to find large non-spherical objects, is due to our understanding of the nature of gravitational force, which keeps large natural objects together. The size is obviously the premier consideration here: as we shall soon discuss, other potentially detectable astroengineering projects have been conceived even at this early stage of human technological thinking. However, for smaller objects on planetary surfaces, such as Pearl-Katar (Figure 8.6) or even a diatom shell, the discussed dilemmas remain, and breakthroughs such as Arnold's remain a tall order.[19]

Thus, the question whether we can do better than Lovecraftian 'colour by analogy' still remains very much open. One pragmatic strategy would be to admit defeat and incorporate some elements of the sceptical view into a revised conceptual framework. This is what gradualists – or at least, those wiser among them – did after the acceptance of plate tectonics and after the Alvarezes' 1980 paper: claiming, for instance, that the relevant timescales are so large that huge catastrophes can effectively happen nowadays, so the 'present is (still) key to the past'. In the SETI case it would lead to relinquishing the idea of *communicating* with extraterrestrial intelligence,[20] even if contemporary and in the 'cabin of contact'; at least, without communication being vague and prone to systematic errors and prejudices. This still does not preclude SETI as an operative and meaningful *observational* research programme. Although, predictably, it would disappoint some of the actual SETI researchers and many more interested non-professionals, it is this moderate, unpretentious context, in which the emphasis is on the search for traces and manifestations, which offers the best prospects for success.

This lack of a universal 'recognition protocol' for strange forms of life and intelligence is perhaps the deepest problem where astrobiology and SETI do intersect with classical philosophical debates in ontology, epistemology and even aesthetics. Many authors have pointed out this issue, including several SETI pioneers, like Sagan and Shklovsky. Again, there is a beautiful example from the realm of literature, and it was again Stanislaw Lem who, in *Fiasco*, offered a haunting scenario of what the existence of viable SETI targets might mean, if coupled with failure in overcoming Wittgenstein's communication problem. I have already mentioned Lem's last masterpiece in the previous chapter, related to Simpson's argument from biological contingency; however, there is much more relevant material in that strange and disturbing book. Although Lem lived almost two decades after the publication of *Fiasco*, with it he was done writing prose – the last and the longest novel of his is, in more than one sense, a literary and philosophical testament; it contains almost all the elements of his long and hard thinking about the problem of extraterrestrial life and intelligence.[21]

Fiasco is a story about being unable to make sense of the alien – in spite of all conceivable advantages. A distant planet Quinta is undeniably inhabited by intelligent beings possessing an advanced technological civilization, including spaceflight and utilization of nuclear energy for both peaceful and military purposes. Although human civilization has evolved by the time of the events described in the novel, to the point of unprecedented physical power and social harmony, enabling long-term interstellar voyages and manipulation of the very fabric of spacetime, the two civilizations – human and Quintan – fail to connect on the level necessary for meaningful communication and dialogue. Humans are willing to undertake an astronomically expensive, complex and difficult interstellar journey in order to establish contact, but all their attempts fail against the extreme biological and cultural differences – and lead straight into delusion, brutality and madness. The details are full of passion and drama (which, especially for Lem afficionados, contains elements and citations of most of his major works) and need no retelling here. But the novel is chock-full of metaphors, symbols and memes relevant for any philosophical thinking of astrobiology, setting the stage for the next century in a similar manner as Wells' *War of the Worlds* set the stage for twentieth-century thinking on the topic. In contrast to Wells' masterpiece, in Lem's novel the villains are both humans and extraterrestrials, except that human guilt is necessarily zoomed-in on and hence more obvious. The failure of the contact is exacerbated by the fact that a man-made advanced artificial intelligence, ironically named DEUS, also miserably fails in the attempt at contact, raising all sorts of questions about the size and stratification of the parameter space describing various forms of intelligence, both extraterrestrial and machine ones.

The depressing ending and the nightmarish *finis* of its last sentence (one of the last lines of prose the master from Krakow ever wrote), 'and as the towering spider webs and the antennas, breaking, fell upon him in flames, he realized that he had seen the Quintans', coupled with the unambiguous title of the novel, leave no doubt that Lem was deeply pessimistic about the prospects that the recognition problem will ever be solved, at least by humans. Even worse, the failure to recognize failure might bring about, Lem warns us, the 'loss of cosmic innocence' tragedy – the story well known from human history and our relationship with poorly understood environments, like the rainforest or Arctic seas. This time, potential moral fiasco (the best interpretation of the title) could have even wider and darker consequences. Although I cannot discuss the ethical ramifications of astrobiology in general, and SETI studies in particular, the ethics and politics of astrobiology are complex and difficult topics that have been barely touched in the literature so far.[22]

The strangeness of astroengineering

If we could detect artefacts over interstellar distances, what should we expect to see? Obviously, there must be some working hypotheses, since otherwise there is no hope for success at all. Big triangular or rectangular objects orbiting and transitting an observed star, as in Arnold's study, constitute one such working hypothesis. Can we expand the inventory?

The crucial ingredient here is an increased awareness of the potentials inherent in astroengineering. The central point was clearly explicated long ago by Freeman Dyson:[23]

> When one discusses engineering projects on the grand scale, one can either think of what we, the human species, may do here in the future, or one can think of what extraterrestrial

species, if they exist, may have already done elsewhere. To think about a grandiose future for the human species ("la nostalgie du future") is to pursue idle dreams, or science fiction. But to think in a disciplined way about what we may now be able to observe astronomically, if it should happen to be the case that technologically advanced species exist in our corner of the universe, is a serious and legitimate part of science... In this way I am able to transpose the dreams of a frustrated engineer into a framework of respectable astronomy.

While we may take a *votum separatum* on some of Dyson's views on future studies (written almost half a century ago), the major point of this passage seems almost self-evident: one of the best ways to ascertain the limits of the technically possible is to discard anthropocentrism and sample a larger volume of space and time, reasonably expecting that somewhere and sometime it has been achieved. Therefore, we need to consider the epistemological and methodological impact of macro-engineering (or astroengineering in this context; I shall use the terms synonymously) on our SETI projects and policies. As we hope to show, this contact between engineering and astrobiology can be encouraging and productive for both sides.

By their fruits ye shall know them. This Biblical proverb neatly encapsulates an unconventional approach to SETI, in which the focus is on a search for traces, manifestations and macro-engineering artefacts, instead of intentional messages. Even more, the metaphor seems particularly apt, since it warns about messages (and efforts to hear them; cf. criticism of actual SETI practice in Chapter 7) being actively misleading in a search for the truth. In 1960, Dyson suggested that the very existence of advanced technological civilizations should provide us with the means of detecting them.[24] Starting from Malthusian assumptions (which were Darwin's and Wallace's starting points as well) and the well-recorded increase in power consumption with the development of technological civilization here on Earth, Dyson concluded that a truly developed society will soon face limits regarding both living space and available energy if constrained only to planetary surfaces. On the contrary, the only way to optimize resources would be a construction of a *Dyson shell*, capturing all energy from the domicile star. This particular paper, hardly longer than a page, has not only motivated manifold subsequent visions and studies in the field of astroengineering, but is also likely to continue to do so for a long time to come. As I have argued in the previous chapter and in more detail elsewhere,[25] Dyson's paper has provided the groundwork for a different sort of SETI from the one historically conducted since the OZMA project in the mid 1960s. The key is his suggestion that infrared signatures of a Dyson shell would be detectable from large distances, and would present a confirmation of the existence of advanced technological societies. This view has been subsequently elaborated by Dyson himself, and in an often-neglected study by Sagan and Walker; recently, there has been a renaissance of interest in Dyson shells.[26]

(It is neither necessary nor desirable for our further considerations to make the notion of advanced technological civilizations more precise. If Steven J. Dick is right in arguing that postbiological evolution is the dominant option for the peaks of astrobiological complexity, such a definitional programme is doomed to failure. The diversity of postbiological evolution is likely at least to match, and probably to dwarf, the diversity of its biological precedent. It is one particular feature – information processing – we assume common for 'mainstream' advanced societies, in accordance with the postbiological paradigm and the Intelligence Principle of Dick. Thus, whether such peaks of the astrobiological landscape can most adequately be described as 'being computers' or 'having computers' is not of key importance for the analysis; we just suppose that in either case the desire for optimization of computations will be one important – if not *the most* important – desire of such advanced entities. It is already clear, from the obviously limited human experience in astronautics, that postbiological evolution offers significant advantages in this respect.[27] However, an operational definition is clear: advanced civilization is the community capable of grand feats of astroengineering. Whether it will actually engage in such huge enterprises is of lesser importance; as in the case of Fermi's paradox, the issue at hand is relative frequency – it is enough that *some* advanced civilizations engage in astroengineering to become detectable over a significant fraction of the GHZ.)

So it was Dyson, 'the eclectic physicist and the frustrated engineer' by his own amusing description in the 1966 paper, who provided the key insight which, unfortunately, almost half a century later has not been sufficiently understood in all its ramifications and consequences. Of course, this 'Dysonian approach' to SETI is not limited to the search for Dyson shells, but to a general class of artefacts, manifestations and traces of the existence of advanced technological societies. Besides Dyson shells, Harris' large-scale antimatter-burning vehicles or industrial plants, and Arnold's artificial transiting objects – which could be many things, from large orbital stations to Tsiolkovsky–O'Neill habitats to stranger options beyond human imagination – there are other astroengineering feats belonging to this potentially detectable category. These include supramundane planets, shell worlds, orbital rings, 'Halo' artefacts and similar circumplanetary constructions, as well as, perhaps, often-discussed space elevators.[28] All these significantly increase the detectability of a SETI target, both during construction (which might span long timescales) and the operational stage. For a small schematic 'primer' of these, see Figure 8.7. It is reasonable to assume that stellar uplifting or stellar rejuvenation operations would cause an increase in detectability, too.[29] The same applies to projects of planetary ecosynthesis, especially if the timescale for detection is increased.[30] At still larger scales, things are less clear and the imagination here fails – except for straightforward extensions, such as Kardashev's star-powered

Figure 8.7. Examples of astroengineering constructs potentially detectable over interstellar distances. A Dyson shell is shown in construction – a phase likely to last a long time by present-day human standards – and the shellworld is in cross-section. (Courtesy of S. Popović.)

Type 3 civilizations. As far as the latter are concerned, the Dysonian approach allows us to reassess an *extragalactic* SETI, in the sense precluded by the orthodox paradigm, since the propagation of intentional signals, not to mention two-way communication at intergalactic distances, is hardly practicable.[31] An additional important, and often ignored, point is that astroengineering – if feasible at all – will *unavoidably* expand the range of astronomically observable phenomena.[32] This may sound trivial, but since the progress of astronomy has been exponential as far as the number of discoveries in the sky is concerned, the conclusion that we are either (i) soon likely to exhaust the spectrum of different observable phenomena, or (ii) begin sampling, intentionally or not, the space of astroengineering constructs, seems inescapable.

The list of both theoretical and observational SETI studies performed so far along Dysonian guidelines is rather short; most of it is contained in the bibliography.[33] It is worth adding that such searches have tangible benefits for macro-engineers, arising from the proposed cooperation with astrobiologists. Here is a fresh outlet for the 'frustration' Dyson wrote about. Different types of macro-engineering projects will have different astrobiological impacts, and will require different detection methods and procedures, thus opening a wide and still almost empty field for theoretical studies and modelling. The same issue has another side: a civilization wishing for some reason to avoid detection will likely refrain from at least some astroengineering projects; whether this can apply to future humanity/posthumanity is for future social sciences and decision-making processes to determine. However, in order to do so, detailed studies of macro-projects' *detectability* need to be performed, exactly of the same kind required by astrobiology. Finally, the very limits of the concept of astroengineering itself, vis-à-vis materials, energy, time and sophistication constraints are possible to probe *only* through the Dysonian approach to SETI; this is similar to the way an astrophysicist learns more about our Sun, its structure, evolution and final destiny by observing billions of other stars in the Milky Way.[34]

Overall, the greatest beneficiary may be the long-term future of intelligences, those sparse peaks of the astrobiological landscape itself. As Konstantin Tsiolkovsky famously wrote in a 1911 letter, 'The Earth is the cradle of the mind, but one cannot live forever in a cradle.' Neither can others.

9

Astrobiology as the neo-Copernican synthesis?

> It is a paradoxical but profoundly true and important principle of life that the most likely way to reach a goal is to be aiming not at that goal itself, but at some more ambitious goal beyond it.
>
> *Arnold J. Toynbee*

This last chapter is not analytic; rather it is hortatory. I shall try to show that the exhortation is called for and, indeed, is quite beneficial. In any astrobiological study, we encounter strands not only of many established scientific disciplines, but also, even more to the point, various styles of research, analytic approaches and ways of thought. While a conservative mind tends to see this as a weakness, I submit it is a source of true strength and promise. What we do need at the start of the third millennium is the original, bold, synthetic thinking unconstrained by outdated philosophical preconceptions and disciplinary fences, structurally accommodating open systems like those encountered 'in the field'. Here I shall sketch the *extended mandate* of astrobiology and its capacity for a wide interdisciplinary and even intercultural synthesis.

The two main strands of this book, Copernicanism and continuity, at a first glance may seem disjoint. After all, Copernicanism has derived its name from the greatest scientific *revolution* of all times, which *severed* the link of humans to the universe at large; the very language alludes to *dis*continuities, rather than the opposite. Even more, if we regard subsequent revolutions, like Darwin's and (more controversially) Freud's, or the one in cognitive science, as further blows to humanity's self-image – as in Freud's oft-cited narrative – we are tempted to regard them as further discontinuities. In fact, as shown long ago by historian Bruce Mazlish in a brilliant essay on 'The Fourth Discontinuity' – the one between man and machine – each of these revolutionary developments in fact served to reinforce the underlying, non-obvious, naturalistic continuity:[1]

In this version of the three historic ego-smashings, man is placed on a continuous spectrum in relation to the universe, to the rest of the animal kingdom, and to himself. He is no longer discontinuous with the world around him. In an important sense, it can be contended, once man is able to accept this situation, he is in harmony with the rest of existence.

Cleverly using examples of Descartes and Pascal, Samuel Butler and Mary Shelley, Mazlish shows how the underlying continuity is tightly related to the naturalistic and evolutionary worldview. By overcoming our parochial and anthropocentric prejudices, we get a glimpse of the underlying universal, cosmic evolution, which spans both the past and future of the observable universe, and an even vaster space of conceptual possibilities. Like Popper, in regard to the power of selection to create diversity, Mazlish invokes an explicitly astrobiological narrative in support of continuity; the continuity has been justified[2]

because modern man perceived that the same scientific concepts help explain the workings of himself and of his machines and that the evolution of matter – from the basic building blocks of hydrogen turning into helium in the distant stars, then fusing into carbon nuclei and on up to iron, and then exploding into space, which has resulted in our solar system – continues on earth in terms of the same carbon atoms and their intricate patterns into the structure of organic life. And now into the architecture of our thinking machines.

The context for Mazlish's study was set by the then nascent AI field; but everything he discussed could be stated even more forcefully in the astrobiological context.[3] The underlying continuity of universal evolution is obscured by various layers of anthropocentrism. By peeling away these layers (which are contingent facts of human history, like geocentrism, belief in special creation, etc.) we are revealing, piece by piece, what observation selection and our disciplinary and cultural barriers keep hidden. 'By forging a true discontinuity in the physical history of intelligence on earth, we may force ourselves to appreciate our own deep embeddedness in nature.'[4]

Why neo-Copernican? The extended mandate is best seen in the multiverse context (as discussed in Chapter 4), although it is present at the lower rungs of Figure 4.3 as well. The use of the adjective 'typical', clearly unavoidable in any discussion, whether referring to the biosphere, Earth, the Solar System, the Galaxy, or the entire universe or values of its parameters, has so far been haphazard and imprecise. For the multiverse, it is clear that in order to achieve completely working anthropic explanations we need to make the preconditions for life and observers precise and expressible in mathematical terms. Some attempts in this direction have recently been made, but it is clear that physics itself cannot give us the whole answer.[5] The reason is that we need elucidation of the link between the properties of living systems, including self-aware or intelligent living systems, and the laws of (effective) physics. While such elucidation is undoubtedly a very tall order in practical terms, the existence of the link itself is a tacit and unquestioned assumption in most practical research. To reject it, would mean either a return to outdated vitalism or accepting the explanatory nihilism of a 'brute fact' variety. In any other case, including various types and shades of 'emergentism', it is meaningful to search for the underlying 'ontological translator', translating the laws of low-energy physics into the allowed properties of living systems. (Of course, this does not mean that it is possible to *explain* those properties by appealing to physics; this would be an example of 'greedy reductionism' of the worst kind. In contrast, to ask for meaningful limits on the range of properties is entirely reasonable and in the best tradition of 'progressive' reductionism.) Such a research programme is Copernican from the start: the very act of invoking the laws of physics, common to Earth, the Milky Way and the entire universe, brings Copernicanism to the forefront. Moreover, since the richness of possible emergent living systems will certainly not be exhausted in one run – or a billion runs – of cosmic evolution, it is the astronomical dimension of astrobiology that gives meaning to the whole project. The same that applies to the rich variety of local conditions under one set of low-energy physical laws – the astrobiological landscape as discussed in Chapter 3 – could be stated, *mutatis mutandis*, in the context of varying (low-energy) physical laws in the multiverse. In this way we shall be able to finally clear out all specifics that blind us to the truly general – the central tenet of Copernicanism.

Why, then, the prefix 'neo'? The reason is simple: classic Copernicanism dealt with selection in tangible and intuitive spaces, notably in the configuration space, or in the real or hypothetical sample of similar objects (e.g., planetary systems in the Galaxy). Hence the conventional designation of the 'Principle of Mediocrity' or something similar which, as I have argued, often muddies the water. That is clearly not enough for the range and scope of the emerging astrobiological synthesis. What is needed is the extension of Copernicanism to more abstract spaces of dynamical parameters, those constituting the astrobiological landscape. Thus, we are dealing with Copernicanism 'with a vengeance' – a very general heuristics which, for the first time, would be based not on philosophical principles, but on the physical understanding of the preconditions for observership. With the increase of our understanding as to what physical, chemical, geological, etc. requirements underlie the complexity of living and intelligent systems, we enter a new phase in which the promise of the Copernican revolution is fulfilled to its logical end. A consequence of such a project is that typicality becomes reductively explainable, in the same manner as, for example, the typicality of hydrogen and helium in our universe is reductively explainable through the theory of primordial nucleosynthesis (and chemical evolution in general). This changed – or, better, *upgraded* – epistemic status of Copernicanism deserves a modest change in designation. The

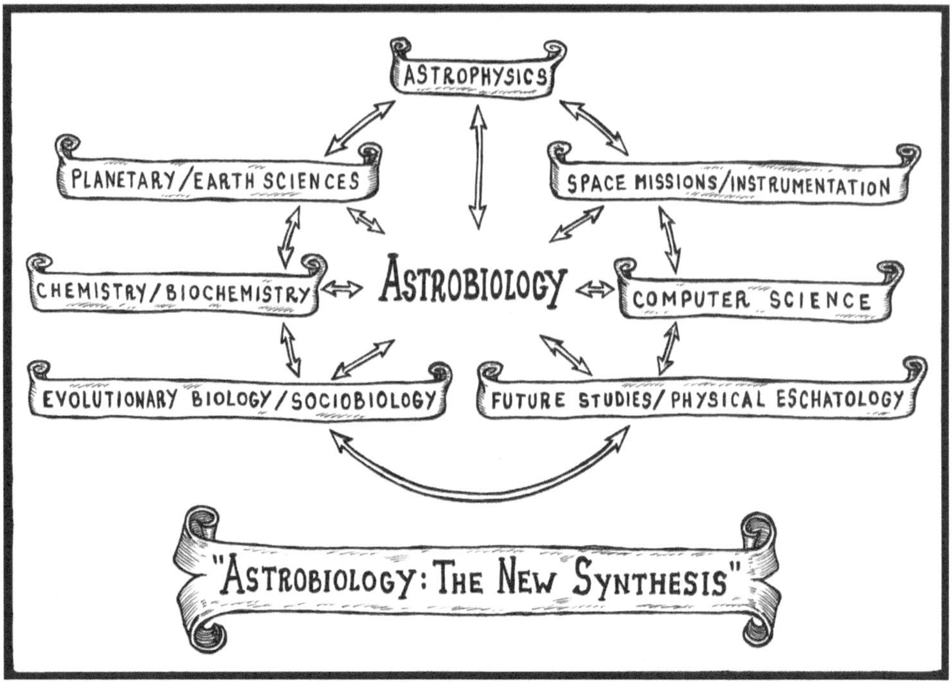

Figure 9.1. A schematic view of formal disciplinary connections mediated by astrobiological revolution and its 'extended mandate'. Of course, the choice of presented fields is highly subjective, but is diverse enough to represent adequately a wide spectrum of the promising synthesis. (Courtesy of S. Popović.)

change symbolizes the formulation of very concrete research programmes, and *disassociation* with the old-fashioned and (often implicitly, however unjustly) disparaging view of philosophical principles.

Why synthesis? Here, a historically important example may help. In 1975, the great biologist Edward O. Wilson published his famous *Sociobiology: The New Synthesis*, in his own words 'a systematic study of the biological basis of all social behavior'. The furore and controversy it caused has been chronicled and described in many places, and are largely irrelevant for our purposes here.[6] What is relevant, though, and deserves paraphrasing the title in Figure 9.1, is the attempt to bridge the gap between the seemingly disparate and often outrightly hostile scientific disciplines or groups of disciplines. Wilson himself expanded upon this synthetic view many years later in *Consilience*.[7] His pleading for the particular role biology needs to play in any such unifying endeavour can be repeated almost verbatim here. However, the most appropriate prefix I would dare to argue in this book is different: it is *astro*biology that offers at present the greatest potential for serious integrative multidisciplinary research.

The conjunction of explanations is a goal, stated or unstated, of all multidisciplinary endeavours. What distinguishes astrobiology is the potential to achieve more than that – to play a role in what Wilson beautifully called 'resuming the Enlightenment quest'. Wilson, of course, applied it to his own consilience programme, built upon sociobiology, of which he was one of the founding fathers. But even if the specific approach is rejected, the general sentiment and even his evocative designation are not only commendable, but become stronger in time. At the beginning of the third millennium, there is a host of issues of both cognitive and practical nature that require novel, provocative, radical, even revolutionary answers; some of them are inherited from the epochs of philosophical speculation ('man's place in the universe' as Wallace would have it), others (future studies, global catastrophic risks) are new in both formulation and implied ways of thinking.

One way of seeing why astrobiology is uniquely positioned in this respect is to survey the pile of specific cases. Among the examples I have mentioned in this book thus far, many possess exactly this particular quality. For instance, take the conjunction of the astrophysical explanation of the origin of polarized light, the chemical explanation of the photodissociation asymmetry between L- and D-enantiomeres, and the planetological explanation of transport of organics to the surface of the early Earth, all comprising a hypothesis for an explanation of the biochemical puzzle of the chirality of molecules in living beings. The specific example of biogenesis on Earth cannot be reasonably accounted for without various assumptions naturally following our understanding of planetary formation, planetesimal properties and evolution, impacts, and the transport of chemical compounds to planets in general and Earth in particular. Ditto for models, not yet built, which will be used to explain possible Martian biogenesis. Let us recall the Areoparadise Lost scenario: the two possibilities – independent biogenesis vs. interplanetary panspermia – could be distinguished only through comparing our general understanding of biogenesis with different physical/chemical/geological inputs in the cases of Earth and Mars. The concept of spatial and temporal correlations encapsulated in the GHZ represents a conjunction of explanatory theories belonging to cosmology, galactic dynamics, stellar and galactic chemical evolution, as well as planetary sciences. The situation in which both Olum's problem and Carter's argument can be resolved by relinquishing old-fashioned gradualism ought to encourage further research along (neo)catastrophist lines. Other such examples of consilience abound. This virtue is fated to remain underestimated as long as multidisciplinary approach is verbally commended rather than specifically supported in research and education, as well as in funding.

The other way to go about arguing for the astrobiological grand synthesis is to look outward at general issues illuminated by the ongoing astrobiological revolution. Of course, this is coupled with the problems traditionally plaguing 'grand narratives', some of them epistemologically serious, others following ephemeral postmodern fads and fallacies. While the detailed treatment certainly requires a separate book, one particular item is often invoked whenever a research programme on the 'deep questions' with profound philosophical implications is concerned. As I have repeatedly emphasized in this book, the 'deep questions' about the relationship of life and the universe has been misused and trivialized by charlatans, mystics and various extrascientific interests. We must not only deny quacks the power to determine the course of the history of ideas, we must take to heart the lesson that 'deep questions' must have strange or even "crazy" answers. How could it be otherwise? As a famous philosopher noticed in a similar context, '[s]omeone who proposes a non-strange answer shows he didn't understand this question'.[8]

Astrobiology selects the self-reflective part of scientific research, but it does that on a deeper level and with a wider justification than previous similar attempts. When leading cosmologists announce that they 'discuss the possibility that the strongest constraint on the number of distinguishable universes may be related not to the properties of the multiverse but to the properties of observers'[9] it is clear that self-reflection, in an evolutionary sense, has become a key ingredient in very practical research. Moreover, the fulfilment of the neo-Copernican programme, as outlined above, will create a satisfactory account of the 'reductionist' side of any unification scheme. Even in the context of Wilsonian consilience, the things taken for granted there (that life emerged on Earth, evolved in a Darwinian manner until sufficient complexity for intelligent observers arose, was decimated by occasional mass extinctions, etc.) naturally demand an explanation in an implied wider explanatory project.

In this book, I have argued that the bridge between cosmology and particle physics on the one side and astrobiology (and certain other 'mundane' disciplines, like artificial life, evolutionary theory, complexity theory, etc.) on the other, can be achieved through an exploration of cosmological parameter space. The Archipelago of Habitability – the set of habitable domains in the multiverse – seems to be a useful metaphor that directs our efforts towards an elucidation of the relationship between astrobiology and cosmology in a wider perspective. It offers directions for tight interdisciplinary collaboration between scientists interested in properties of life and those interested in a grand-scale structure of physical reality. It also offers great opportunities for philosophers – notably, the opportunity to generalize classical anthropic reasoning to the entire Archipelago, where all teleological and other unnecessary baggage is explicitly discarded.

It seems mandatory for the Enlightenment quest to be self-reflecting and perhaps, ultimately, self-explaining. This circumstance makes the candidature of astrobiology as the best available *vehicle* of the Enlightenment quest even more persuasive on multiple levels. On the basic level, perennial questions of self-knowledge, questions containing the delicate 'we', *Where do we come from? Where are we going?* are just simplified, vernacular renditions of the Canonical Three. However, on a subtler level, it is the complementary understanding of the different, of Otherness, which gives weight to our cognitive effort. We have not really understood our planet – in the purely astronomical sense – until we have observed and understood the properties of other planets in our neighbourhood, the Solar System. What holds for the configuration space of astronomy is even more likely to hold for the more abstract spaces of biological evolution and cognitive phenomena.

An inspiring recent attempt to achieve unification on the higher level has been made in the abovementioned work of Eric Chaisson, whose framework (or proto-theory) of cosmic evolution represents one of the most valiant attempts in this direction:[10]

> Although the majority of those relatively few workers seeking unification in science today embrace aspects of quantum gravity . . . other, less exotic roads toward unification exist. To be sure, when looking out of the window, we see no elementary strings, no 11 dimensions and no multiple universes – principal features of currently fashionable efforts to merge the very big and the very small. Instead, we sense mainly, and especially when aided by technology, the very complex. We observe a single universe of four dimensions, populated with, details aside, galaxies, stars, planets and life forms.

Chaisson's own work is an example of such synthetic endeavours, to which actual astrobiological research offers foundations and new meaning. I shall briefly discuss another such example below, the one of Gardner's 'Selfish Biocosm' hypothesis; some further examples can easily be identified in the literature. While none of those presented so far will, probably, be around for long, they are important and necessary steps in the general direction in a field that is *even younger* that it is usually presented: even younger because the comprehension of the extended mandate makes the task we face larger and, consequently, our attained level of understanding smaller in comparison. Such bold hypotheses might be entirely wrong, but we still need to hope that one or more of them will play the catalysing and energizing role of the classical steady state theory in physical cosmology.

There are empirical indications that this wider synthesis is, actually, taken seriously in many circles and already exerts quite a beneficial influence. As with all great intellectual movements, the signs are there, but they are not always easy to spot. When one looks at the titles and authors of research papers both in astrobiology journals and in specialized journals in astronomical, space-, life-, or geo-sciences, one is pleasantly surprised to find more and more interdisciplinary topics discussed by scientists with wildly heterogeneous affiliations and career profiles. The new spirit is exemplified by the approach taken, for instance, by the Astrobiology Institute, envisaged by NASA as a virtual entity, 'rather than being a traditional bricks-and-mortar facility'.[11] Astrobiology is a discipline for the age of globalization and cyberspace, for quite some time already taking advantage of resources such as the Second Life, former SETI@home, global databases, virtual observatories, virtual conferences and GRID-like distributed computing.[12] Again and again, *we are at the very beginning*.

Finally, a true temporal mirror is offered to us by SETI and other parts of the astrobiological enterprise attempting to address the third of the Canonical Three: they are offering a glimpse of our own future. The underlying assumption is again Copernicanism, since to argue that the generic evolutionary trajectory of intelligent species in the Galaxy applies to humanity as well is equivalent to arguing that we are, in a broad enough sense, typical in the set of Galactic intelligent species. Of course, the limits of such reasoning also have to be understood. First, there might not be one generic evolutionary trajectory of intelligent species, but their distribution could be bimodal or something more complex. Second, even if we were entirely typical, the trajectory would still be only a probabilistic inference, not an unchangeable destiny – if the distribution is broad enough, many outliers are likely to exist, and we may worry or hope that humanity will be among those outliers. Nevertheless, even the limitations testify to the potential reach and significance of the interface between astrobiology and future studies.

While the detailed consideration of this relationship is certainly beyond the scope of the present book, a few notes are in order. By future studies, I here call the entire array of related topics and disciplines dealing with the future evolution of humanity and its ecological and, even more widely understood, astrophysical surroundings. Some of these have specific names, often of very recent origin, like *physical eschatology* – the study of the future evolution of astrophysical objects.[13] Others are recognized as particular themes within a wider discipline; this applies, for example, to economic, strategic and climate forecasts. Some themes have the potential to emerge eventually as distinct disciplines – most notably a cluster of ethical, technological, social and even political questions, like those discussed by Tofler, Bostrom, Hughes and Kurzweil.[14] Finally, some of the research topics have been overlapping and often confused with a specific set of ideas or intellectual movements such as transhumanism. (All these I lump together here under the name of future studies, while fully understanding that it is a very heterogeneous set, with members sometimes antithetical to one another. This is also distinct from, though of course related to, the historical meaning for which the term 'future studies' has been reserved, namely economic and cultural forecasting.)[15]

Both astrobiology and future studies yearn for the resumption of the Enlightenment quest, and for yet another reason elucidated by Wilson in both *Consilience* and his subsequent writings, especially those on environmental issues. At present, humanity faces an unprecedented array of global risks from both natural and anthropogenic sources.[16] Global risk analysis touches the domain of astrobiology at several points: a part of the risk spectrum obviously has played an important role in shaping the astrobiological landscape (asteroidal/cometary impacts, high-energy astrophysical explosions, supervolcanism on habitable planets), while other risks might do so at the peaks of astrobiological complexity (warfare, misuse of advanced technologies, in particular, AI risks).[17] In addition, global risk management encompasses content from complexity theory, system engineering, Bayesian decision theory and organizational studies, all of relevance to practical astrobiological research. Finally, as emphasized by many authors, the resolution of one big astrobiological puzzle, Fermi's paradox, will give us an important insight into the future of humanity as well.[18] It is impossible to say whether, as species and civilization, we shall succeed in the management of the global risks we are facing. But we clearly have an obligation to try; thus, research in astrobiology might provide additional very practical uses along this route as well.

No promise in the entire history of ideas completely lacks an element of the Faustian bargain. By finishing the unfinished business of the Copernican revolution, astrobiology offers to deprive us of not only self-aggrandizing and hubristic concoctions and relics of a pre-scientific age, but also of concepts built into the foundations of our human exceptionalism. This human exceptionalism runs much deeper than conventional anthropocentrism: today, everyone will profess Copernicanism, and yet human exceptionalism reigns supreme in all of our institutions – in domains ranging from ecology to moral philosophy to law- and policy-making where important issues for the future are decided. Debates such as the one on animal rights are hampered from the start by unabashed human exceptionalism. The astrobiological enterprise systematically erases not only stale anthropocentrisms, but also many a cozy illusion and perhaps even a supporting beam of our sanity when faced with Pascal's 'eternal silence of infinite space' and even those kingdoms that know us not.

At the very end of this journey, let me try to explore, in an overoptimistic and overspeculative way, the possibility of annihilating some of the constraints used throughout. In Chapter 4, I have argued that a useful bridge between cosmology and particle physics on one side, and astrobiology on the other, can be achieved immediately through an exploration of cosmological parameter space. The Archipelago of Habitability – the set of habitable domains in the multiverse – appears to be a useful operational concept that directs our efforts towards a comprehensive elucidation of the true relationship between cosmology and astrobiology. Suppose that in some distant future moment we map it out completely; will it be the end of the road? As an astrobiologist pointedly warned:[19]

> If immortality is limited to having the same thoughts and experiences over and over again forever, it does truly seem pointless. However, if immortality is combined with progress, then we can imagine living in a state of perpetual novelty, always learning or doing something new and exciting. The trouble is, what for? When human beings embark on a project for a purpose, they have in mind a specific goal. If the goal is not achieved, the project will have failed (though the experience may not necessarily be valueless). On the other hand, if the goal is attained, the project will be completed and the activity will then cease. Can there be a true purpose in a project that is *never* completed? Can existence be meaningful if it consists of an unending journey toward a destination that is never reached?

Therefore, it is meaningful, from a philosophical if not from a practical point of view, to ask about the limits of any quest, including the one enabled by the extended mandate of astrobiology.

Beyond it is the rejection of the perennial dichotomy discussed in Chapter 8, the dichotomy between 'natural' and 'artificial'. There is no reason why speculation following the Rennaissance ideal of *vita activa* should be less bold than those inspired by *vita contemplativa.* In recorded human history, navigation has been followed by colonization; relevant are the cases of colonization of uninhabited islands, like Cape Verde or the Falklands. The increase of human knowledge (in the final analysis, *physics* knowledge) has been inexorably followed by adjusting even hostile conditions for human economic use and habitation. The same general pathway applies to currently discussed plans for permanent manned bases on the Moon, orbital habitat stations and – in the long run – terraforming other planets and satellites. In essence, all this reduces to the extension of the habitable part of the parameter space *within* our actual vaccuum physics; but why should this be the end? The Enlightenment quest does not end in understanding – its inseparable part is the increase in humanity's control of nature. So the elucidation of the theoretical concept of habitability leads naturally to ways and techniques for extending the habitable realm.

Visitors to St Paul's cathedral in London may marvel at the inscription on the floor, marking the grave of the astronomer and architect of the cathedral, Sir Christopher Wren: *Si monumentum requiris, circumspice!* (Roughly: *If you are searching for the monument, look around!*) This inscription, made by the great man's son, can serve as a metaphor for an active large-scale view of the role of evolved intelligence in physical reality. Information about an individual actor (or creator in the non-supernaturalist sense) that we expect to find on a conventional tomb or monument – or whose absence is an element of symbolic strangeness, as in Magritte's *The Beyond* (Figure 8.2) – may in fact be widely distributed and embedded in the environment. In particular, there is no obvious reason why it could not be distributed over cosmological temporal and spatial scales. Although it would be extremely difficult to detect such cases, there have already been some steps in this direction.

The 'Selfish Biocosm' hypothesis of James N. Gardner is another bold attempt in the right, synthetic direction.[20] Its essence is that the universe we inhabit is in the process of becoming pervaded with increasingly intelligent life – but since it is eminently neo-Copernican, not necessarily human or even human-successor life. Under this hypothesis, the emergence of life and increasingly competent intelligence are not meaningless accidents in a hostile, largely lifeless cosmos but at the very heart of the vast machinery of creation, cosmological evolution and cosmic replication. According to Gardner, the peculiarly life-friendly laws and constants that prevail in our universe – which we perceive as empirical fine-tunings discussed in Chapters 3 and 4 – serve a function equivalent to that of DNA in living creatures on Earth, providing a recipe for development and a blueprint for the construction of offspring. In other words, the capacity for the universe to generate life and to evolve ever more capable intelligence is encoded as a hidden subtext to the basic laws and constants of nature, stitched like the finest embroidery into the very fabric of our universe. A corollary – and a key falsifiable implication of the 'Selfish Biocosm' hypothesis – is that we are likely not to be alone in the universe, but are probably part of a vast, yet undiscovered extraterrestrial community of lives and intelligences spread across billions of galaxies and megaparsecs. We share a possible common fate with that hypothesized community – to help shape the future of the universe and transform it from a collection of lifeless particles into a vast, transcendent mind. While Gardner's hypothesis is certainly worth further investigation and development, it is not due to the likelihood of its being correct in the present form, but as a blueprint for a new kind of thought motivated by the explosion of astrobiological activity.

Where could this end in the future accorded to us – and other intelligent observers capable of managing existential risks – by physical eschatology? Could the rules be changed and, in the fullness of time, parts of the landscape be modified by intentional influence, like the building of artificial islands in today's world? Fortunately, at least two visionary pioneers strode with giant steps in that direction: Olaf Stapledon and Stanislaw Lem. In the last quarter section of Stapledon's *Star Maker*, we are offered an 'internal' perspective on the creation of universes.[21] As the narrator accesses larger and larger spatial and temporal scale in Stapledon's strange novel, the distinction between untamed nature and the 'artifice of eternity' dissolves. 'The cosmos which he now created was that which contains the readers and the writer of this book.' The creation described therein is naturalistic, but it is human language that, indeed, fails in this limit of physical capacities: ‘To speak thus of the universal creative spirit is almost childishly anthropomorphic.’ Some of the textual description is surprisingly fresh, considering its early-twentieth century vintage: ‘This most subtle medium the Star Maker now rough-hewed into the general form of a cosmos. Thus he fashioned a still indeterminate space-time, as yet quite ungeometrized; an amorphous physicality with no clear quality or direction, no intricacy of physical laws . . . ’ It incorporates the emergence of diverse conceivable observers (as we expect in the Archipelago):[22]

> I can, indeed, say of all these creations that, like our own cosmos, they were immensely capacious, immensely subtle; and that, in some alien manner or other, every one of them had both a physical and a mental aspect; though in many the physical, however crucial to the spirit's growth, was more transparent, more patently phantasmal than in our own cosmos. In some cases this was true equally of the mental, for the beings were often far less deceived by the opacity of their individual mental processes, and more sensitive to then-underlying unity.

Stapledon perhaps did not entirely succeed in conveying such a grand vision, but it is a glorious failure, and one still awaiting its proper inheritor; perhaps s/he will belong to the realm of physics, rather than – or in addition to – the one of poetry.

The ‘external’ perspective is given in Lem's mind-boggling story/essay/paper, *The New Cosmology*, originally written in 1971.[23] In contrast to the timid Victorian observer of Stapledon's novel, Lem's narrator is a modern-day Nobel-prize winner, a hard-nosed and somewhat cynical physicist who first describes his ‘trivial’ solution of Fermi's paradox: *Where is Everybody?* Why don't we find artefacts and engineering feats of supercivilizations billions of years older than ours?[24] But, he says, we do: the laws of physics themselves are products of their engineering, of the Great Game, played on the largest possible spatial and temporal scales. The Players' pool of knowledge is so vast and their capacities so overwhelming, that again *only* ‘by their fruits ye shall know them’ – and those are the (effective) laws of physics:[25]

> If one considers "artificial" to be that which is shaped by an active Intelligence, then the entire Universe that surrounds us is already *artificial*... Instrumental technologies are required only by a civilization still in embryonic stage, like Earth's. A billion-year-old civilization employs none. Its tools are what *we* call the Laws of Nature. Physics itself is the "machine" of such civilizations!

He continues to explain the hierarchical levels of various ‘physicses’ and the strategies of reconstructing, using game-theory approaches, the evolutionary pathway of our own effective laws. However, his main point is clear: ‘[T]heoretically, if the energy Earth's science invests in elementary particle research were to be multiplied 10^{19} times, that research as a *discovering* of the state of things would turn into a *changing* that state! Instead of examining the laws of Nature, we would be imperceptibly altering them.' And, again, in the anecdote, told by the narrator at the finale of his speech:[26]

> I should like to quote . . . Professor Ernst Ahrens, my teacher. Many years ago, when, still a youth, I went to him with my first drafts containing the conception of the Game, to ask him his opinion, Ahrens said: "A theory? A theory, yet? Maybe it is not a theory. Mankind is going to the stars, yes? Then, even if there is nothing to it, this thing, maybe what we have here is a blueprint, maybe it will all come to pass someday, just so!"

So, Lem concludes his splendid essay; we could hope to become a Player one day and strive to increase our own habitable domain, to reclaim new lands for our island in the Archipelago.[27] (Or, in a more general sense, to engineer it for whatever purpose is important for superadvanced Players.)

How could such a 'physics engineering' be performed? An analogy with condensed matter physics points to the reheating of a region and subsequent forcing of primordial symmetry to be broken down in a different stable vacuum configuration. Thus are synthetic diamonds produced from charcoal. Either in the form of creating 'basement universes' (Linde's term) in the laboratory,[28] or finding and manipulating pre-existing bubbles in the Planckian substrate, the net effect is the same: creating new universes in stable vacuum states with convenient effective physics. If 'convenient' is construed as habitable, this will lead to an increase in the measure of the Archipelago of Habitability. Thus, the capacities of Lem's Players (analogous to the reclaiming of habitable land from the sea by Dutch farmers or Japanese macro-engineers; compare Figure 8.6) represent the answer to the question with which we started – the ultimate limit of what is possible in physics – reshaping and transforming the Archipelago.

The road from resuming the Enlightenment quest to the reshaping of physical reality itself is a long and winding one, possibly requiring astronomical timescales to traverse. However, there is no reason why we should refrain from contemplating it right here, right now. To shun speculation and censor ourselves at the threshold of the yet unknown would be to betray what is best in science – and in our lives. About a century ago, H. G. Wells called for the establishment of 'departments of foresight' and corresponding academic specialities. Among many contemporary scientific endeavours, none is more appropriate than astrobiology to prepare present and future generations for fulfilling this Wellsian role.

Notes

Introduction

1 See http://www.nasa.gov/kepler for continuous updates and summaries on the *Kepler* mission (last accessed on 1 December 2011).
2 For reviews of the subject matter that by necessity becomes obsolete quickly, see Mason (2008) and the popular book of Kasting (2010); for putative extragalactic planets discovered by pixel lensing, see Ingrosso *et al.* (2009, 2011).
3 Dick (1996).
4 Fry (2000).
5 Eddington (1952), p. 125.
6 Gould (2003), pp. 132–134.
7 Percy Bysshe Shelley, *Prometheus Unbound* 4:418–423 (e.g., the electronic edition of Professor Jack Lynch at http://andromeda.rutgers.edu/~jlynch/Texts/prometheus.html; last accessed on 1 February 2012).

1 Astrobiology: 'The Colour Out of Space?'

Quote: text in De Chirico's self-portrait (1911); image reproduced in *'De Chirico' (Great Modern Masters)*, Jose Maria Faerna. Harry N. Abrams Inc., New York, 1995.
1 Wells [1898] (2006), p. 11.
2 Wayne (1978). A dedicated website with a pile of information is http://www.thewaroftheworlds.com/ (last accessed on 27 July 2011).
3 Gould (1996). *The Musical Version*, produced in the immediate aftermath of the extremely successful *Viking* 1 and 2 missions to Mars, adds a disturbing coda about *our* visits to the Red Planet, which might also be interpreted in the light of worries about planetary contamination.
4 For an additional, completely different, aspect of Wells' thriller relevant to contemporary philosophy, see Gere (2004).
5 http://www.oosa.unvienna.org/oosa/SpaceLaw/outerspt.html (last accessed on 25 July 2011).
6 For the best comprehensive reviews see Des Marais and Walter (1999); Chyba and Hand (2005).
7 In *Jacobellis v. Ohio*, 378 U.S. 184, 197 (1964); see http://laws.findlaw.com/us/378/184.html (last accessed on 25 July 2011).
8 Typical undergraduate textbooks are Gilmore and Sephton (2004); Bennett and Shostak (2011).
9 E.g., Cleland and Chyba (2002); Tsokolov (2009); Benner (2010); Jagers op Akkerhuis (2010); Tirard, Morange and Lazcano (2010).
10 Schrödinger (1944).
11 Joyce (1994).
12 Cohen and Stewart (2002).
13 Des Marais *et al.* (2008); for all versions, see http://astrobiology.arc.nasa.gov/roadmap/ (last accessed on 27 July 2011).
14 Bertka (2009), p. 1. In his superb popular book on astrobiology, *Lonely Planets*, David Grinspoon gives his version as: 'How often new biospheres get started on planets? How likely is life, once started, to develop intelligence? What might intelligence become?' (Grinspoon 2003,

p. 135). That there exist many versions of the Canonical Three clearly indicates that there is something deep and profound captured by them.

15 McKay *et al.* (1996); see also Schulze-Makuch and Darling (2010).

16 An extremely small violation of parity in the world of elementary particles is usually considered irrelevant for the problem of biological homochirality; see Keszthelyi (1984) for the important arguments.

17 Palladino (1990); Fry (2000). But see Hengeveld (2011).

18 E.g., Engel and Macko (1997); Bailey *et al.* (1998); Clark (1999); Hazen and Sholl (2003).

19 Chaisson (2001, 2003).

20 Crow (1986); Dick (1996).

21 Gale (2009).

22 Wallace (1903); I shall analyse Wallace's confluence of astrobiology and cosmology in some detail in the next chapter.

23 The newest and most fascinating results are in Schmitt-Kopplin *et al.* (2010).

24 This especially pertains to the aftermath of the Barnard Star's affair, in which the Dutch–American astronomer Piet Van de Kamp, on the basis of observations at the Sproul Observatory spanning more than 40 years, claimed the astrometric discovery of two planetary companions of sub-Jupiter masses to this nearby red dwarf star. However, later searches failed to replicate Van de Kamp's results, and the subsequently formed consensus assumed that the signal was due to instrumental errors at the Sproul Observatory 60-cm refractor. See Gatewood and Eichhorn (1973); Van de Kamp (1982); a fine history of the affair can be found at http://weblore.com/richard/barnard%27s_star.htm.

25 Simpson (1964).

26 More on Simpson's argument and related issues in Chapter 7.

27 Tipler (1980, 1981a); Ulvestad (2002).

28 Robson (2005); some comments on astrobiological relevance of the novel is given in my review of the novel, Ćirković (2007).

29 Benner (2010), p. 1029.

30 Ward (2005), p. 237.

31 Shanahan (2004), p. 1.

32 Often called the Principle of Mediocrity in both popular and research literature (e.g., Davies 1995). This locution, however, is misleading since it diverts the attention from the fact that typicality is entirely a function of the context of problem-situation; and a function that needs to be *calculated* in the course of research. In Chapter 3, I shall introduce the concept of astrobiological landscape, offering prospects to at least partially mitigate this confusion.

33 Note that this does not mean that our locations in these spaces are *random*; we shall return to this point in Chapter 3.

34 The extension of the concept of the circumstellar habitable zone to bodies possessing internal energy sources has been excellently discussed by Cockell in his contribution to Horneck and Rettberg (2007); see also Chapter 4 in Gilmour and Sephton (2004).

35 Hull (2002), p. 124.

36 Lovecraft (2000), pp. 236–271.

37 Joshi (2004); Lovecraft (2005).

38 Joshi and Schultz (2001), p. 42.

39 For some of the many examples in this new and fascinating 'genre' of theoretical physics, see Aguirre (2001); Harnik, Kribs and Perez (2006); Adams (2008). We shall return to this important topic in more detail in Chapter 4.

40 H. P. Lovecraft, letter to J. F. Morton, 14. III 1924. (see Lovecraft 1965, p. 327).

41 Leiber (2001).

2 Cosmology, life and duration of the past

Quote: Homer, *The Odyssey* (8th Century BC).

1 Hofstadter (2000). This magnificent book has been, for more than 30 years now, an exquisite prototype of what science writing ought to be – a text from which both an expert and a layperson could invariably learn something new at each re-reading.

2 Dick (1996, 1997, 2003).
3 The publication of Dick's book in 1996 roughly corresponds to the onset of the astrobiological revolution, so it gives the best comprehensive account of the pre-1995 view.
4 Wallace (1903); the biography in question is Shermer (2002).
5 E.g., Shanahan (2004), pp. 250–253.
6 Wallace (1907).
7 See Paul (1993) for a fine summary of Kapteyn's life and work.
8 The degree of progress in the understanding of our astronomical environment can be, ironically, gauged from the fact that Kapteyn in this respect was in sharp *disagreement* with Thomas Wright, the first man who identified the Milky Way as our planar stellar system, who in 1750 concluded that the Solar System must be located on the outskirts of the Galaxy. Namely, Wright derived this apparently correct conclusion from his premise that the centre of our stellar system must be populated by 'the most perfect beings', coupled with rather solid empirical evidence of the terrestrial inhabitants' imperfect nature (Belkora 2002). What a gold mine for, until recently, fashionable debates about 'Whiggish' and 'priggish' history of science (e.g., Harrison 1987; Mayr 1990)!
9 Kapteyn (1913); Jeans (1923). That Wallace was entirely competent with the astronomy of his day is testified by a 'note added in proof' on page 306 in which he mentions the new (March 1903) study by Luigi d'Auria, putting the Solar System at less than 48 parsecs from the centre of the Milky Way.
10 GHZ is an annular ring in the thin disc of the Milky Way in which we expect to find Earth-like planets, habitable for the kind of life we know. See Gonzalez, Brownlee and Ward (2001); Lineweaver, Fenner and Gibson (2004); Gonzalez (2005); for an early version, see Marochnik (1983).
11 Wallace (1903), pp. 308–309.
12 Gould (1987a), p. 392.
13 Wallace (1903), p. 306.
14 It is no accident that A. R. Wallace was also the author of an essay on the need for diversity of scholarship, something approaching heresy in academia to this day (written by a young man of 20, but published more than 60 years later; see http://people.wku.edu/charles.smith/wallace/S001.htm, last accessed 1 December 2011).
15 Heffernan (1978), p. 91.
16 Einstein (1917); for the Boltzmann–Zermelo debate, see Boltzmann (1895); Steckline (1983).
17 The best comprehensive histories are Kragh (1996, 2007) and North (1965, 1994).
18 Namely, Olbers' paradox (the thermodynamical disequilibrium of the visible universe) and the Hubble expansion, together with 'a half fact' of (observationally insecure) isotropy of the distribution of galaxies. Also, one could add a controversial datum of the age of the Earth, which was in conflict with the then reigning values of the Hubble constant, giving strong impetus to the steady state and other unorthodox cosmological models. For a modern overview, see Nussbaumer and Bieri (2009).
19 For comprehensive reviews of the field in this period, see Weinberg (1972); Harrison (1973).
20 Perlmutter *et al.* (1999); Peacock (1999); Weinberg (2008).
21 The English translation of Professor Jasper Hopkins can be found on his website: http://jasper-hopkins.info/ (last accessed 1 February 2012).
22 Guthrie (1979), p. 29.
23 E.g., North (1965).
24 Eddingtom (1928), p. 85. It is interesting to note that these words of Eddington preceded for more than two decades the coining of the expression 'Big Bang', so they should not be interpreted as a critique of a particular model (after all, the first model which could, in a loose sense, be called a Big Bang model, was constructed by Lemaître only in 1931), but as rejection of the general concept of the origination of the world at a finite moment of time.
25 Steckline (1983); Price (1996), and references therein.
26 Kragh (1996).
27 Davies (1978).
28 Ellis, Maartens and Nel (1978).

29 This biographical detail is not of primary importance, but is still telling; we shall return to this at the end of this chapter.
30 Tipler (1982).
31 Barrow and Tipler (1986), pp. 601–608.
32 For a sketch from the pen of a 'contact pessimist', see Tipler (1981b).
33 E.g., Hawkins (1971); Munitz (1986); Grünbaum (1991); Balashov (1994); Kragh (1996). The present section is partially based upon the discussions in Ćirković (2004c).
34 It seems clear that Empedocles held a sort of absolutist (substantialist) theory of the nature of time. In particular, the fragment B 16 of the Diels collection reads (according to the translation of Burnet 1908): 'For of a truth they (Strife and Love) were aforetime and shall be; nor ever, methinks, will boundless time be emptied of that pair.'
35 See the excellent discussion of Empedocles' cosmology in O'Brien (1969).
36 Another pioneering contribution of Empedocles lies exactly in the separation (the earliest one in Western thought) of physical nature and artefacts of human cognizance. See, for instance, the Diels' fragment B 8, reading (in Burnet's translation): 'There is no coming into being of aught that perishes, nor any end for it in baneful death; but only mingling and change of what has been mingled. Coming into being is but a name given to these by men.' Even more telling along the same lines are fragments B 11 and B 15.
37 Balashov (1994).
38 Guthrie (1979).
39 Fragment B 21 in Diels (1983), translation by Burnet (1908). Similar content can be found in B 20.
40 *Ecc.*, 1:10–11.
41 See, for instance, Davies (1994).
42 Brown (1984) comes to a similar conclusion by arguing that the Empedoclean cosmology does not support personal immortality *via* eternal recurrence.
43 Gregory (2007), p. 2.
44 Ibid., p. 95.
45 In *The Dream Quest of Unknown Kadath*; Lovecraft (1999), p. 485.
46 Thucydides (1972), I, 1.
47 Spengler (1918); see also the controversial, but immensely inspiring book by Francis Cornford (1965).
48 This is clear, for instance, from the fragment A 10 in Diels (1983), preserved by Plutarch, in which it is explicitly asserted that the formation and destruction of many worlds occurs within the global temporal infinity. In the continuation of the very same excerpt from *Stromateis*, an evolutionary doctrine is attributed to Anaximandros: '. . . Farther he says that at the beginning man was generated from all sorts of animals, since all the rest can quickly get food for themselves, but man alone requires careful feeding for a long time; such a being at the beginning could not have preserved his existence.' (Fairbanks 1898) Hyppolites quotes Anaximandros as emphasizing the nature of *apeiron* as not only everlasting, but truly eternal (B 2), obviously in opposition to mankind, which has a fixed beginning in time. Even more intriguing is the doctrine ascribed to Anaximandros by Cicero: 'It was the opinion of Anaximandros that gods have a beginning, at long intervals rising and setting, and that they are the innumerable worlds. But who of us can think of god except as immortal?' Did he have in mind essentially what we today denote as supercivilizations or Kardashev's Type 3 extraterrestrial communities or even 'basement universe' creators? More on this in Chapters 8 and 9.
49 Guthrie (1979).
50 This would be a historical analogue to the Eddington–Lemaître cosmological model, considered on p. 45 below.
51 In translation of William E. Leonard, available via WWW Project Gutenberg at http://www.gutenberg.org.dirs/7/8/785/785-h/785-h.htm: last accessed 1 February 2012.
52 Tipler (1982).
53 Hutton (1788), p. 304.
54 Hutton (1788), pp. 294–295.

55 Which was standard in Hutton's time, even more so than in Wallace's; cf. Fontenelle [1686] (1990), and the above-cited monographs of Crow and Dick.
56 Haldane (1945); emphasis by the present author.
57 E.g., Eddington (1952).
58 Bok (1946); Kragh (1996).
59 Bondi (1961), p. 118.
60 E.g., Hinckfuss (1975); Newton-Smith (1980); Arsenijević (1986).
61 This is entirely distinct from the fact that the characteristic timescale for the onset of inhomogeneity is finite (it is roughly equal to $(4\pi G\rho)^{-1/2}$, where ρ is the density of homogeneous matter), which had been known for quite some time before the Eddington–Lemaître universe; in fact, this result predates Einstein's general relativity by some years (e.g., Peebles 1993).
62 The fragment B 64 of Diels suggests that Anaxagoras endorsed a version of what were to become the absolutist or Newtonian (or Platonist in Newton-Smith's parlance) theory of time. In this description of Anaxagoras' cosmology, preserved in Simplicius' commentaries on Aristotle's *Physics*, it is emphasized that the world itself was created at the beginning of motion, but that *before that time* everything was static. As in the Eddington–Lemaître universe, the time axis can be formally extended backwards to $-\infty$.
63 Of course, this interpretation of the Eddington–Lemaître universe is not mandatory. It may as well be said that stars and galaxies have existed indefinitely as such before the expansions starts. In this case the anthropic argument becomes valid at any epoch in the finite past, even before the start of the expansion, since it reduces to the application in the static Einstein universe. However, there are additional arguments against such an interpretation of the Eddington–Lemaitre model. For instance, as shown in an instructive study by Pegg (1971), the model with an indefinite past containing stars can be rejected as a consequence of a very grave form of Olbers' paradox.
64 Newton-Smith (1980), p. 44.
65 Barrow and Tipler (1978).
66 Misner (1969).
67 Is such a rescaling just a mathematico-philosophical perversion lacking any relevance for the physical world? It seems that the answer is firmly in the negative. While the elaboration lies beyond the scope of the present study, it is enough to point out that the famous 'biological scaling hypothesis' of Freeman Dyson is just one guise of the rescaling of time in the relationist context (Dyson 1979). Many results in the nascent discipline of physical eschatology depend on Dyson's hypothesis, and it is obvious that its being true or not has very real, physical consequences.
68 Bondi and Gold (1948); Hoyle (1948).
69 Hoyle (1949).
70 Balashov (1994).
71 Cf. Milne (1940).
72 Quoted according to Kragh (1996).
73 Sciama (1955); Harwit (1961).
74 E.g., Bok (1946).
75 I use 12 Gyr as the rough age of the Milky Way thin disc, i.e., that particular Galaxy component where we expect to find habitable planets.
76 Annis (1999b).
77 Linde (1988, 1990); Linde, Linde and Mezhlumian (1994); Vilenkin (1992, 1995).
78 Narlikar (1984); Hoyle (1992).
79 Linde *et al.* (1994).
80 I shall discuss a particularly indicative instance of Toulmin and Goodfield [1966] (1982) in Chapter 7; some others are listed in the histories of Kragh (1996, 2004).
81 For early 'cosmographic' speculations such as Soddy's or Nernst's, see Kragh (2004).
82 Weinberg [1977] (1993).
83 Kragh (2005); see also Kragh (1996), pp. 136–138.
84 One could argue that some of the astrobiological sectors, notably observational searches for habitable planets or calculations of circumstellar habitable zones, or analysis of organics from

meteorites, etc. are factual enough (of course, the term needs not to be taken as referring to purely empirical work; Gamow, who after all was a theoretician himself, did not use it in such a way either). On the other hand, the same could be said for some bits and pieces of cosmology even in Gamow's time (e.g., developing techniques for more precise measuring of the Hubble parameter or Zwicky's prescient study of the potential usage of gravitational lenses on cosmological scales).

85 At the very least, it did require further weakening of the teleological stranglehold on science.

86 For historical overviews see, apart from Dick's book, studies of Wolfe (2002); Raulin-Cerceau (2010).

87 Who used it in a fine review (Wilczek 1993) of the fellow Nobel laureate Steven Weinberg's popular book *Dreams of the Final Theory* (Weinberg 1992) in order to emphasize that even hard-nosed scientists with self-professed disinterest in philosophy cannot really avoid it when they turn to 'big' or 'final' questions of science.

88 Bondi (1955).

89 Beer *et al.* (2004).

90 Literal crossover examples include Almár (1992); Ćirković and Bostrom (2000).

91 Dicke (1961). The density of matter sets both the expansion and the matter creation rates in the steady state cosmology and it is unconnected to the properties of stars or planets.

92 For Hoyle's life and science, see Hoyle (1994); Gregory (2005); Mitton (2005).

93 Hoyle (1957).

94 E.g., Hoyle (1994); Hoyle and Wickramasinghe (1999, 2000); Hoyle, Wickramasinghe, and Watkins (1986).

95 Hoyle (1994); Wickramasinghe, in Gough (2005), pp. 58–92.

96 Wallis and Wickramasinghe (2004); Napier (2004).

97 E.g., Lineweaver *et al.* (1996); Lineweaver (1998); Griffiths and Lineweaver (2004).

98 Lineweaver and Davis (2002); Lineweaver and Egan (2007). Seminal results on GHZ are in Lineweaver (2001); Lineweaver, Fenner, and Gibson (2004); see also Robles *et al.* (2008).

99 Further examples of such a crossover between astrobiological and cosmological themes could be found in the careers of other distinguished scientists of the last half-century: Sir Martin Rees, Freeman Dyson, Paul C.W. Davies, Max Tegmark, Nikolai Kardashev, Brandon Carter, Eric Chaisson, Chris Impey, etc. Even without any of the structural arguments sketched above (and highlighted in the further course of this book), this would constitute an interesting correlation, especially when the modest numerical sizes of both disciplines is taken into account.

3 Cosmology, life and selection effects

Quote: political epigram. Sir John Harington (1561–1612).

1 Hogan (2000); Bostrom (2002).

2 For a pre-2009 exception to this, the famous Gliese 581 system, see, e.g., Udry *et al.* (2007).

3 There is even a website devoted to this famous newspaper error: http://www.deweydefeatstruman.com/ (last accessed 1 February 2012).

4 E.g., Ferland *et al.* (1998); for the main resources, see http://www.nublado.org/wiki (last accessed 1 February 2012).

5 As we do; see, e.g., Chambers (2010).

6 Lineweaver and Grether (2003), p. 1358.

7 Malmquist (1920); Eddington (1940).

8 This is luminosity over the whole electromagnetic spectrum, the so-called bolometric luminosity. Proxima is much fainter when we restrict ourselves to visible light – only 0.000056 solar luminosities!

9 Teerikorpi (1997) and the extensive list of references therein.

10 A digression for afficionados of the history of science: Malmquist bias has been implicitly present in the massive effort by the Dutch astronomer Jacobus Kapteyn, mentioned in the previous chapter, who was the last and most diligent in an array of distinguished astronomers trying to complete the Herschellian programme to determine the shape of our stellar system – the Galaxy – from star counts in standardized patches of the sky. Lacking knowledge of the

intrinsic luminosity function of stars and having no representative volume-limited sample to rely upon, Kapteyn assumed that stellar luminosities are distributed according to a Gaussian distribution around a mean value (for a discussion, see Kapteyn 1913; Eddington 1914). With this assumption, Malmquist bias led him to severely underestimate the number of intrinsically faint stars at large distances, leading to the small and erroneous 'Kapteyn universe'. As discussed in Chapter 2, an idiosyncratic version of the Kapteyn universe was a major ingredient in the proto-astrobiological theory of Alfred Russel Wallace of 1903.

11 Signor and Lipps (1982), p. 295.

12 The Dell was closed and demolished in 2001 after 103 years (!) of service, and 'The Saints' now play at St Mary's stadium.

13 For lists of particular fine-tuned constants and parameters, see, e.g., Carr and Rees (1979); Barrow and Tipler (1986); Hogan (2000).

14 Lineweaver and Egan (2007).

15 Dirac (1937, 1974); Jordan (1955).

16 Dicke (1961); a rejoinder is Dirac (1961).

17 Hellings *et al.* (1983); Damour, Gibbons and Taylor (1988).

18 Palladino (1990).

19 Cf. Arsenijević (1986).

20 E.g., Callender (2001, 2004).

21 Some of it has been contained in various musings of John A. Wheeler; e.g., Wheeler (1975). For an unabashedly anthropocentric vision of this kind from a pen of a famous playwright, see Frayn (2006).

22 Baxter (2000); Bostrom (2003); Holt (2004); Walker and Ćirković (2006). The unpublished 'Peer Hypothesis' of Paul Birch also belongs to the same broad category (http://www.paulbirch.net/PeerHypothesis.pdf; (last accessed 1 February 2012).

23 Especially the review paper of Tegmark (2003) to which I shall return in the next chapter.

24 For the question of prevalence of L-enantiomeres could, in principle, be answered by postulating that both L- and D-forms of life exist with equal frequency throughout the universe, preserving the *global* symmetry; see Gleiser, Thorarinson and Walker (2008). If there is an appropriate physical mechanism how the initial small asymmetry can be amplified and spread – as in the absorption of polarized radiation hypothesis – we have successfully embedded our particular problem into a wider whole. After that, it is essentially a semantic issue ('how we shall call *our* stereoisomer?') that completes the explanatory account.

25 The title of the seminal book on observation selection by Bostrom (2002).

26 Ćirković, Sandberg and Bostrom (2010).

27 Cohen and Stewart (2002), p. 100.

28 Ironically, Cohen and Stewart are distinguished *opponents* of anthropic reasoning (as elaborated, for instance, in a detailed interview in Cohen and Stewart 2009). Such is the widespread confusion surrounding simple observation selection!

29 Gould (1987b), p. 205 – it is contained in a review of Dyson's intellectual autobiography, *Disturbing the Universe*, which does indeed have pronounced teleological elements.

30 Smolin (2007).

31 Susskind (2006), pp. 170–172.

32 Olum (2004).

33 Note that the definition of 'large' in terms of spatial extent is primary; although Olum does not explain it. Otherwise, one of his own solutions, based on the discrepancy between spatial extent and number of observers, ceases to make sense (Section 2.9 of his study; see the solutions listed below).

34 Gott (1993); for the Doomsday Argument, see also Bostrom (2001, 2002).

35 In particular, those of WMAP: Bennett *et al.* (2003); Schwarz and Terrero-Escalante (2004).

36 E.g., Billyard *et al.* (2000).

37 Ćirković (2006).

38 Olum's problem is related to Fermi's paradox (Brin 1983; Webb 2002), the famous question asking why we do not find extraterrestrial intelligences or traces of their activity in our past light cone, as well as to Davies–Tipler argument outlined in Chapter 2. However, there is an all-important difference: in most expositions of Fermi's paradox there is no mention of the number of observers. In fact, as Frank Tipler (e.g., Tipler 1980) and other SETI sceptics have

repeatedly emphasized, Fermi's paradox is worsened because it makes sense for a civilization to launch self-reproducing robotic probes ('von Neumann probes'), which need not be considered intelligent at all! Thus, Fermi's paradox is of a more general nature; the price paid is, obviously, higher precision of Olum's problem, where we can use quantitative machinery of anthropic reasoning (in the form developed, for instance, by Bostrom 2002). I shall return to issues closely related to Olum's problem in Chapters 7 and 8.

39 Or, in a less merciful Nietzschean term, 'mummificationist' view. In a beautiful passage in *Twilight of the Idols*, philosophers are rebuked for 'their lack of historical sense, their hatred of even the idea of becoming . . . They think they are doing a thing *honor* when they dehistoricize it, *sub specie aeternitatis* – when they make a mummy of it.' (Nietzsche 1968, p. 351.)

40 There is a set of necessary conditions for universal 'cosmic time' to be well defined. They are satisfied in all currently favoured cosmological models, and in all the inflationary models that Olum considers. They are not satisfied, for instance, in the celebrated rotating universe of Kurt Gödel (e.g., Stein 1970), which contains closed time-like curves.

41 Olum (2004), p. 1.

42 Lineweaver, Fenner and Gibson (2004).

43 For reviews, see e.g., Hartmann (1999); Kumar and Piran (2000); Piran (2000); Mészáros (2002).

44 Leonard and Bounell (1998); Annis (1999a); Gonzalez, Brownlee and Ward (2001); Scalo and Wheeler (2002); Vukotić (2010).

45 This is unrealistic, but is more in agreement with Olum's scenario and the continuity thesis, to be discussed in Chapter 6. We shall see below how this assumption may also be relativized with respect to the evolution of the universe in time.

46 Of course, some very weak form of the principle of indifference is necessary here for the said conclusion. Nevertheless, this option is quite implausible on other grounds, as Olum points out.

47 Non-exclusivity means that we need a solution that will act uniformly in space and time over vast scales without the need for 'conspiracy' or fine-tuning; see especially Brin (1983).

48 Carter (1983, 2008); Maynard Smith and Szathmary (1997); Knoll and Bambach (2000); Watson (2008).

49 Ho and Monton (2005).

50 In addition, Ho and Monton provide interesting remarks about the decrease of credence in a theory due to a particular improbable observation. While they agree that anthropic reasoning as used by Olum decreases our credence in the best cosmological theory (eternal inflation), they argue that our credence in it should still be very high, lacking further contrary evidence.

51 Darwin (1859), pp. 489–490.

52 Although many authors make a big deal of the problems about defining complexity (e.g., Tsokolov 2009), I find this issue a typical red herring. Critics tend to forget, time and again, that theories define concepts, so the fact that we still do not have a *general* theory of complexity does not preclude our intuitive grasp of it.

53 Gonzalez, Brownlee and Ward (2001); Gonzalez (2005).

54 Cf. Bracewell (1975).

55 E.g., Kaltenegger, Fridlund and Kasting (2002); Summers *et al.* (2002); Edwards (2004).

56 This also gives another instance of the usefulness of pop-cultural references for astrobiological research. Heretical as it may sound, many of these astrobiological histories have been first explored in the recreational context of SF literature; while, of course, this cannot substitute for quantitative research, this circumstance should be an inspiration for building specific models.

57 Adapted from Ćirković and Vukotić (2008), p. 540.

58 Gould (1985).

59 Vukotić and Ćirković (2007).

60 The landscape is temporally additive as well, in the sense that we can take snapshots of it at particular epochs and co-add them to obtain the landscape over an extended period of time.

4 Cosmology, life and the Archipelago

Quote: from Springer and American Institute of Physics, September 1997, *At Home in the Universe (Masters of Modern Physics)*, John Archibald Wheeler; with kind permission from Springer Science and Business Media B.V.

1 The story has been recently told in fabulous detail by Levy (2011).
2 Gott *et al.* (2005).
3 There is some additional subtlety here. The SDSS Great Wall, similar to other parts of large-scale cosmological structure, is a physical association – as contrasted with chance groupings of objects in the sky – and thus we are entitled to regard it as a legitimate object of (astro)physical inquiry without falling into a trap of reification error. However, this does not mean that such structures are *bound* systems, like the netbook I use for writing this sentence, or a chair, or Earth, or any of us, or any of the hypothetical extraterrestrial observers in the Milky Way, or the Milky Way itself. In everyday life, we tend to call bound objects 'things' and to allocate a definite physical size to them. A set of billiard balls is not a thing with definite size, although it is obviously an interacting physical system – indeed, strong interaction between its components is the key to any snooker game! Large-scale structure is not bound and will dissipate and cease to exist, according to the new standard cosmological model, under the inexorable accelerating expansion of the universe, driven by dark energy. The changes are, of course, so slow that even the huge look-back times to these great walls and superclusters of galaxies are smaller than the dissipation timescale and we can treat our surveys essentially as snapshots (and, consequently, assign definite physical size and other properties to them).
4 This chapter is loosely based upon Ćirković (2008). It relies on slightly more complex technical issues in fundamental physics and cosmology, so it may be somewhat more exacting to the casual reader. Although it plays an important role in the overall 'ascent-and-descent' plan of the book, it is optional in the thematic sense: astrobiology is well defined in the single universe and the discussion in all the other chapters is essentially independent of the universe–multiverse dichotomy. By skipping the present chapter, a reader may lose some richness of perspective, but will not lose any coherence of the narrative.
5 No pun intended here. Tegmark (2003), in what is rapidly becoming a classic paper, lists four levels of the multiverse: (I) extensions of infinite spacetimes outside of our cosmological horizon, (II) 'bubble universes' created by cosmological inflation, (III) Everettian 'many worlds' of quantum mechanics, and (IV) a Platonic multiverse of totality of mathematical structure. Each level has many different realizations – it is easiest to see at Level I, when we take into account that there are many spatially infinite cosmological models – they are partially overlapping, and perhaps further classifications according to explanatory roles are forthcoming.
6 Smolin (1997, 2007).
7 Deutsch (1998). While Deutsch uses the term 'multiverse' in the sense of Everettian quantum mechanics, it joins smoothly with the Tegmark's classification and other ways of thinking about the totality of existence.
8 For further literature on the multiverse see, at the popular level, Rees (1997, 2001); Vilenkin (2006); Susskind (2006); Manly (2011); more advanced discussions are Garriga and Vilenkin (2001); Linde (1990); Freivogel and Susskind (2004); Ellis, Kirchner, and Stoeger (2004); Carroll (2006); Carr (2007); Aguirre, Johnson, and Shomer (2007); for a philosophical version, see Lewis (1986).
9 While this is a huge topic in itself, some entry points into the literature are Lloyd (2000); Fredkin (2003); Chaitin (2006).
10 The term promoted (if not truly invented) in popular culture by the contemporary British novelist Michael Moorcock; see Moorcock (1995), especially the new Introduction.
11 Susskind (2003, 2006); Bousso and Polchinski (2004).
12 Tegmark (1998).
13 Cf. Houellebecq (2007).
14 Weinberg (1987).
15 Carroll (2006).
16 Susskind (2003).
17 Cf. Hamkins and Montero (2000); Tegmark (2008).
18 E.g., Silk (2000); Saitoh *et al.* (2008).
19 In comparison, the New Standard Cosmology (after 1998) gives the empirical ratio as about three.
20 Susskind (2006), p. 356.

21 Shapiro and Feinberg (1982); Dyson (2003, 2007); Benner (2009).
22 See, for example, the review of Bedau (2003).
23 E.g., Ray (2003).
24 Even more precisely, Tegmark uses the term 'self-aware substructure' in order to pose the problem as mathematical, since from the Platonist viewpoint an observer is nothing else than part of a wider mathematical structure. But the same general reasoning about the criteria applies if we (i) regard observers as physical systems only approximated by mathematical descriptions, and (ii) have a particular physical mechanism to introduce variations (like spontaneous symmetry breaking in inflationary cosmologies).
25 Cohen and Stewart (2002), p. 5. This popular book is, as already mentioned in Chapter 1, one of the best examples of criticism of astrobiology 'from the left', to use an outdated political cliché. Among its many rants against mainstream astrobiology being too conservative and unimaginative, there are some useful and thought-inspiring gems.
26 For classical reviews of such evidence, see Barrow and Tipler (1986); Balashov (1991); Hogan (2000).
27 Penrose (1989); Ćirković (2003b).
28 Tegmark and Rees (1998).
29 Aguirre (2001).
30 Harnik, Kribs and Perez (2006).
31 Chavanis (2007); Adams (2008).
32 Bostrom (2002).
33 Starkman and Trotta (2006).
34 Trotta (2007).
35 The documentary is Alex Gibney's *Enron: The Smartest Guys in the Room* (2005), and the name of genecentrism-savvy CEO is Jeffrey Skilling. In April 2011, at the time of this book's writing, the US 5th Circuit Court has upheld the 2006 conviction against Skilling, who is currently serving a 24-year, 4-month prison sentence, after being found guilty of 19 criminal counts, including conspiracy, securities fraud, false statements, and insider trading.
36 Bostrom (2002).
37 Which is, by the way, quantified through the Observation Equation on page 173 of Bostrom (2002).
38 A paradigmatic example is Klee (2002); ironically enough, that particular study, which attempts to list numerical imprecisions of various fine-tuning arguments, contains serious numerical errors itself; see a critique in Walker and Ćirković (2006). The very fact that this promptly written critique took four years and countless toing-and-froing with referees to be published (and even then only as a part of a wider study), testifies how strongly a part of the philosophical community feels about anthropic reasoning – to the point of condoning even numerical blunders if they are made in an 'ideologically correct' tract.
39 Dyson, Kleban and Susskind (2002).
40 Page (2006).
41 Carlip (2007).
42 Vilenkin (2007).
43 Boltzmann (1895), p. 414.
44 Ellis and Brundrit (1979).
45 Harrison (1987), p. 213.
46 Pagels (1998), p. 181.
47 E.g., Mosterín (2000, 2004), after many others. This is not just the 'usual' well-intentioned philosophical hair-splitting, but an aggressive and systematic attempt to elevate a semantic pseudo-problem to a serious issue in metaphysics and the philosophy of science. Being an astronomer, I fear the day some philosopher will discover that the very bread and butter of my trade consists of misnomers: 'stellar magnitudes', 'planetary nebulae', 'forbidden spectral lines', 'phases of interstellar gas', 'Hubble's constant', 'metallicity' – to give just a few glaring examples!
48 For some of those applications, see, e.g., Bostrom (2002); Ćirković, Sandberg and Bostrom (2010).

49 Lineweaver and Davis (2002); Carter (2008); for Fontenelle's early anthropic argument about comets, see Ćirković (2002).

5 Astrobiology as a natural extension of Darwinism

Quote: taken from Teilhard de Chardin. *The Phenomenon of Man* 2nd edition. Harper and Row, New York (1965).

1 Misner, Thorne and Wheeler (1973), pp. 24–25.
2 For some of the recent contributions in this long-standing debate, see, e.g., Brandon (1997); Cartwright (1997); Hull (2002); Elgin (2006); Fodor and Piattelli-Palmarini (2010).
3 Popper (1972, 1974).
4 Popper (1974), p. 136.
5 Ruse (1977).
6 Besides the historical fact that no extrasolar planets were known to exist with certainty at the time of Popper's writings. It was still conceivable at the time – although not very likely in the view of most astronomers – that the formation of the Solar System was a rare fluke, occurring as a consequence of an incredibly rare stellar near-collision or some other event of very small probability measure.
7 For a refreshingly clear and optimistic view of the prospects for present life on Mars, see Schulze-Makuch and Darling (2010).
8 Ball (1985); Rospars (2011).
9 Shanahan (2004).
10 Dawkins (1983); Gould (1989, 2002).
11 Lem [1964] (1973).
12 Lem [1974] (1977).
13 Gould (1996).
14 Ward (2005).
15 E.g., Fodor and Piatelli-Palmarini (2010).
16 Note that it is quite a distinct issue from the question whether this skewing was 'random' or 'deterministic'. I am speaking here about epistemic (or 'Bayesian' or 'subjective') probabilities without prejudicing the deeper (meta)physical issue as to whether relevant physical, chemical or biochemical processes are deterministic or not.
17 Gould (1989), pp. 48–50.
18 Gould (2002), esp. Chapters 6, 11, 12. In an eerie episode of the tragic history of biology in the Soviet Union, the famous geneticist Nikolai Ivanovich Vavilov was, immediately before his arrest, subjected to brutal investigation by the 'Grand Inquisitor' of Soviet life sciences, Trofim Denisovich Lysenko. One item he was accused of was writing that evolution can sometimes lead to simplification, a view apparently at odds with the Stalinist ideology of inexorable progress throughout the universe. As Lysenko put it succinctly and, in the context of the times, threateningly: 'Yet in Chapter 4 of the history of the Party it says that evolution is increase in complexity . . . ' Vavilov ultimately paid for his multiple 'heresies' with his life; see Medvedev (1971).
19 Conway Morris (2003); Beatty (2006); Turner (2011).
20 There is a distinct possibility that the *ceteris paribus* clause is untenable on physical grounds. In this particular case, it would mean that, due to some not-yet-understood cosmogonical reason, the existence of a massive cometary disc precludes the formation of a habitable planet in the same planetary system. Or that, even if a habitable planet is formed, it is predominantly likely to migrate outside of the habitable zone due to some physical reason connected to the existence of the massive cometary belt. On the other hand, it could be argued that even with an order of magnitude increase, the mass of the cometary belt in the τ Ceti system is still quite small compared to the characteristic mass scales for dynamically important planets. We simply do not understand the formation of planetary systems and their early evolution well enough to decide either way. I shall neglect the possibility for the purpose of this particular scenario, but shall return to the whole class of 'unphysical *ceteris paribus*' arguments in some detail when dealing with the controversial 'Rare Earth hypothesis' in Chapter 6.

21 Greaves *et al.* (2004).
22 For the age of τ Ceti, see Mamajek and Hillenbrand (2008).
23 Of course, in accordance with the Bayesian approach to observation selection sketched in the previous chapters, there could occur a probability shift if there are many observers throughout the Galaxy and some weak version of the principle of indifference holds. If, for example, there are 10^9 planets with observers in the Galaxy in a sufficiently broad spatial distribution (not necessarily homogeneous), even a very rare star would with reasonable probability be within a couple of parsecs from *some* observers. But, the premise is almost certainly untrue – even the fiercest SETI-optimists do not think that intelligent beings are that common; the distance to τ Ceti is, for any reasonable density of planets with observers, much smaller than the average distance between such observer-inhabited planets.
24 Conway Morris (1998, 2003).
25 Of course, this is entirely naturalistic reading of Conway Morris, who in any case devotes very little attention to biogenesis itself. Less charitably, one could argue that a more adequate interpretation of Conway Morris is that terrestrial biogenesis is not just contingently improbable, as famously argued, for example, by Jacques Monod, but supernaturalistic and unique. This is so especially in light of the last chapter of *Life's Solution* and some passages in his previous book, notably the rhetorical question: '[i]s not much of our disenchantment with the barbarity of much recent architecture due to this banishment of the numinous?' (Conway Morris 1998, p. 11) While supernaturalism does not imply uniqueness – indeed, as persuasively shown by the historian of science Michael J. Crow, almost all pluralists roughly between Lucretius and Svante Arrhenius were supernaturalists – the converse thesis needs only weak additional assumptions to hold true (Crow 1986). In a spatially and temporally finite universe, very low probability of even clearly naturalistic biogenesis would ensure uniqueness, below some critical a-priori probability. If we relax either spatial or temporal finiteness, it is much more difficult to maintain uniqueness.
26 Conway Morris (2003), p. 282.
27 Ibid., p. 136.
28 The term 'galactic convergence' is from Sterelny (2005). Add to this a rather bizarre consequence of Conway Morris' account (if not taken as a joke) that the convergence in the subspace of flowering plants is weaker than convergence in the subspace of intelligent humanoids' cultural patterns, e.g., greetings!
29 Compare this with the classical hypothesis for resolving Fermi's paradox suggested by Shklovsky, von Hoerner and some other mainly disgruntled SETI pioneers: intelligent species are rare in any particular epoch of Milky Way history because they destroy themselves soon after inventing nuclear weapons (or biotech weapons, or military nanotechnology, or whatever infernal weapons of mass destruction are on the horizon). Such a tremendously strong convergence – millions of civilizations on millions of planets exhibiting the same self-destructive behaviour, without any physical correlations possible – does sound more like mystical 'fate' than naturalistic science. Seemingly paradoxical, such extreme convergentionism has much in common with the long-abandoned formalist theories of orthogenesis, which purported to explain biological change through the innate and unchangeable tendencies of each morphological type. See Grene (1958), Bowler (1992), Gould (2002, esp. Chapter 5), and for an example specifically related to SETI, see Ćirković, Dragićević and Berić-Bjedov (2005).
30 Di Giulio (2010).
31 Briggs and Fortey (2005), pp. 108–109.
32 Perhaps by using quantum computing in the scenario depicted in Robert Charles Wilson's (2003) novel *Blind Lake*.
33 Although, incredibly enough, some beautiful proposals have already been made in that direction; see, for example, Turnbull *et al.* (2006); Beckwith (2008); Kaltenegger and Sasselov (2010).
34 Consider, for example, the wonderful results of modelling of desert formation in Daisyworld in Ackland, Clark and Lenton (2003).
35 McShea (1994, 1996).
36 Vermeij (2006), p. 1808.
37 Sterelny (2005).
38 E.g., Zdravković (2008), ch. 6.

39 The situation might be different if life based on nuclear, instead of electromagnetic, force exists somewhere in the universe.
40 Gold (1999).
41 Davies *et al.* (2009).
42 Chela-Flores (2003).
43 In Impey (2010), p. 162.
44 Gould (1989), pp. 309–311.
45 Ward and Brownlee (2002), p. 144.
46 Caldeira and Kasting (1992); see also Bounama, Franck, and von Bloh (2001).
47 Of course, these models do not include any intentional influence upon the Earth system. This is important for two related, but distinct reasons: (i) contemporary warnings about anthropogenic global warming clearly demonstrate the capacity of human civilization to adversely influence climate on the global scale in short periods of time, so the end of the biosphere could be immensely quickened by irresponsible human actions, and (ii) it is conceivable that human or posthuman civilization on Earth – or, to be logically complete, an advanced extraterrestrial civilization visiting the Solar System in the future – will be able to postpone or avoid Earth's dire fate by acts of macro-engineering. In fact, some of the proposals along these lines have already been put forward in order to mitigate the effects of anthropogenic global warming. This is a small part of the complex problematic of macro-engineering in its general context, obviously of huge importance for the third of the 'Canonical Three' and in particular for SETI studies; I shall return to this topic in Chapters 7 and 8.
48 See Bostrom and Ćirković (2008), esp. the Introduction, pp. 1–29.
49 Turner (2011), p. 65.
50 In passing, one could mention that, apart from the topic of biogenesis, which by its very nature presents a bridge connecting biochemistry and molecular biology with astrobiology, there are some intriguing steps in different directions as well, benefitting from the increase of scope offered by astrobiological research. An interesting example is the recent criticism of the panspermia hypotheses based on the genetic properties of the supposed last universal common ancestor (Di Giulio 2010).
51 Okasha (2006), p. 3.
52 Race, M. S. in Bertka (2009), pp. 205–219.
53 Rospars (2011).
54 Voiced recently, for example, by Fodor and Piattelli-Palmarini (2010); see a critical account in Sober (2010).

6 Rare Earths and the continuity thesis

Quote: The Book of Job, King James Bible.

1 Impey (2010). In cosmology, there has been a similar volume, *Origins: The Lives and Worlds of Modern Cosmologists*, edited by Alan Lightman and Roberta Brawer (1990), containing interviews with 27 distinguished cosmologists. Besides offering another instance of 'parallel evolution' of thinking about the physical and biological universes, the differences are also interesting. By the time of *Origins*, cosmology had grown out of its birth pangs and even adolescent unrest, so Lightman and Brawer's task was easier, at least formally – they posed a set of rather similar questions to their interviewees. Impey, on the other hand, could not enjoy such luxury; even lumping together 'big themes' into separate parts could not pre-empt the need for rather disparate, finely tuned and personalized questions. This is partly a reflection of the real diversity of astrobiological research; but a part of it is, as I attempt to argue throughout this book, the consequence of immaturity, unclear philosophical and methodological infrastructure, and an over-defensive attitude widespread in any multidisciplinary field, all of which act to impede and slow down the emergence of a truly common 'astrobiological language'.
2 A few years ago, I received a referee's report which somewhat timidly asked if I was aware that some claims in my submitted manuscript were contrary to 'Rare Earth theory' (!) and the book by Conway Morris (2003) was explicitly invoked by the concerned reviewer.
3 Ward and Brownlee (2000), p. 283.

4 Ibid., Chapter 12.
5 Except for the helium-hydride ion HeH^{+}; Roberge and Dalgarno (1982).
6 Havel (1999).
7 Havel's paper is interesting from another point of view related to anthropic reasoning: it explicitly treats the 'implied observer' problem in conceived worlds.
8 Ferguson (1999), p. 83.
9 Ibid., p. 18.
10 Some of the entry points, apart from Ferguson's excellent anthology, are Hawthorne (1991) and Cowley (1999). An amusing popular book by Comins (2010) shows how the counterfactual approach works in astronomy.
11 Horner and Jones (2008, 2009).
12 Horner and Jones (2009), p. 75.
13 Part of the problem with REH is also that (4) (and analogous premises in other requirements) is uncertain – as explicated by Ward and Brownlee themselves. In other places (e.g., pp. 186–189) they claim that *some* impacts are in fact necessary for diversification of the biosphere. Obviously, the risk and habitability functions are more complex than intuitively assumed.
14 This is an instance of the general and quite tough metaphysical problem of *transworld identity*: can there be the same objects in different worlds, and if so, how do we establish their identity? Perhaps the best introduction is http://plato.stanford.edu/entries/identity-transworld/ (last accessed 22 November 2011). See also Quine (1976), Lewis (1986) and Mackie (2006).
15 Chyba and Hand (2005), pp. 47–48.
16 Benn (2001); Waltham (2004). See also Néron de Surgy and Laskar (1997), especially their Figure 1.
17 E.g., Kitcher (1985).
18 Darling (2001), Chapter 6.
19 That Conway Morris' (2003) book is quite religion-friendly is seen at many layers. At the bottom are those instances in which the usual portrayal of conflicts between science and religious fundamentalism is a target of witticisms and sarcasms (e.g., p. 321). All this is intermingled with approving quotes from distinguished and, on the subject of religion, hardly dispassionate sources such as C. S. Lewis and G. K. Chesterton.
20 The paper on GHZ is Gonzalez, Brownlee and Ward (2001); the book is Gonzalez and Richards (2004).
21 Valencia, O'Connell and Sasselov (2007).
22 Bounama, von Bloh and Franck (2007).
23 Waltham (2004).
24 See, e.g., Ćirković (2004a, 2009); Ćirković and Vukotić (2008); McCabe and Lucas (2010); Forgan and Rice (2010); Vukotić (2010).
25 Cohen and Stewart (2002), p. 6.
26 Ward and Brownlee (2000), p. 282.
27 Haldane (1927), p. 286.
28 Fry (1995, 2000, 2011).
29 Fry (1995), p. 389.
30 Eigen (1992).
31 Hoyle and Wickramasinghe (1981, 1999). Hoyle (1983), p. 19 gives the probability as $10^{-40\,000}$, which is usually quoted in the literature (and even became known as Hoyle's fallacy, see http://en.wikipedia.org/wiki/Hoyle%27s_fallacy, last accessed May 10, 2011), while his long-term collaborator in fascinating recollections quotes the smaller value given above; Wickramasinghe (2005), p. 143. There is also some amusing confusion about the exact type of airliner; while Hoyle and most commentators mention a Boeing 747, Maynard Smith (1986), p. 49, and Wikramasinghe give it as the smaller and older Boeing 707. (While the 'Jumbo Jet' is certainly a more appealing and iconic version, the difference in the number of parts is unsubstantial in this context, as is the difference between the two hyperastronomically small numbers.)
32 Barrow and Tipler (1986), p. 21.
33 Chaisson (2001), pp. 220–221.

34 Tsokolov (2009).
35 de Chardin [1955] (1959). The related 'noosphere', as a parallel to 'geosphere' and 'biosphere', was used previously, in the 1920s, by both Teilhard and the Russian geologist and philosopher Vladimir Vernadsky. Like Haldane, Vernadsky, one of the major figures of the philosophical movement known as *cosmism*, has also built a system of thought based on the continuity of various forms of evolution.
36 Parenthetically, a similar neologism for the origination of complex metazoans seems to be missing.
37 Haldane (1924, 1927); see also Adams (2000).
38 There seems not to be an adequate adjective here, since 'observable universe' conveys a different *kind* of quality from 'habitable universe'. 'Habitable universe' – which gives rise to the Archipelago of Habitability – means 'the universe capable of supporting life', while the 'observable universe' does not mean, in its conventional usage, 'the universe capable of supporting observers', but 'the part of the universe accessible to our (astronomical) observations' instead. While one might insist on a particular construal of an 'observable universe' in the present context, I still felt that being entirely consequential form-wise was simply not worth the effort.
39 E.g., Borges (1999).

7 SETI and its discontents

Quote: Heraclitus of Ephesus, pre-Socratic Greek philosopher, 6th Century BC.
1 For Fermi's paradox, the best reviews are the classic review of Brin (1983) and the recent popular book of Webb (2002); see also Ćirković (2009). Carter's argument has been presented in Carter (1983, 1993). The argument from biological contingency, although it could be traced back to Wallace, was clearly formulated by George Gaylord Simpson in 1964, and subsequently elaborated (in wildly varying styles) by Barrow and Tipler (1986); Mayr (1993); Ward and Brownlee (2000); Tipler (2003).
2 Jones (1985).
3 Simpson (1964); see also the discussion in Conway Morris (2003), esp. Chapter 9.
4 This should not augment our pride, since we understand intelligence and capacity for communication in a strictly operational sense, without endowing them with any values.
5 Gould (1987a).
6 Benton (2003); for the end-Permian extinction, see also Erwin (1990, 1993, 2006). For EQ history, see Russell (1983, 1995).
7 Lem (1987); see also Lem [1974] (1977, Chapter 3). I shall return to another important aspect of this novel for astrobiology in Chapter 8.
8 Consider, for example, the 'Introdus' of Greg Egan's *Diaspora*: the mass uploading of future humans in virtual 'city states' where uploads are immortal and mostly devoted to art or abstract mathematics (Egan 1997).
9 Carter (1983); see also Barrow and Tipler (1986).
10 Strictly speaking, this is just a half of Carter's argument in his fascinating 1983 paper. The rest concerns the issue of the number of 'crucial' (or 'critical') steps in the emergence of intelligent observers on Earth and possibly elsewhere. Barrow and Tipler (1986) and Wilson (1994) devoted much attention to 'anthropic' prediction of this number.
11 Maddox (1984); McKay (1996); Bostrom (2002); Barrow (2002); Lineweaver and Davis (2002); Davies (2003); Duric and Field (2003).
12 Tipler (1988, 2003).
13 Livio (1999).
14 A more detailed discussion is in Ćirković, Vukotić and Dragićević (2008).
15 Kirschvink *et al.* (2000); Hoffman *et al.* (1998)
16 However, for a view ascribing even the 'Snowball' glaciations to our astrophysical environment, see Pavlov *et al.* (2005).
17 Gonzalez (2005).
18 A minor additional argument to the same effect may come from the panspermia hypotheses, which, although quite speculative, have experienced recent resurgence. Thus, Napier (2004), as well as Wallis and Wickramasinghe (2004), have constructed working panspermia models.

19 Russell (1983, 1995).
20 McKay (1996).
21 Wilson (1994).
22 E.g., Rampino (1997); Byakov, Stepanov and Stepanova (1997, 2001); Leitch and Vasisht (1998); Stozhkov, Okhlopkov and Pokrevsky (2001); Shaviv (2002a,b); Carslaw *et al.* (2002); Stozhkov (2003); Melott *et al.* (2004); Pavlov *et al.* (2005); Gies and Helsel (2005).
23 Raup and Valentine (1983); Maher and Stevenson (1988); Oberbeck and Fogleman (1989).
24 We do not presume any special understanding of evolutionary biology here; the formulation is applicable to the case of a stable, only very slowly changing biological environment – a single macro-evolutionary regime in terms of Jablonski (1986, 1989) – in which Barrow and Tipler's metaphor of a 'monotonic approach to intelligence' might perhaps work. Such a world is not the real world, but its juxtaposition with the real world can teach us some important lessons.
25 Annis (1999a). A similar suggestion was made earlier by Clarke (1981), although his model was entirely qualitative and incorrectly incorporated a physical mechanism (Galactic core outbursts) for global regulation.
26 Thorsett (1995); Crutzen and Brühl (1996); Dar (1997); Sergi *et al.* (2001); Dar and De Rújula (2002); Scalo and Wheeler (2002); Hartmann, Kretschmer and Diehl (2002); Gehrels *et al.* (2003); Smith, Scalo and Wheeler (2004); Thomas *et al.* (2005, 2008); Galante and Horvath (2007).
27 Dar (1997).
28 Scalo and Wheeler (2002).
29 It is important to understand that the GRB-mechanism is just one of several possible physical processes hypothesized for 'resetting astrobiological clocks', whose main advantage is that its secular evolution is constrained through cosmological observations (e.g., Bromm and Loeb 2002). Any catastrophic mechanism that operates (1) on sufficiently large scales and (2) exhibits secular evolution can play a similar role. There is no dearth of such mechanisms; some of the bolder ideas proposed in the literature are cometary-impact-causing 'Galactic tides' (Clube and Napier 1984, 1990; Torbett 1986; Matese and Whitman 1990; Asher *et al.* 1994; Matese and Whitmire 1996; Rampino 1997), cosmic-ray jets (Dar, Laor and Shaviv 1998), neutrino irradiation (Collar 1996), clumpy cold dark matter (Abbas and Abbas 1998), or climate changes induced by spiral-arm crossings (Leitch and Vasisht 1998; Zank and Frisch 1999; Shaviv 2002a,b). All such effects are cumulative: the total risk function of global regulation is the sum of all risk functions of individual catastrophic mechanisms.
30 Melott *et al.* (2004).
31 Eldredge and Gould (1972); for a detailed elaboration and synthetic view, see Gould (2002).
32 For the historical context of pluralism and its discontents, see Crow (1986), Dick (1996).
33 Tipler (1994).
34 Mayr (1993).
35 In fact, there is a fourth type of criticism, which I will entirely skip here, the one suggesting that, although SETI is, in fact, well founded epistemologically, we still should not engage in it, since it might be physically dangerous or at least might cause adverse societal or psychological consequences. There could be much of importance here, but since this type of criticism touches upon quite complex topics in risk analysis and applied ethics there is no possibility to discuss it in the framework of the present book; see Tough (1990); Diamond (1999); Carrigan (2006); Baum, Haqq-Misra and Domagal-Goldman (2011).
36 Heraclitus, fragment B18 (Diels 1983).
37 Mash (1993); Basalla (2006); Pigliucci (2010), pp. 33–40.
38 Some of the other players on the same side of the philosophical fence are McMullin (1980); Rescher (1985); Kukla (2001); Ulvestad (2002).
39 Of course, as is well known (e.g., Dick 1996), this project was cancelled less than a year later. The amount of ink spilt on the cost of SETI projects – quite modest in terms of contemporary science funding – is an amazing topic for STS studies, which could shed, perhaps, some light on the mixture of jealousy, zeal and anthropocentrism behind this particular obsession. I have yet to encounter a research paper in science or philosophy that would start a criticism of some *other* controversial direction of research – string theory, artificial intelligence, climate modelling, etc. – by listing prices of particle accelerators, running costs of supercomputers, geoscience satellite equipment cost in dollars per kg payload, etc. involved in such research. (Of course, this does

not apply to various institutional reports and 'white papers'.) In the specific case of NASA's *High Resolution Microwave Survey*, the historical fact is that at roughly about the same time other, much more ambitious scientific projects were being cancelled, the most (in)famous being the Superconducting SuperCollider.

40 Mash (1993), p. 210.

41 In spite of the fact that it has subsequently been cited approvingly in even stronger and less cautious terms by Kukla (2001).

42 Is this now scholastic hair-splitting on my behalf? Not necessarily, since no less philosophical a VIP than Georg Wilhelm Friedrich Hegel famously argued (in his *Lectures on the History of Philosophy*) that, contra Parmenides and almost all philosophical tradition since Eleatics, there is no contradiction in claiming that being (*das Sein*) and nothing (*das Nichts*) are one and the same *unless we have some additional specification* of the situation (Hegel [1832] 1971). This highlights very deep ontological difficulties associated with any attempt to achieve perfect abstraction away from all empirical knowledge; see also Nozick (1981), esp. Chapter 2.

43 Mash (1993), pp. 217–218.

44 E.g., Izakov (1997); Chaisson (2001, 2003).

45 Kukla (2001), p. 49.

46 Earman (1995), p. 269.

47 Mash (1993), *op. cit.*, p. 212.

48 Ibid., p. 221.

49 E.g., Franck *et al.* (2001); Franck, von Bloh and Bounama (2007); Bounama, von Bloh and Franck (2007); Ćirković and Vukotić (2008); Forgan and Rice (2010); Vukotić (2010); McCabe and Lucas (2010). A classical blueprint for such numerical studies is the paper of Newman and Sagan (1981).

50 By the way, this passage of Mash repeats the rhetorical device of division into 'believers' and 'sceptics' with regard to SETI, implying that 'sceptics' are somehow closer to what is 'true science'; I suggest, as discussed in relation to REH, that the tables could equally well be turned, and Mash and company could justifiably be called 'believers' – in the uniqueness of humans – while Sagan, Drake, Tarter, etc., could be regarded as 'sceptics'.

51 One of the main tenets of the present book is that such a distinction is hard to make, so Basalla may be on the right track here; but one of the SETI sceptics' favourite strategies (as we have seen while discussing REH) has been to strongly separate the two. As long as we do not doubt the naturalistic origin of intelligent life *on Earth*, the onus of proof lies with those postulating a sharp separation between the two in the wider cosmic context.

52 Basalla (2006), p. 102. I dare not joke with such 'arguments' – at least not having already, in Chapter 5, cited Lysenko's merciless inquisition of Vavilov's scientific 'sins'. I guess the temptation to straighten out academic opponents in this manner, abusing innocent personal data like the contents of vanity licence plates, is present in democratic societies as well.

53 Ibid., p. 124.

54 Ibid., p. 14.

55 Basalla (2006), p. xiii. The sentence is given in the Introduction, and not as a preview of subsequent conclusions of the exploration, which is also indicative of the largely *ex cathedra* and paternalizing relations towards the scientists involved in SETI projects.

56 Even more to the point, the discussion of the continuity thesis in the previous chapter is something entirely different from the ideas of the continuity of the physical and biological world held by ancient thinkers, such as Anaximandros and Empedocles.

57 Toulmin and Goodfield [1965] (1982).

58 Kragh (1996), p. 244.

59 For a detailed calculation, see Lineweaver (2001).

60 One is tempted to say that Pigliucci's justifications for putting other items in this 'middle kingdom' are weak as well – this particularly pertains to string theory and the multiverse concept. On the other hand, one should not take these weaknesses as important, since they are far removed from the book's main concerns (which are quite laudable anyway).

61 Remember Oscar Wilde's quip following the invention of the radio – when people finally are able to be heard by many, nobody has anything of importance to say. A key early precursor to SETI,

Nikola Tesla, expressed, in the early years of the 20th century, his hopes that the new electronic media (he contributed to the invention of both radio and television) will annihilate illiteracy all over the planet in a matter of years... Sadly enough, the age of innocence seems to be over.

62 E.g., Walters, Hoover and Kotra (1980); Wilson (1984); Olson (1988); Chyba (1997); Ćirković (2004b).
63 Excellent criticism is given by Burchell (2006).
64 E.g., Tarter (2001); Wilson (2001).
65 Shklovskii (1962), Chapters 25, 26; Sagan (1973); Dyson (1960, 1966, 2003); Freitas (1985).
66 Harris (1986, 2002).
67 Ćirković and Bradbury (2006); Ćirković (2006); Ćirković and Vukotić (2008).
68 Pigliucci (2010), pp. 39–40.
69 Brin (1983); Lem [1974] (1977), Chapter III.
70 Using Fermi's lunchtime comments as the authoritative formulation of the paradox is about as appropriate as the hypothetical usage of Lou Gehrig's sport biography in a study of pathophysiology of the Lou Gehrig disease – it may look like a flashy rhetorical strategy and it may impress the readership, but it is obviously of very limited cognitive value. At the very best it can give us a *particular* instance, which one then confuses with the *general* issue at one's own peril. Fermi's remarks are part of the history of astrobiology, not of astrobiology itself. As in Basalla's case, Pigliucci tends to conflate the history of science with science, with the same result of oversimplifying SETI. See also Almár (2011).
71 Rose and Wright (2004).
72 Calvino (1993).
73 By the way, that particular parenthetical question is similar to the question, 'whoever decided that all functions of a real variable have defined values at points corresponding to natural numbers: 1, 2, 3...?' Of course, nobody did – there are infinitely many real functions whose values are not defined at those particular points. In practice, most real functions, and almost all *interesting* real functions, will have such values – but not because such points are somehow exceptional or special or – *pace* Leopold Kronecker's famous quip – God-given. It is because those points are just there, embedded in the dense set of all real numbers.
74 It also is fairly indicative that the entire section on SETI as 'almost science' in Pigliucci's book does not quote a single reference, if one does not count two websites, one being the famous SETI@home, and the other having an online calculator for the Drake equation. In contrast, the following section on evolutionary psychology, although significantly shorter than the section on SETI, cites eight fully academic bibliographical references. Not really a problem for the book as such (after all, to paraphrase Pigliucci's own comments on Fermi's lunch, its author was just casually chatting during *hors d'oeuvre* before passing to the main courses of creationism, postmodernism and the public perception of science), but even a cursory survey of the literature would show that things are not that simple, as the vignette on SETI attempts to demonstrate.
75 Especially Ward (2005).
76 Drake (1976).

8 Natural and artificial: the cosmic *Domain of Arnheim*

Quote: from a letter to Robert Graves, quoted in *J.B.S.: The Life and Work of J.B.S. Haldane*, Ronald William Clark, Oxford University Press, Oxford (1984).

1 E.g., Bischoff (2005); Klingsör-Leroy (2009).
2 Among exceptions, see Almár (1984); Rubtsov (1991); Michaud (2007).
3 Wittgenstein (1953).
4 This is not to deny that they are tightly connected. Not only is 'the medium the message', as the cliché goes; in the high days of behaviourism, as Fodor and Piatelli-Palmarini (2010) remind us, some argued that, in principle, we could teach worms to speak English. That such a position is today considered mindless (no pun intended!) extremism, useful only for comic relief, it should give us food for thought. Since semiotics in the astrobiological context has been largely neglected so far, I separate the two here just in 'zeroth approximation'.

5 Raup (1992).
6 Alvarez *et al.* (1980); for a detailed account of the whole revolutionary period, see, e.g., Raup (1991, 1999); Glen (1994).
7 Raup (1999).
8 Cases in which the distinction cannot be clearly made are always part of strange and even bizarre contexts, and are often misused for extrascientific purposes. The most famous case is, of course, the whole UFO phenomenon, or at least its subset concerning the misidentification of natural phenomena. The endless debates about many witnesses confusing planet Venus with a UFO (or things like meteor showers, ball lightnings, etc.) testify upon the power of the strange context and preconceived notions to scramble that particular aspect of our visual perception. However, even at the height of UFO sightings, confusion of man-made – that is, manifestly artificial – objects like aeroplanes, weather balloons and satellites with alleged extraterrestrial spacecraft happened significantly more often (cf. Condon and Gillmor 1969; Klass 1974).
9 What I shall not attempt to do is to invoke the tangled mass of complex philosophical lingo usually associated with the dichotomy of natural vs. artificial. There has been a great deal of effort invested, since Descartes and Vico, in discussing concepts like 'natural kind' or Nelson Goodman's distinction of 'genuine' vs. 'artificial'. While undoubtedly this work could shed some light on the issues involved, it is still unexplored how it could connect. The same applies to the rapidly growing work going under the name of biosemiotics; while some of the critics pointed out its for SETI studies (e.g., Ulvestad 2002), its relationship to actual research remains to be elaborated.
10 E.g., Paquet (2006); for different versions, see a fine article of Richard Matteson at http://www.mattesonart.com/magritte-and-poe-strange-brew.aspx (last accessed May 2, 2011). The story is in Poe [1847] (1912), pp. 31–44.
11 Ibid., p. 35.
12 Which is also why we deny pseudoscientific interpretations of the Cydonia geological formation as the (human) 'face on Mars'. With good reason, too, although it is not always clear to the general public, which usually (a relict of positivism?) tends to diminish the role of *interpretation* in science.
13 Note that we cannot strictly say that the eggs shown on the sill are eggs of an eagle and not a sparrow – this would require much more detailed insight into the relevant ornithology and morphology (the insight we are not ascribing to the painter either!). *We* are making that connection, which reflects *our* particular epistemic and biosemiotic background.
14 Thompson (1917).
15 The story has been beautifully told by Gould (1983), p. 77.
16 Walker and Ćirković (2006). Of course, in order for this example to be useful, we need to overcome well-publicized conceptual difficulties in defining planets in a way applicable to extrasolar planetary systems.
17 For a serious recent anthology on this hot topic, see Badescu, Cathcart, and Schuiling (2006).
18 Arnold (2005).
19 Among many examples from the science-fiction domain, Greg Bear's novel, *Queen of Angels*, contains a particularly poignant subplot describing this sort of confusion and incorrect attribution when a human interstellar probe visits a distant Earth-like planet and observes its surface features (Bear 1990).
20 What is occasionally dubbed CETI, from *Communication with ExtraTerrestrial Intelligence*, especially in the older literature; e.g., Shklovskii and Sagan (1966).
21 Lem (1987). Lem died in 2006, but except for collections of essays and feuilletons, he did not publish any prose after *Fiasco* and *Peace on Earth*, also written in 1987.
22 For some exceptions, see Hargrove (1986); McKay and Marinova (2001); Cockell (2008); contributions of Mark Lupisella and Kelly Smith in Bertka (2009).
23 Dyson (1966), p. 642.
24 Dyson (1960).
25 Ćirković and Bradbury (2006); Bradbury, Ćirković and Dvorsky (2011).
26 Dyson (1966); Sagan and Walker (1966); Badescu (1995); Badescu and Cathcart (2000); Badescu, Cathcart and Schuiling (2006); Carrigan (2009).

27 Parkinson (2005).

28 Birch (1983, 1991); Roy *et al.* (2004). For a cross-section of the already massive literature on space elevators see, for instance, Fawcett, Laine and Nugent (2006); the proceedings of the 1999 NASA Space Elevator Conference can be found at http://www.spaceelevator.com/docs/elevator.pdf, and a nice summary of the 2009 meeting can be found at http://hplusmagazine.com/2009/08/31/notes-space-elevator-conference-august-13–16-2009/ (last accessed 6 May 2011) – the decadal progress is clearly visible.

29 Criswell (1985); Beech (1990, 2007).

30 E.g., Fogg (1995); McKay and Marinova (2001); Beech (2009).

31 Extragalactic SETI has not been considered very seriously so far (for exceptions, see Gott 1982; Wesson 1990; Annis 1999b). The reason is, perhaps, the same old comforting prejudice that we should expect target-specific (and most conveniently radio) signals. Since these are not likely to be forthcoming over intergalactic distances and the two-way communication desired by some of the pioneers is practicably impossible, there is no point in even thinking seriously about extragalactic SETI. Such a view is fallacious: when we remove the cozy assumption of target-specific signals (together with the second-order assumption of their radio nature), it collapses. On the contrary, extragalactic SETI would enable us to probe an enormously larger part of physical space, as well as the space of possible evolutionary histories of advanced societies. (Of course, part of what we get ensemble-wise, we lose time- and resolution-wise.) In fact, the definition of Kardashev's Type 3 civilization (i.e., those managing energy resources of its entire home galaxy; Kardashev 1964) should prompt us to consider it more carefully, at least for a sample of nearby galaxies, visible at epochs significantly closer to us than the 1.8-Gyr difference between the average of Lineweaver (2001) and the age of Earth, which is about 4.56 Gyr. In fact, it could be argued (although it is beyond the scope of the present study) that the null result of extragalactic SETI observations so far represents a strong argument against the viability of Kardashev's Type 3 civilizations.

32 Almár (1984); Beech (1990).

33 Searches for artificial objects near Earth and anomalous spectral lines in stellar spectra performed by Robert A. Freitas, Jr. and Francisco Valdes in the 1980s (Freitas and Valdes 1980; Valdes and Freitas 1983, 1986). The Japanese programme of searches for Dyson shells around nearby stars has been presented by Jugaku, Noguchi and Nishimura (1995) and Jugaku and Nishimura (2003); a recent survey by Carrigan (2009) used the IRAS database to put serious constrains on the number of Dyson shells in the temperature range between 100 K and 600 K in a volume of space roughly 300 pc from the Sun. Some further proposals for observational or archival searches for Dyson shells or related astroengineering projects are given by Slysh (1985); Tilgner and Heinrichsen (1998); Timofeev *et al.* (2000). The detailed study of Sandberg (2000) fruitfully links information-processing to macro-engineering. Investigation of γ-ray signatures of antimatter burning by advanced Galactic societies was performed by Harris (1986, 2002). The analysis of archival extragalactic data by Annis (1999b) suggests the absence of star-powered Kardashev Type 3 civilizations in nearby galaxies.

34 There are further philosophically interesting issues which I cannot afford the necessary space for discussing here: e.g., this approach works under the assumption that evolutionary trajectories of advanced civilizations are essentially ergodic; cf. Vermeij (2006). Further research on properties of postbiological evolution is thus highly desirable from the point of view of both astrobiology and future studies.

9 Astrobiology as the neo-Copernican synthesis?

Quote: attributed to Arnold J. Toynbee (1889–1975).

1 Mazlish (1967), p. 3.

2 Ibid., p. 14.

3 For similar thoughts inspired by *modern* development of computing, see Dyson (1997).

4 Gould (1987b), p. 215.

5 Hartle and Srednicki (2007); Page (2008).

6 Wilson (1975); for the most detailed – though not quite impartial – account of the controversy, see Segerstråle (2000).
7 Wilson (1998).
8 Nozick (1981), p. 116.
9 Linde and Vanchurin (2010), p. 1.
10 Chaisson (2003), p. 91.
11 Ehrenfreund *et al.* (2004), p. 454; see also Pierson (2006).
12 For SETI@home, see http://setiathome.ssl.berkeley.edu (last accessed 27 July 2011); for a detailed primer on GRID, see Sciabà (2011).
13 Adams and Laughlin (1997); for a survey of literature up to 2003, see Ćirković (2003a).
14 Tofler (1990); Kurzweil (2000); Bostrom (2009); Hughes (2004).
15 Perhaps the only common ground can be found in the (in)famous quip of the former US president George W. Bush: *I think we agree, the past is over.* (Reported in *Dallas Morning News*, 10 May 2000.)
16 Bostrom and Ćirković (2008), especially the Introduction.
17 Rampino (1997, 2002); Baum (2010).
18 Notably Brin (1983); Lem [1974] (1977); Kardashev (1997).
19 Davies (1994), pp. 154–155.
20 Gardner (2004, 2007); for some earlier thinking on the same topic, see Harrison (1995); Dyson (2004); Crane (2010), whose original proposal for a 'meduso-anthropic' principle was published as a preprint in Crane (1994).
21 Stapledon (1937).
22 Ibid., p. 242.
23 Lem [1971] (1993), pp. 197–229.
24 Like Fermi, Lem did not know much about the age distribution of habitable planets, which we now know, thanks to the seminal studies of Lineweaver and collaborators. But the magnitude of Lem's genius can be gauged from the very fact that he did not shun from the then radical speculation about billion-year-old supercivilizations – on the contrary, he accepted it provisionally and then followed it to logical conclusions.
25 Ibid., p. 208, emphasis in the original.
26 Ibid., pp. 228–229.
27 In the mentioned (endnote 14 to Chapter 2) essay on the diversity of knowledge, Alfred Russel Wallace wrote a paragraph which, when interpreted in a naturalistic manner, contains the same thought: 'Yet our horizon ever widens, the limits to our advance seem more distant than ever, and there seems nothing too noble, too exalted, too marvellous, for the ever-increasing knowledge of future generations to attain to. Is it not fitting that, as intellectual beings with such high powers, we should each of us acquire a knowledge of what past generations have taught us, so that, should the opportunity occur, we may be able to add somewhat, however small, to the fund of instruction for posterity?... And, lastly, can any reflecting mind have a doubt that, by improving to the utmost the nobler faculties of our nature in this world, we shall be the better fitted to enter upon and enjoy whatever new state of being the future may have in store for us?' (see http://people.wku.edu/charles.smith/wallace/S001.htm; last accessed 1 December 2011).
28 Linde (1988; 1990); Gott and Li (1998); cf. Holt (2004).

References

Abbas, S. and Abbas, A. (1998). Volcanogenic dark matter and mass extinctions. *Astropart. Phys.* **8**, 317–320.

Ackland, G. J., Clark, M. A. and Lenton, T. M. (2003). Catastrophic desert formation in Daisyworld. *J. Theor. Biol.* **223**, 39–44.

Adams, F. C. (2008). Stars in other universes: stellar structure with different fundamental constants. *J. Cosmol. Astropart. Phys.* **08**, 010.

Adams, F. C. and Laughlin, G. (1997). A dying universe: the long-term fate and evolution of astrophysical objects. *Rev. Mod. Phys.* **69**, 337–372.

Adams, M. B. (2000). Last judgement: the visionary biology of J. B. S. Haldane. *J. Hist. Biol.* **33**, 457–491.

Aguirre, A. (2001). The cold big-bang cosmology as a counter-example to several anthropic arguments. *Phys. Rev. D* **64**, 083508(12).

Aguirre, A., Johnson, M. C. and Shomer, A. (2007). Towards observable signatures of other bubble universes. *Phys. Rev. D* **74**, 063509(17).

Almár, I. (1984). The rate of cosmic discoveries and the influence of ETI on it. *Acta Astronaut.* **11**, 731–733.

Almár, I. (1992). Analogies between Olbers' paradox and the Fermi paradox. *Acta Astronaut.* **26**, 253–256.

Almár, I. (2011). SETI terminology: do we interpret SETI terms correctly? *Acta Astronaut.* **68**, 351–357.

Alvarez, L. W., Alvarez, W., Asaro, F. and Michel, H. V. (1980). Extraterrestrial cause for the Cretaceous-Tertiary boundary extinction. *Science* **208**, 1095–1108.

Annis, J. (1999a). An astrophysical explanation for the Great Silence. *J. Br. Interplanet. Soc.* **52**, 19–22 (preprint astro-ph/9901322).

Annis, J. (1999b). Placing a limit on star-fed Kardashev type III civilisations. *J. Br. Interplanet. Soc.* **52**, 33–36.

Arnold, L. F. A. (2005). Transit light-curve signatures of artificial objects. *Astrophys. J.* **627**, 534–539.

Arsenijević, M. (1986). *Prostor, Vreme, Zenon.* Belgrade, Serbia: Filozofsko društvo Srbije (in Serbian).

Asher, D. J., Clube, S. V. M., Napier, W. M. and Steel, D. I. (1994). Coherent catastrophism. *Vistas Astron.* **38**, 1–27.

Badescu, V. (1995). On the radius of Dyson's sphere. *Acta Astronaut.* **36**, 135–138.

Badescu, V. and Cathcart, R. B. (2000). Stellar engines for Kardashev's type II civilization. *J. Br. Interplanet. Soc.* **53**, 297–306.

Badescu, V., Cathcart, R. B. and Schuiling, R. D. (2006). *Macro-Engineering: A Challenge for the Future*. New York: Springer.

Bailey, J. *et al.* (1998). Circular polarization in star-formation regions: implications for biomolecular homochirality. *Science* **281**, 672–674.

Balashov, Yu. V. (1991). Resource letter AP-1: the anthropic principle. *Am. J. Phys.* **59**, 1069–1076.

Balashov, Yu. (1994). Uniformitarianism in cosmology: background and philosophical implications of the steady-state theory. *Stud. Hist. Philos. Sci.* **25B**, 933–958.

Ball, J. A. (1985). Universal aspects of biological evolution. In *The Search for Extraterrestrial Life: Recent Developments*, ed. by M. D. Papagiannis. Dordrecht: IAU, Reidel Publishing Co., pp. 251–254.

Barrow, J. D. (2002). *The Constants of Nature: From Alpha to Omega*. London: Jonathan Cape.

Barrow, J. D. and Tipler, F. J. (1978). Eternity is unstable. *Nature* **276**, 453–459.

Barrow, J. D. and Tipler, F. J. (1986). *The Anthropic Cosmological Principle*. New York: Oxford University Press.

Basalla, G. (2006). *Intelligent Life in the Universe*. Oxford: Oxford University Press.

Baum, S. D. (2010). Is humanity doomed? Insights from astrobiology. *Sustainability* **2**, 591–603.

Baum, S. D., Haqq-Misra, J. D. and Domagal-Goldman, S. D. (2011). Would contact with extraterrestrials benefit or harm humanity? A scenario analysis. *Acta Astronaut.* **68**, 2114–2129.

Baxter, S. (2000). The planetarium hypothesis: a resolution of the Fermi paradox. *J. Br. Interplanet. Soc.* **54**, 210–216.

Bear, G. (1990). *Queen of Angels*. New York: Warner Books.

Beatty, J. (2006). Replaying life's tape. *J. Philos.* **103**, 336–362.

Beckwith, S. V. W. (2008). Detecting life-bearing extrasolar planets with space telescopes. *Astrophys. J.* **684**, 1404–1415.

Bedau, M. A. (2003). Artificial life: organization, adaptation and complexity from the bottom up. *Trends Cogn. Sci.* **7**, 505–512.

Beech, M. (1990). Blue stragglers as indicators of extraterrestrial civilisations? *Earth, Moon, Planets* **49**, 177–186.

Beech, M. (2007). *Rejuvenating the Sun and Avoiding Other Global Catastrophes*. New York: Springer.

Beech, M. (2009) *Terraforming: The Creating of Habitable Worlds*. New York: Springer.

Beer, M. E., King, A. R., Livio, M. and Pringle, J. E. (2004). How special is the Solar System? *Mon. Not. R. Astron. Soc.* **354**, 763–768.

Belkora, L. (2002). *Minding the Heavens: The Story of Our Discovery of the Milky Way*. Bristol: Institute of Physics Publishing.

Benn, C. R. (2001). The Moon and the origin of life. *Earth, Moon, Planets* **85/86**, 61–66.

Benner, S. (2009). *Life, the Universe... and the Scientific Method*. Gainesville, FL: The FfAME Press.

Benner, S. A. (2010). Defining life. *Astrobiology* **10**, 1021–1030.

Bennett, C. L. *et al.* (2003). First year Wilkinson Microwave Anisotropy Probe (WMAP) observations: preliminary maps and basic results. *Astrophys. J. Suppl.* **148**, 1–27.

Bennett, J. O. and Shostak, S. (2011). *Life in the Universe*. 3rd edition. San Francisco: Benjamin Cummings.

Benton, M. J. (2003). *When Life Nearly Died: The Greatest Mass Extinction of All Time*. London: Thames and Hudson.

Bertka, C. M. (ed.) (2009). *Exploring the Origin, Extent, and Future of Life*. Cambridge: Cambridge University Press.

Billyard, A. P., Coley, A. A., Lidsey, J. E. and Nilsson, U. S. (2000). Dynamics of M-theory cosmology. *Phys. Rev. D* **61**, 043504.

Birch, P. (1983). Orbital ring systems and Jacob's ladders: III. *J. Br. Interplanet. Soc.* **36**, 231–238.

Birch, P. (1991). Supramundane planets. *J. Br. Interplanet. Soc.* **44**, 169–182.

Bischoff, U. (2005). *Max Ernst*. Cologne: Taschen.

Bok, B. J. (1946). The time-scale of the universe. *Mon. Not. R. Astron. Soc.* **106**, 61–75.

Boltzmann, L. (1895). On certain questions of the theory of gases. *Nature* **51**, 413–414.

Bondi, H. (1955). Fact and inference in theory and in observation. *Vistas Astron.* **1**, 155–162.

Bondi, H. (1961). *Cosmology*. 2nd edition. London: Cambridge University Press.

Bondi, H. and Gold, T. (1948). The steady-state theory of the expanding universe. *Mon. Not. R. Astron. Soc.* **108**, 252–270.

Borges, J. L. (1999). *Collected Fictions*. Translated by A. Hurley. New York: Penguin.

Bostrom, N. (2001). The Doomsday Argument, Adam & Eve, UN++ and Quantum Joe. *Synthese* **127**, 359–387.

Bostrom, N. (2002). *Anthropic Bias: Observation Selection Effects in Science and Philosophy*. New York: Routledge.

Bostrom, N. (2003). Are you living in a computer simulation? *Philos. Q.* **53**, 243–55.

Bostrom, N. (2009). The future of humanity. In *New Waves in Philosophy of Technology*, ed. by J.-K. B. Olsen, E. Selinger and S. Riis. New York: Palgrave Macmillan, pp. 186–216.

Bostrom, N. and Ćirković, M. M. (eds.) (2008). *Global Catastrophic Risks*. Oxford: Oxford University Press.

Bounama, C., Franck, S. and von Bloh, W. (2001). The fate of Earth's oceans. *Hydrol. Earth Sci.* **5**, 569–575.

Bounama, C., von Bloh, W. and Franck, S. (2007). How rare is complex life in the Milky Way? *Astrobiology* **7**, 745–755.

Bousso, R. and Polchinski, J. (2004). The string theory landscape. *Sci. Am.* **291**, 60–69.

Bowler, P. J. (1992). *The Eclipse of Darwinism*. Baltimore, MD: The Johns Hopkins University Press.

Bracewell, R. N. (1975). *The Galactic Club: Intelligent Life in Outer Space*. San Francisco: W. H. Freeman.

Bradbury, R. J., Ćirković, M. M. and Dvorsky, G. (2011). Considerations for a Dysonian approach to SETI: a fruitful middle ground? *J. Br. Interplanet. Soc.* **64**, 156–165.

Brandon, R. N. (1997). Does biology have laws? The experimental evidence. *Philos. Sci.* **64**, S444–S457.

Briggs, D. E. G. and Fortey, R. A. (2005). Wonderful strife: systematics, stem groups, and the phylogenetic signal of the Cambrian radiation. *Paleobiol.* **31**, 94–112.

Brin, G. D. (1983). The 'Great Silence': the controversy concerning extraterrestrial intelligence. *Q. J. R. Astron. Soc.* **24**, 283–309.

Bromm, V. and Loeb, A. (2002). The expected redshift distribution of gamma-ray bursts. *Astrophys. J.* **575**, 111–116.

Brown, G. (1984). The cosmological theory of Empedocles. *Apeiron* **18**, 97–101.

Burchell, M. J. (2006). W(h)ither the Drake equation? *Int. J. Astrobiol.* **5**, 243–250.

Burnet, J. (1908). *Early Greek Philosophy*. London: Adam and Charles Black.

Byakov, V. M., Stepanov, S. V. and Stepanova, O. P. (1997). Quasi-regular staying of Solar System in supernova remnants and natural Earth history. *Radiat. Phys. Chem.* **49**, 299–305.

Byakov, V. M., Stepanov, S. V. and Stepanova, O. P. (2001). Role of ionizing radiation in the natural history of Earth. *Radiat. Phys. Chem.* **60**, 297–301.

Caldeira, K. and Kasting, J. F. (1992). The life span of the biosphere revisited. *Nature* **360**, 721–723.
Callender, C. (2001). Taking thermodynamics (too) seriously. *Stud. Hist. Philos. Mod. Phys.* **32**, 539–553.
Callender, C. (2004). Measures, explanation and the past: should 'special' initial conditions be explained? *Br. J. Philos. Sci.* **55**, 195–217.
Calvino, I. (1993). *Le Cosmicomiche*. Milan: Arnoldo Mondadori Editore S.p.A.
Carlip, A. (2007). Transient observers and variable constants, or repelling the invasion of the Boltzmann's brains. *J. Cosmol. Astropart. Phys.* **6**, 001.
Carr, B. (ed.) (2007). *Universe or Multiverse?* Cambridge: Cambridge University Press.
Carr, B. J. and Rees, M. J. (1979). The anthropic principle and the structure of the physical world. *Nature* **278**, 605–612.
Carrigan, R. A., Jr. (2006). Do potential SETI signals need to be decontaminated? *Acta Astronaut.* **58**, 112–117.
Carrigan, R. A. (2009). IRAS-based whole-sky upper limit on Dyson spheres. *Astrophys. J.* **698**, 2075–2086.
Carroll, S. M. (2006). Is our universe natural? *Nature* **440**, 1132–36.
Carslaw, K. S. *et al.* (2002). Cosmic rays, clouds, and climate. *Science* **298**, 1732–1737.
Carter, B. (1983). The anthropic principle and its implications for biological evolution. *Philos. Trans. R. Soc. Lond.* **A310**, 347–363.
Carter, B. (1993). The anthropic selection principle and the ultra-Darwinian synthesis. In *The Anthropic Principle, Proceedings of the Second Venice Conference on Cosmology and Philosophy*, ed. by F. Bertola and U. Curi. Cambridge: Cambridge University Press, pp. 33–66.
Carter, B. (2008). Five- or six-step scenario for evolution? *Int. J. Astrobiol.* **7**, 177–182.
Cartwright, N. (1997). Models: the blueprints for laws. *Philos. Sci.* **64**, S292–303.
Ćirković, M. M. (2002). On the first anthropic argument in astrobiology. *Earth, Moon, Planets* **91**, 243–254.
Ćirković, M. M. (2003a). Resource letter PEs-1: physical eschatology. *Am. J. Phys.* **71**, 122–133.
Ćirković, M. M. (2003b). The thermodynamical arrow of time: reinterpreting the Boltzmann–Schuetz argument. *Found. Phys.* **33**, 467–490.
Ćirković, M. M. (2004a). Earths: rare in time, not space? *J. Br. Interplanet. Soc.* **57**, 53–59.
Ćirković, M. M. (2004b). On the temporal aspect of the Drake equation and SETI. *Astrobiology* **4**, 225–231.
Ćirković, M. M. (2004c). The anthropic principle and the duration of the cosmological past. *Astron. Astrophys. Trans.* **23**, 567–597.
Ćirković, M. M. (2006). Too early? On the apparent conflict of astrobiology and cosmology. *Biol. Philos.* **21**, 369–379.
Ćirković, M. M. (2007). Natural history, by Justina Robson. *J. Evol. Technol.* **16**, 167–170 (http://www.jetpress.org/v16/cirkovic.html).
Ćirković, M. M. (2008). Sailing the Archipelago. *Collapse* **5**, 292–329.
Ćirković, M. M. (2009). Fermi's paradox: the last challenge for Copernicanism? *Serb. Astron. J.* **178**, 1–20.
Ćirković, M. M. and Bostrom, N. (2000). Cosmological constant and the final anthropic hypothesis. *Astrophys. Space Sci.* **274**, 675–687.
Ćirković, M. M. and Bradbury, R. J. (2006). Galactic gradients, postbiological evolution and the apparent failure of SETI. *New Astron.* **11**, 628–639.
Ćirković, M. M. and Vukotić, B. (2008). Astrobiological phase transition: towards resolution of Fermi's paradox. *Orig. Life Evol. Biosph.* **38**, 535–547.

Ćirković, M. M., Dragićević, I. and Berić-Bjedov, T. (2005). Adaptationism fails to resolve Fermi's paradox. *Serb. Astron. J.* **170**, 89–100.

Ćirković, M. M., Vukotić, B. and Dragićević, I. (2008). Galactic 'punctuated equilibrium': how to undermine Carter's anthropic argument in astrobiology. *Astrobiology* **9**, 491–501.

Ćirković, M. M., Sandberg, A. and Bostrom, N. (2010). Anthropic shadow: observation selection effects and human extinction risks. *Risk Analysis* **30**, 1495–1506.

Chaisson, E. J. (2001). *Cosmic Evolution: The Rise of Complexity in Nature*. Cambridge, MA: Harvard University Press.

Chaisson, E. J. (2003). A unifying concept for astrobiology. *Int. J. Astrobiol.* **2**, 91–101.

Chaitin, G. (2006). From Leibniz to Ω: epistemology as information theory. *Collapse* **I**, 27–51.

Chambers, J. E. (2010). Stellar elemental abundance patterns: implications for planet formation. *Astrophys. J.* **724**, 92–97.

Chavanis, P.-H. (2007). White dwarf stars in *D* dimensions. *Phys. Rev. D* **76**, 023004.

Chela-Flores, J. (2003). Testing evolutionary convergence on Europa. *Int. J. Astrobiol.* **2**, 307–312.

Chyba, C. F. (1997). Catastrophic impacts and the Drake equation. In *Astronomical and Biochemical Origins and the Search for Life in the Universe*, ed. by C. B. Cosmovici, S. Bowyer and D. Werthimer. Bologna: Editrice Compositori, pp. 157–164.

Chyba, C. F. and Hand, K. (2005). Astrobiology: the study of the living universe. *Annu. Rev. Astron. Astrophys.* **43**, 31–74.

Clark, S. (1999). Polarized starlight and the handedness of life. *Am. Sci.* **87**, 336–343.

Clarke, J. N. (1981). Extraterrestrial intelligence and Galactic nuclear activity. *Icarus* **46**, 94–96.

Cleland, C. E. and Chyba, C. F. (2002). Defining 'life'. *Orig. Life Evol. Biosph.* **32**, 387–393.

Clube, S. V. M. and Napier, W. M. (1984). The microstructure of terrestrial catastrophism. *Mon. Not. R. Astron. Soc.* **211**, 953–968.

Clube, S. V. M. and Napier, W. M. (1990). *The Cosmic Winter*. Oxford: Basil Blackwell.

Cockell, C. S. (2008). An essay on extraterrestrial liberty. *J. Br. Interplanet. Soc.* **61**, 255–275.

Cohen, J. and Stewart, I. (2002). *What Does a Martian Look Like? The Science of Extraterrestrial Life*. Hoboken, NJ: John Wiley & Sons.

Cohen, J. and Stewart, I. (2009). Alien science (interview). *Collapse* **V**, 225–290.

Collar, J. I. (1996). Biological effects of stellar collapse neutrinos. *Phys. Rev. Lett.* **76**, 999–1002.

Comins, N. F. (2010). *What If the Earth Had Two Moons*? New York: St. Martin's Press.

Condon, E. U. and Gillmor, D. S. (1969). *Final Report of the Scientific Study of Unidentified Flying Objects*. New York: Dutton.

Conway Morris, S. (1998). *The Crucible of Creation*. Oxford: Oxford University Press.

Conway Morris, S. (2003). *Life's Solution: Inevitable Humans in a Lonely Universe*. Cambridge: Cambridge University Press.

Cornford, F. (1965). *Thucydides Mythistoricus*. New York: Greenwood Press Publishers.

Cowley, R. (ed.) (1999). *What If? The World's Foremost Military Historians Imagine What Might Have Been*. New York: Bantam.

Crane, L. (1994). Possible implications of the quantum theory of gravity. Preprint, http://arxiv.org/abs/hep-th/9402104.

Crane, L. (2010). Possible implications of the quantum theory of gravity: an introduction to the Meduso-anthropic principle. *Found. Sci.* **15**, 369–373.

Criswell, D. (1985). Solar System industrialization: implications for interstellar migration. In *Interstellar Migration and the Human Experience*, ed. by B. Finney and E. Jones. Berkeley, CA: University of California Press, pp. 50–87.

Crow, M. J. (1986). *The Extraterrestrial Life Debate 1750–1900*. Cambridge: Cambridge University Press.

Crutzen, P. J. and Brühl, C. (1996). Mass extinctions and supernova explosions. *Proc. Natl. Acad. Sci. USA* **93**, 1582–1584.

Damour, T., Gibbons, G. W. and Taylor, J. H. (1988). Limits on the variability of G using binary-pulsar data. *Phys. Rev. Lett.* **61**, 1151–1154.

Dar, A. (1997). Life extinctions by neutron star mergers. In *Very High Energy Phenomena in the Universe; Morion Workshop*, ed. by Y. Giraud-Heraud and J. Tran Thanh Van. Paris: Editions Frontieres, pp. 379–386.

Dar, A. and De Rújula, A. (2002). The threat to life from Eta Carinae and gamma-ray bursts. In *Astrophysics and Gamma Ray Physics in Space*, ed. by A. Morselli and P. Picozza, Frascati Physics Series Vol. XXIV. Rome: Istituto Nazionale di Fisica Nucleare, Frascati, pp. 513–523.

Dar, A., Laor, A. and Shaviv, N. J. (1998). Life extinctions by cosmic ray jets. *Phys. Rev. Lett.* **80**, 5813–5816.

Darling, D. (2001). *Life Everywhere: The Maverick Science of Astrobiology*. New York: Basic Books.

Darwin, C. (1859). *On the Origin of Species by Means of Natural Selection*. London: John Murray.

Davies, P. C. W. (1978). Cosmic heresy? *Nature* **273**, 336–337.

Davies, P. C. W. (1994) *The Last Three Minutes*. New York: Basic Books.

Davies, P. C. W. (1995). *Are We Alone? Philosophical Implications of the Discovery of Extraterrestrial Life*. New York: Basic Books.

Davies, P. C. W. (2003). Does life's rapid appearance imply a Martian origin? *Astrobiology* **3**, 673–679.

Davies, P. C. W. *et al.* (2009). Signatures of a shadow biosphere. *Astrobiology* **9**, 241–249.

Dawkins, R. (1983). Universal Darwinism. In *Evolution from Molecules to Men*, ed. by D. S. Bendall. Cambridge: Cambridge University Press, pp. 403–425.

de Chardin, T. [1955] (1959). *The Phenomenon of Man*. Translated by B. Wall. London: Collins.

Des Marais, D. J. and Walter, M. R. (1999). Astrobiology: exploring the origins, evolution, and distribution of life in the universe. *Annu. Rev. Ecol. Syst.* **30**, 397–420.

Des Marais, D. J. *et al.* (2008). The NASA Astrobiology Roadmap. *Astrobiology* **8**, 715–730.

Deutsch, D. (1998). *The Fabric of Reality*. New York: Penguin Books.

Di Giulio, M. (2010). Biological evidence against the panspermia theory. *J. Theor. Biol.* **266**, 569–572.

Diamond, J. (1999). To whom it may concern. *The New York Times Magazine*, 5 December, 68–69.

Dick, S. J. (1996). *The Biological Universe: The Twentieth-Century Extraterrestrial Life Debate and the Limits of Science*. Cambridge: Cambridge University Press.

Dick, S. J. (1997). The biophysical cosmology: the place of bioastronomy in the history of science. In *Astronomical and Biochemical Origins and the Search for Life in the Universe*, ed. by C. B. Cosmovici, S. Bowyer and D. Werthimer, IAU Colloquium No. 161. Bologna: Editrice Compositori, pp. 785–788.

Dick, S. J. (2003). Cultural evolution, the postbiological universe and SETI. *Int. J. Astrobiol.* **2**, 65–74.

Dicke, R. H. (1961). Dirac's cosmology and Mach's principle. *Nature* **192**, 440–441.

Diels, H. (1983). *Presocratic Fragments*. Zagreb, Croatia: Naprijed (in Croatian).

Dirac, P. A. M. (1937). The cosmological constants. *Nature* **139**, 323.

Dirac, P. A. M. (1961). Dirac replies [to Dicke's letter]. *Nature* **192**, 441.

Dirac, P. A. M. (1974). Cosmological models and the large numbers hypothesis. *Proc. R. Soc.* **A338**, 439–446.

Drake, F. (1976). On hands and knees in search of Elysium. *MIT Technol. Rev.* **78** (June), 22–29.

Duric, N. and Field, L. (2003). On the detectability of intelligent civilizations in the Galaxy. *Serb. Astron. J.* **167**, 1–10.

Dyson, F. J. (1960). Search for artificial stellar sources of infrared radiation. *Science* **131**, 1667–1668.

Dyson, F. J. (1966). The search for extraterrestrial technology. In *Perspectives in Modern Physics*, ed. by R. E. Marshak. New York: Interscience Publishers, pp. 641–655.

Dyson, F. J. (1979). Time without end: physics and biology in an open universe. *Rev. Mod. Phys.* **51**, 447–460.

Dyson, F. J. (2003). Looking for life in unlikely places: reasons why planets may not be the best places to look for life. *Int. J. Astrobiol.* **2**, 103–10.

Dyson, F. J. (2004). *Infinite in All Directions*. New York: Harper Perennial.

Dyson, F. J. (2007). *A Many-Colored Glass: Reflections on the Place of Life in the Universe*. Charlottesville, VA: University of Virginia Press.

Dyson, G. B. (1997). *Darwin Among the Machines: The Evolution of Global Intelligence*. New York: Basic Books.

Dyson, L., Kleban, M. and Susskind, L. (2002). Disturbing implications of a cosmological constant. *J. High Energy Phys.* **10**, 011.

Earman, J. (1995). Recent work on time travel. In *Time's Arrow Today*, ed. by S. Savitt. Cambridge: Cambridge University Press, pp. 268–310.

Eddington, A. S. (1914). *Stellar Movements and the Structure of the Universe*. London: Macmillan and Co.

Eddington, A. S. (1928). *The Nature of the Physical World*. London: Cambridge University Press.

Eddington, A. S. (1940). The correction of statistics for accidental error. *Mon. Not. R. Astron. Soc.* **100**, 354–361.

Eddington, A. S. [1933] (1952). *The Expanding Universe*. Cambridge: Cambridge University Press.

Edwards, H. G. M. (2004). Raman spectroscopic protocol for the molecular recognition of key biomarkers in astrobiological exploration. *Orig. Life Evol. Biosph.* **34**, 3–11.

Egan, G. (1997). *Diaspora*. London: Orion/Millennium.

Ehrenfreund, P. *et al.* (eds.) (2004). *Astrobiology: Future Perspectives*. Dordrecht: Kluwer Academic Publishers.

Eigen, M. (1992). *Steps towards Life*. Oxford: Oxford University Press.

Einstein, A. (1917). Kosmologische Betrachtungen zur allgemeinen Relativitätstheorie. *Sitzungsberichte der Königlich Preußischen Akademie der Wissenschaften* (Berlin), 142–152.

Eldredge, N. and Gould, S. J. (1972). Punctuated equilibria: an alternative to phyletic gradualism. In *Models in Paleobiology*, ed. by T. M. Schopf. San Francisco: Freeman, Cooper, pp. 82–115.

Elgin, M. (2006). There may be strict empirical laws in biology, after all. *Biol. Philos.* **21**, 119–134.

Ellis, G. F. R. and Brundrit, G. B. (1979). Life in the infinite universe. *Q. J. R. Astron. Soc.* **20**, 37–41.

Ellis, G. F. R., Maartens, R. and Nel, S. D. (1978). The expansion of the universe. *Mon. Not. R. Astron. Soc.* **184**, 439–465.

Ellis, G. F. R., Kirchner, U. and Stoeger, W. R. (2004). Multiverses and physical cosmology. *Mon. Not. R. Astron. Soc.* **347**, 921–936.

Engel, M. H. and Macko, S. A. (1997). Isotopic evidence for extraterrestrial non-racemic amino acids in the Murchison meteorite. *Nature* **296**, 837–840.

Erwin, D. H. (1990). The end-Permian mass extinction. *Annu. Rev. Ecol. Syst.* **21**, 69–91.

Erwin, D. H. (1993). *The Great Paleozoic Crisis: Life and Death in the Permian*. New York: Columbia University Press.

Erwin, D. H. (2006). *Extinction: How Life on Earth Nearly Ended 250 Million Year Ago*. Princeton, NJ: Princeton University Press.

Fairbanks, A. (1898). *The First Philosophers of Greece*. London: Kegan Paul, Trench & Trubner.

Fawcett, B., Laine, M. and Nugent, T., Jr. (2006). *Liftport: Opening Space to Everyone*. Decatur, IL: Meisha Merlin Publishing.

Ferguson, N. (1999). *Virtual History: Alternatives and Counterfactuals*. New York: Basic Books.

Ferland, G. J. *et al.* (1998). CLOUDY 90: numerical simulation of plasmas and their spectra. *Publ. Astron. Soc. Pac.* **110**, 761–778.

Fodor, J. and Piattelli-Palmarini, M. (2010). *What Darwin Got Wrong*. New York: Farrar, Straus, and Giroux.

Fogg, M. J. (1995). *Terraforming: Engineering Planetary Environments*. Warrendale, PA: SAE International.

Fontenelle, B. B. [1686] (1990). *Conversation on the Plurality of the Worlds*. Translated by H. A. Hargreaves. Berkeley, CA: University of California Press.

Forgan, D. H. and Rice, K. (2010). Numerical testing of the Rare Earth hypothesis using Monte Carlo realization techniques. *Int. J. Astrobiol.* **9**, 73–80.

Franck, S., Block, A., von Bloh, W., Bounama, C. and Schellnhuber, H.-J. (2001). Planetary habitability: estimating the number of Gaias in the Milky Way. In *Proceedings of the First European Workshop on Exo- and Astro-Biology*, ed. by P. Ehrenfreund, O. Angerer and B. Battrick. ESA SP-496, pp. 73–78.

Franck, S., von Bloh, W. and Bounama, C. (2007). Maximum number of habitable planets at the time of Earth's origin: new hints for panspermia and the mediocrity principle. *Int. J. Astrobiol.* **6**, 153–157.

Frayn, M. (2006). *The Human Touch: Our Part in the Creation of a Universe*. London: Faber and Faber.

Fredkin, E. (2003). An introduction to digital philosophy. *Int. J. Theor. Phys.* **42**, 189–247.

Freitas, R. A., Jr. (1985). Observable characteristics of extraterrestrial technological civilizations. *J. Br. Interplanet. Soc.* **38**, 106–112.

Freitas, R. A., Jr. and Valdes, F. (1980). A search for natural or artificial objects located at the Earth–Moon libration points. *Icarus* **42**, 442–447.

Freivogel, B. and Susskind, L. (2004). Framework for the string theory landscape. *Phys. Rev. D* **70**, 126007(22).

Fry, I. (1995). Are the different hypotheses on the emergence of life as different as they seem? *Biol. Philos.* **10**, 389–417.

Fry, I. (2000). *The Emergence of Life on Earth: A Historical and Scientific Overview*. New Brunswick, NJ: Rutgers University Press.

Fry, I. (2011). The role of natural selection in the origin of life. *Orig. Life Evol. Biosph.* **41**, 3–16.

Galante, D. and Horvath, J. E. (2007). Biological effects of gamma-ray bursts: distances for severe damage on the biota. *Int. J. Astrobiol.* **6**, 19–26.

Gale, J. (2009). *Astrobiology of Earth: The Emergence, Evolution, and Future of Life on a Planet in Turmoil*. Oxford: Oxford University Press.

Gardner, J. N. (2004). The physical constants as biosignature: an anthropic retrodiction of the Selfish Biocosm hypothesis. *Int. J. Astrobiol.* **3**, 229–236.

Gardner, J. N. (2007). *The Intelligent Universe: AI, ET, and the Emerging Mind of the Cosmos*. Pompton Plains, NJ: New Page Books.

Garriga, J. and Vilenkin, A. (2001). Many worlds in one. *Phys. Rev. D* **64**, 043511(5).

Gatewood, G. and Eichhorn, H. (1973). An unsuccessful search for a planetary companion of Barnard's Star (BD +4 °3561). *Astron. J.* **78**, 769–776.

Gehrels, N. *et al.* (2003). Ozone depletion from nearby supernovae. *Astrophys. J.* **585**, 1169–1176.

Gere, C. (2004). Brains-in-vats, giant brains and world brains: the brain as metaphor in digital culture. *Stud. Hist. Philos. Biol. & Biomed. Sci.* **35**, 351–366.

Gies, D. R. and Helsel, J. W. (2005). Ice age epochs and the Sun's path through the Galaxy. *Astrophys. J.* **626**, 844–848.

Gilmour, I. and Sephton, M. A. (2004). *An Introduction to Astrobiology*. Cambridge: Cambridge University Press.

Gleiser, M., Thorarinson, J. and Walker, S. I. (2008). Punctuated chirality. *Orig. Life Evol. Biosph.* **38**, 499–508.

Glen, W. (ed.) (1994). *The Mass-Extinction Debates: How Science Works in a Crisis*. Palo Alto, CA: Stanford University Press.

Gold, T. (1999). *The Deep Hot Biosphere: The Myth of Fossil Fuels*. New York: Springer-Verlag.

Gonzalez, G. (2005). Habitable zones in the universe. *Orig. Life Evol. Biosph.* **35**, 555–606.

Gonzalez, G. and Richards, J. (2004). *The Privileged Planet: How Our Place in the Cosmos is Designed for Discovery*. Washington DC: Regnery Publishing.

Gonzalez, G., Brownlee, D. and Ward, P. (2001). The Galactic Habitable Zone: Galactic chemical evolution. *Icarus* **152**, 185–200.

Gott, J. R. (1982). Cosmology and life in the universe. In *Extraterrestrials – Where Are They*? ed. by B. Zuckerman and M. H. Hart. Elmsford, NY: Pergamon Press, pp. 122–134.

Gott, J. R. (1993). Implications of the Copernican principle for our future prospects. *Nature* **363**, 315–319.

Gott III, J. R. and Li, L.-X. (1998). Can the universe create itself? *Phys. Rev. D* **58**, 023501(43).

Gott III, J. R. *et al.* (2005). A map of the universe. *Astrophys. J.* **624**, 463–484.

Gough, D. (ed.) (2005). *The Scientific Legacy of Fred Hoyle*. Cambridge: Cambridge University Press.

Gould, S. J. (1983). *Hen's Teeth and Horse's Toes*. New York: W. W. Norton.

Gould, S. J. (1985). The paradox of the first tier: an agenda for paleobiology. *Paleobiol.* **11**, 2–12.

Gould, S. J. (1987a). *The Flamingo's Smile: Reflections in Natural History*. New York: W. W. Norton.

Gould, S. J. (1987b). *An Urchin in the Storm: Essays about Books and Ideas*. New York: W. W. Norton.

Gould, S. J. (1989). *Wonderful Life: The Burgess Shale and the Nature of History*. New York: W. W. Norton.

Gould, S. J. (1996). *Full House: The Spread of Excellence from Plato to Darwin*. New York: Three Rivers Press.

Gould, S. J. (2002). *The Structure of Evolutionary Theory*. Cambridge, MA: Belknap Press.

Gould, S. J. (2003). *The Hedgehog, the Fox, and the Magister's Pox: Mending and Minding the Misconceived Gap Between Science and the Humanities*. London: Vintage.

Greaves, J. S., Wyatt, M. C., Holland, W. S. and Dent, W. R. F. (2004). The debris disc around τ Ceti: a massive analogue to the Kuiper Belt. *Mon. Not. R. Astron. Soc.* **351**, L54–L58.

Gregory, A. (2007). *Ancient Greek Cosmogony*. London: Duckworth.

Gregory, J. (2005). *Fred Hoyle's Universe*. Oxford: Oxford University Press.

Grene, M. (1958). Two evolutionary theories (I). *Br. J. Philos. Sci.* **9**, 110–127.

Griffiths, L. M. and Lineweaver, C. H. (2004). Testing the cosmic microwave background data for systematic effects. *Astrophys. J.* **603**, 371–382.

Grinspoon, D. (2003). *Lonely Planets: The Natural Philosophy of Alien Life*. New York: HarperCollins.

Grünbaum, A. (1991). Creation as a pseudo-explanation in current physical cosmology. *Erkenntnis* **35**, 233–254.

Guthrie, W. K. C. (1979). *A History of Greek Philosophy: II. The Presocratic Tradition from Parmenides to Democritus*. London: Cambridge University Press.

Haldane, J. B. S. (1924). *Daedalus, or Science and the Future: A Paper Read to the Heretics, Cambridge on February 4th, 1923*. London: Kegan Paul, Trench, Trubner & Co.

Haldane, J. B. S. (1927). *Possible Worlds and Other Essays*. London: Harper and Brothers.

Haldane, J. B. S. (1945). A quantum theory of the origin of the Solar System. *Nature* **155**, 133–135.

Hamkins, J. D. and Montero, B. (2000). Utilitarianism in infinite worlds. *Utilitas* **12**, 91–96.

Hargrove, E. C. (ed.) (1986) *Beyond Spaceship Earth: Environmental Ethics and the Solar System*. San Francisco: Sierra Club Books.

Harnik, R., Kribs, G. and Perez, G. (2006). A universe without weak interactions. *Phys. Rev. D* **74**, 035006(15).

Harris, M. J. (1986). On the detectability of antimatter propulsion space-craft. *Astrophys. Space Sci.* **123**, 297–303.

Harris, M. J. (2002). Limits from CGRO/EGRET data on the use of antimatter as a power source by extraterrestrial civilizations. *J. Br. Interplanet. Soc.* **55**, 383–393.

Harrison, E. R. (1973). Standard model of the early universe. *Annu. Rev. Astron. Astrophys.* **11**, 155–186.

Harrison, E. (1987). Whigs, prigs and historians of science. *Nature* **329**, 213–214.

Harrison, E. R. (1995). The natural selection of universes containing intelligent life. *Q. J. R. Astron. Soc.* **36**, 193–203.

Hartle, J. B. and Srednicki, M. (2007). Are we typical? *Phys. Rev. D* **75**, 123523(6).

Hartmann, D. H. (1999). Afterglows from the largest explosions in the universe. *Proc. Natl. Acad. Sci. USA* **96**, 4752–4755.

Hartmann, D. H., Kretschmer, K. and Diehl, R. (2002). Disturbance ecology from nearby supernovae. In *Proceedings of the 11th Workshop on Nuclear Astrophysics, Ringberg Castle, 2/11–16, 2002*, ed. by W. Hillebrandt and E. Müller. Garching: Max-Planck-Institut, pp. 154–163.

Harwit, M. (1961). Can gravitational forces alone account for galaxy formation in a steady-state universe? *Mon. Not. R. Astron. Soc.* **122**, 47–50.

Havel, I. M. (1999). Living in conceivable worlds. *Found. Sci.* **3**, 375–394.

Hawkins, D. (1971). On the self consistency of the steady-state cosmology. *Philos. Sci.* **38**, 273–279.

Hawthorn, G. (1991). *Plausible Worlds: Possibility and Understanding in History and the Social Sciences*. Cambridge: Cambridge University Press.

Hazen, R. M. and Sholl, D. S. (2003). Chiral selection on inorganic crystalline surfaces. *Nat. Mater.* **2**, 367–374.

Heffernan, W. C. (1978). The singularity of our inhabited world: William Whewell and A. R. Wallace in dissent. *J. Hist. Ideas* **39**, 81–100.

Hegel, G. W. F. [1832] (1971). *Vorlesungen über die Geschichte der Philosophie I–III*. Frankfurt: Suhrkamp Verlag.

Hellings, R. W. *et al.* (1983). Experimental test of the variability of G using Viking lander ranging data. *Phys. Rev. Lett.* **51**, 1609–1612.

Hengeveld, R. (2011). Definitions of life are not only unnecessary, but they can do harm to understanding. *Found. Sci.* **16**, 323–325.

Hinckfuss, I. (1975). *The Existence of Space and Time*. Oxford: Clarendon Press.

Ho, B. and Monton, B. (2005). Anthropic reasoning does not conflict with observation. *Analysis* **65**, 42–45.

Hoffman, P. F., Kaufman, A. J., Halverson, G. P. and Schrag, D. P. (1998). A neoproterozoic snowball Earth. *Science* **281**, 1342–1346.

Hofstadter, D. R. (2000). *Gödel, Escher, Bach: An Eternal Golden Braid*. 20th-anniversary edition. London: Penguin Books.

Hogan, C. J. (2000). Why the universe is just so. *Rev. Mod. Phys.* **72**, 1149–61.

Holt, J. (2004). The big lab experiment: was our universe created by design? *Slate*, http://slate.msn.com/id/2100715.

Horneck, G. and Rettberg, P. (eds.) (2007). *Complete Course in Astrobiology*. Weinheim, Germany: Wiley-VCH.

Horner, J. and Jones, B. W. (2008). Jupiter friend or foe? I. The asteroids. *Int. J. Astrobiol.* **7**, 251–261.

Horner, J. and Jones, B. W. (2009). Jupiter friend or foe? II. The Centaurs. *Int. J. Astrobiol.* **8**, 75–80.

Houellebecq, M. (2007). *The Possibility of an Island*. Translated by G. Bowd. New York: Vintage.

Hoyle, F. (1948). A new model for the expanding universe. *Mon. Not. R. Astron. Soc.* **108**, 372–382.

Hoyle, F. (1949). On the cosmological problem. *Mon. Not. R. Astron. Soc.* **109**, 365–371.

Hoyle, F. (1957). *The Black Cloud*. London: William Heinemann.

Hoyle F. (1983) *The Intelligent Universe*. London: Michael Joseph.

Hoyle, F. (1992). Mathematical theory of the origin of matter. *Astrophys. Space Sci.* **198**, 195–230.

Hoyle, F. (1994). *Home Is Where the Wind Blows: Chapters from a Cosmologist's Life*. Mill Valley, CA: University Science Books.

Hoyle, F. and Wickramasinghe, N. (1981). *Evolution from Space*. London: J. M. Dent.

Hoyle, F. and Wickramasinghe, N. (1999). The universe and life: deductions from the weak anthropic principle. *Astrophys. Space Sci.* **268**, 89–102.

Hoyle, F. and Wickramasinghe, N. (2000). Cometary impact and ice-ages. *Astrophys. Space Sci.* **275**, 367–376.

Hoyle, F., Wickramasinghe, N. C. and Watkins, J. (1986). *Viruses from Space*. Cardiff: University College Cardiff Press.

Hughes, J. (2004). *Citizen Cyborg: Why Democratic Societies Must Respond to the Redesigned Human of the Future*. New York: Basic Books.

Hull, D. L. (2002). Recent philosophy of biology: a review. *Acta Biotheor.* **50**, 117–128.
Hutton, J. (1788). Theory of the Earth; or an investigation of the laws observable in the composition, dissolution, and restoration of land upon the globe. *Trans. R. Soc. Edinb.* **1**, 209–304.
Impey, C. (ed.) (2010). *Talking About Life: Conversations on Astrobiology*. Cambridge: Cambridge University Press.
Ingrosso, G. *et al.* (2009). Pixel lensing as a way to detect extrasolar planets in M31. *Mon. Not. R. Astron. Soc.* **399**, 219–228.
Ingrosso, G. *et al.* (2011). Search for exoplanets in M31 with pixel-lensing and the PA-99-N2 event revisited. *Gen. Relativ. Gravit.* **43**, 1047–1060.
Izakov, M. N. (1997). Self-organization and information for planets and ecosystems. *Phys. Uspekhi* **40**, 1035–1042.
Jablonski, D. (1986). Background and mass extinctions: the alternation of macroevolutionary regimes. *Science* **231**, 129–133.
Jablonski, D. (1989). The biology of mass extinction: a palaeontological view. *Philos. Trans. R. Soc. Lond.* **B325**, 357–368.
Jagers op Akkerhuis, G. A. J. M. (2010). Towards a hierarchical definition of life, the organism, and death. *Found. Sci.* **15**, 245–262.
Jeans, J. (1923). Obituary Notices: Associates: Kapteyn, Jacobus Cornelius. *Mon. Not. R. Astron. Soc.* **83**, 250–255.
Jones, E. M. (1985). 'Where is everybody?' An account of Fermi's question. Los Alamos preprint, http://lib-www.lanl.gov/la-pubs/00318938.pdf.
Jordan, P. (1955). *Schwerkraft und Weltall*. Braunschweig, Germany: Vieweg.
Joshi, S. T. (2004). *H. P. Lovecraft: A Life*. Warwick: Necronomicon Press.
Joshi, S. T. and Schultz, D. E. (2001). *An H. P. Lovecraft Encyclopedia*. New York: Hippocampus.
Joyce, G. F. (1994). Foreword. In *Origins of Life: The Central Concepts*, ed. by D. W. Deamer and G. Fleischaker. Boston: Jones and Bartlett.
Jugaku, J. and Nishimura, S. (2003). A search for Dyson spheres around late-type stars in the solar neighborhood. In *Bioastronomy 2002: Life Among the Stars, Proceedings of IAU Symposium 213*, ed. by R. Norris and F. Stootman. San Francisco: ASP Conference Series, pp. 437–438.
Jugaku, J., Noguchi, K. and Nishimura, S. (1995). A search for Dyson spheres around late-type stars in the solar neighborhood. In *Progress in the Search for Extraterrestrial Life*, ed. by G. Seth Shostak. San Francisco: ASP Conference Series, pp. 381–385.
Kaltenegger, L. and Sasselov, D. (2010). Detecting planetary geochemical cycles on exoplanets: atmospheric signatures and the case of SO_2. *Astrophys. J.* **708**, 1162–1167.
Kaltenegger, L., Fridlund, M. and Kasting, J. (2002). Review on habitability and biomarkers. In *Earth-like Planets and Moons. Proceedings of the 36th ESLAB Symposium, 3–8 June 2002, ESTEC, Noordwijk, The Netherlands*, ed. by B. Foing and B. Battrick. Noordwijk, The Netherlands: ESA Publications Division, pp. 277–282.
Kapteyn, J. C. (1913). The structure of the universe. *Science* **38**, 717–724.
Kardashev, N. S. (1964). Transmission of information by extraterrestrial civilizations. *Sov. Astron.* **8**, 217–220.
Kardashev, N. S. (1997). Cosmology and civilizations. *Astrophys. Space Sci.* **252**, 25–40.
Kasting, J. (2010). *How to Find a Habitable Planet*. Princeton, NJ: Princeton University Press.

Keszthelyi, L. (1984). Parity violation as a source of chirality in nature. *Orig. Life Evol. Biosph.* **14**, 375–382.

Kirschvink, J. L. *et al.* (2000). Paleoproterozoic snowball Earth: extreme climatic and geochemical global change and its biological consequences. *Proc. Natl. Acad. Sci. USA* **97**, 1400–1405.

Kitcher, P. (1985). *Vaulting Ambition: Sociobiology and the Quest for Human Nature.* Cambridge, MA: MIT Press.

Klass, P. J. (1974). *UFOs Explained.* New York: Random House.

Klee, R. (2002). The revenge of Pythagoras: how a mathematical sharp practice undermines the contemporary design argument in astrophysical cosmology. *Br. J. Philos. Sci.* **53**, 331–354.

Klingsör-Leroy, C. (2009). *Surrealism.* Cologne: Taschen.

Knoll, A. H. and Bambach, R. K. (2000). Directionality in the history of life: diffusion from the left wall or repeated scaling of the right? In *Deep Time: Paleobiology's Perspective*, ed. by D. H. Erwin and S. L. Wing. Lawrence, KS: The Paleontological Society, pp. 1–14.

Kragh, H. (1996). *Cosmology and Controversy.* Princeton, NJ: Princeton University Press.

Kragh, H. (2004). *Matter and Spirit in the Universe: Scientific and Religious Preludes to Modern Cosmology.* London: Imperial College Press.

Kragh, H. (2005). George Gamow and the 'factual approach' to relativistic cosmology. In *The Universe of General Relativity: Einstein Studies, Volume 11*, ed. by A. J. Cox and J. Eisenstaedt. Boston: Birkhäuser, pp. 175–188.

Kragh, H. (2007). *Conceptions of Cosmos from Myth to the Accelerating Universe: A History of Cosmology.* Oxford: Oxford University Press.

Kukla, A. (2001). SETI: on the prospects and pursuitworthiness of the Search for Extraterrestrial Intelligence. *Stud. Hist. Philos. Sci.* **32**, 31–67.

Kumar, P. and Piran, T. (2000). Energetics and luminosity function of gamma-ray bursts. *Astrophys. J.* **535**, 152–157.

Kurzweil, R., (2000). *The Age of Spiritual Machines: When Computers Exceed Human Intelligence.* New York: Penguin.

Leiber, F., Jr. (2001). A literary Copernicus. In *Discovering H. P. Lovecraft*, ed. by D. Schweitzer. Holicong, PA: Wildside Press, pp. 7–16.

Leitch, E. M. and Vasisht, G. (1998). Mass extinctions and the Sun's encounters with spiral arms. *New Astron.* **3**, 51–56.

Lem, S. [1964] (1973). *The Invincible.* London: Sidgwick and Jackson.

Lem, S. [1974] (1977). *Summa Technologiae.* Belgrade: Nolit (in Serbian).

Lem, S. (1987). *Fiasco.* Orlando, FL: Harcourt Brace Jovanovich.

Lem, S. [1971] (1993). The new cosmology. In *A Perfect Vacuum*, ed. by S. Lem and M. Kandel. Evanston, IL: Northwestern University Press, pp. 197–227.

Lem, S. [1968] (1999). *His Master's Voice.* Evanston, IL: Northwestern University Press.

Leonard, P. J. T. and Bonnell, J. T. (1998). Gamma-ray bursts of doom. *Sky & Telesc.* (February), 28–34.

Levy, B. (2011). *River of Darkness: Francisco Orellana's Legendary Voyage of Death and Discovery Down the Amazon.* New York: Bantam.

Lewis, D. (1986). *On the Plurality of Worlds.* Oxford: Blackwell.

Lightman, A. and Brawer, R. (1990). *Origins: The Lives and Worlds of Modern Cosmologists.* Cambridge, MA: Harvard University Press.

Linde, A. D. (1988). Life after inflation. *Phys. Lett. B* **211**, 29–31.

Linde, A. D. (1990). *Inflation and Quantum Cosmology.* San Diego, CA: Academic Press.

Linde, A. D. and Vanchurin, V. (2010). How many universes are in the multiverse? *Phys. Rev. D* **81**, 083525(11).

Linde, A. D., Linde, D. and Mezhlumian, A. (1994). From the Big Bang theory to the theory of a stationary universe. *Phys. Rev. D* **49**, 1783–1826.

Lineweaver, C. H. (1998). The cosmic microwave background and observational convergence in the omega–m-omega–lambda plane. *Astrophys. J.* **505**, L69–L73.

Lineweaver, C. H. (2001). An estimate of the age distribution of terrestrial planets in the universe: quantifying metallicity as a selection effect. *Icarus* **151**, 307–313.

Lineweaver, C. H. and Davis, T. M. (2002). Does the rapid appearance of life on Earth suggest that life is common in the universe? *Astrobiology* **2**, 293–304.

Lineweaver, C. H. and Egan, C. A. (2007). The cosmic coincidence as a temporal selection effect produced by the age distribution of terrestrial planets in the universe. *Astrophys. J.* **671**, 853–860.

Lineweaver, C. H. and Grether, D. (2003). What fraction of Sun-like stars have planets? *Astrophys. J.* **598**, 1350–1360.

Lineweaver, C. H. *et al.* (1996). The dipole observed in the COBE DMR 4 year data. *Astrophys. J.* **470**, 38–42.

Lineweaver, C. H., Fenner, Y. and Gibson, B. K. (2004). The Galactic Habitable Zone and the age distribution of complex life in the Milky Way. *Science* **303**, 59–62.

Livio, M. (1999). How rare are extraterrestrial civilizations, and when did they emerge? *Astrophys. J.* **511**, 429–431.

Lloyd, S. (2000). Ultimate physical limits to computation. *Nature* **406**, 1047–1054.

Lovecraft, H. P. (1965). *Selected Letters, Volume 1*. Sauk City, WI: Arkham House.

Lovecraft, H. P. (1999). *The H. P. Lovecraft Omnibus: I. At the Mountains of Madness*. London: Harper Collins.

Lovecraft, H. P. (2000). *The H. P. Lovecraft Omnibus: III. The Haunter of the Dark*. London: Harper Collins.

Lovecraft, H. P. (2005). *Collected Essays, Volume 3: Science*, ed. by S. T. Joshi. New York: Hippocampus Press.

Lucretius (1997). *On the Nature of Things*. Translated by William E. Leonard, e-text version. Urbana, IL: Project Gutenberg.

Mackie, P. (2006). *How Things Might Have Been: Individuals, Kinds, and Essential Properties*. Oxford: Oxford University Press.

Maddox, J. (1984). New twist for the anthropic principle. *Nature* **307**, 409.

Maher, K. A. and Stevenson, D. J. (1988). Impact frustration of the origin of life. *Nature* **331**, 612–614.

Malmquist K. G. (1920). A study of the stars of spectral type A. *Lund Medd. Ser. II* **22**, 1–39.

Mamajek, E. E. and Hillenbrand, L. A. (2008). Improved age estimation for solar-type dwarfs using activity-rotation diagnostics. *Astrophys. J.* **687**, 1264–1293.

Manly, S. (2011). *Visions of the Multiverse*. Pompton Plains, NJ: New Page Books.

Marochnik, L. S. (1983). On the origin of the Solar System and the exceptional position of the Sun in the Galaxy. *Astrophys. Space Sci.* **89**, 61–75.

Mash, R. (1993). Big numbers and induction in the case for extraterrestrial intelligence. *Philos. Sci.* **60**, 204–222.

Mason, J. W. (ed.) (2008). *Exoplanets: Detection, Formation, Properties, Habitability*. Chichester: Springer Praxis.

Matese, J. J. and Whitman, P. G. (1992). A model of the Galactic tidal interaction with the Oort comet cloud. *Celest. Mech. & Dyn. Astron.* **54**, 13–35.

Matese, J. J. and Whitmire, D. (1996). Tidal imprint of distant Galactic matter on the Oort comet cloud. *Astrophys. J. Lett.* **472**, L41–L43.

Maynard Smith, J. (1986). *The Problems of Biology*. Oxford: Oxford Paperbacks.

Maynard Smith, J. and Szathmary, E. (1997) *The Major Transitions in Evolution*. Oxford: Oxford University Press.

Mayr, E. (1990). The myth of non-Darwinian revolution. *Biol. Philos.* **5**, 85–92.

Mayr, E. (1993). The search for intelligence. *Science* **259**, 1522–1523.

Mazlish, B. (1967). The fourth discontinuity. *Technol. Cult.* **8**, 1–15.

McCabe, M. and Lucas, H. (2010). On the origin and evolution of life in the Galaxy. *Int. J. Astrobiol.* **9**, 217–226.

McKay, C. P. (1996). Time for intelligence on other planets. In *Circumstellar Habitable Zones. Proceedings of The First International Conference*, ed. by L. R. Doyle. Menlo Park, CA: Travis House Publications, pp. 405–419.

McKay, C. P. and Marinova, M. M. (2001). The physics, biology, and environmental ethics of making Mars habitable. *Astrobiology* **1**, 90–104.

McKay, D. S. *et al.* (1996). Search for past life on Mars: possible relic biogenic activity in Martian meteorite ALH84001. *Science* **273**, 924–930.

McMullin, E. (1980). Persons in the universe. *Zygon* **15**, 69–89.

McShea, D. W. (1994). Mechanisms of large-scale evolutionary trends. *Evol.* **48**, 1747–1763.

McShea, D. W. (1996). Metazoan complexity and evolution: is there a trend? *Evol.* **50**, 477–492.

Medvedev, Z. A. (1971). *The Rise and Fall of T. D. Lysenko*. Garden City, NY: Doubleday & Co.

Melott, A. L. *et al.* (2004). Did a gamma-ray burst initiate the late Ordovician mass extinction? *Int. J. Astrobiol.* **3**, 55–61.

Mészáros, P. (2002). Theories of gamma-ray bursts. *Annu. Rev. Astron. Astrophys.* **40**, 137–169.

Michaud, M. A. G. (2007). *Contact with Alien Civilizations: Our Hopes and Fears about Encountering Extraterrestrials*. New York: Copernicus Books.

Milne, E. A. (1940). Cosmological theories. *Astrophys. J.* **91**, 129–158.

Misner, C. W. (1969). Absolute zero of time. *Phys. Rev.* **186**, 1328–1333.

Misner, C. W., Thorne, K. S. and Wheeler, J. A. (1973). *Gravitation*. San Francisco: W. H. Freeman.

Mitton, S. (2005). *Conflict in the Cosmos: Fred Hoyle's Life in Science*. Washington DC: Joseph Henry Press.

Moorcock, M. (1995). *The Eternal Champion*. Clarkston, GA: White Wolf.

Mosterín, J. (2000). The anthropic principle in cosmology: a critical review. *Acta Inst. Philos. Aesthet.* (Tokyo, Japan) **18**, 111–139.

Mosterín, J. (2004). Anthropic explanations in cosmology. In *Proceedings of the 12th International Congress of Logic, Methodology and Philosophy of Science*, ed. by P. Hajek, L. Valdés-Villanueva and D. Westerstahl. Amsterdam: North-Holland Publishing.

Munitz, M. (1986). *The Cosmic Understanding*. Princeton, NJ: Princeton University Press.

Napier, W. M. (2004). A mechanism for interstellar panspermia. *Mon. Not. R. Astron. Soc.* **348**, 46–51.

Narlikar, J. V. (1984). Nonconservation of baryons in cosmology – revisited. *Astron. & Astrophys.* **5**, 67–68.

Néron de Surgy, O. and Laskar, J. (1997). On the long term evolution of the spin of the Earth. *Astron.& Astrophys.* **318**, 975–989.

Newman, W. I. and Sagan, C. (1981). Galactic civilizations: population dynamics and interstellar diffusion. *Icarus* **46**, 293–327.

Newton-Smith, W. H. (1980). *The Structure of Time*. London: Routledge and Kegan Paul.

Nietzsche, F. (1968). *Twilight of the Idols*. Translated by R. J. Hollingdale. Baltimore: Penguin.

North, J. (1965). *The Measure of the Universe: A History of Modern Cosmology*. London: Oxford University Press.

North, J. (1994). *The Fontana History of Astronomy and Cosmology*. London: Fontana Press.

Nozick, R. (1981). *Philosophical Explanations*. Cambridge, MA: Harvard University Press.

Nussbaumer, H. and Bieri, L. (2009). *Discovery of the Expanding Universe*. Cambridge: Cambridge University Press.

Oberbeck, V. R. and Fogleman, G. (1989). Estimates of the maximum time required to originate life. *Orig. Life Evol. Biosph.* **19**, 549–560.

O'Brien, D. (1969). *Empedocles' Cosmic Cycle*. Cambridge: Cambridge University Press.

Okasha, S. (2006). The levels of selection debate: philosophical issues. *Philos. Compass* **1**, PS 001, 1–12.

Olson, E. C. (1988). *N* and the rise of cognitive intelligence on Earth. *Q. J. R. Astron. Soc.* **29**, 503–509.

Olum, K. (2004). Conflict between anthropic reasoning and observation. *Analysis* **64**, 1–8.

Page, D. N. (2006). The lifetime of the universe. *J. Korean Phys. Soc.* **49**, 711–14 (preprint at arXiv:hep-th/0510003).

Page, D. N. (2008). Typicality derived. *Phys. Rev. D* **78**, 023514(7).

Pagels, H. R. (1998). A cozy cosmology. In *Modern Cosmology & Philosophy*, ed. by J. Leslie. Amherst, MA: Prometheus Books, pp. 180–86.

Palladino, P. (1990). Stereochemistry and the nature of life: mechanist, vitalist, and evolutionary perspectives. *Isis* **81**, 44–67.

Paquet, M. (2006). *Magritte*. Cologne: Taschen.

Parkinson, B. (2005). The carbon or silicon colonization of the universe? *J. Br. Interplanet. Soc.* **58**, 111–116.

Paul, E. R. (1993). The life and works of J. C. Kapteyn. *Space Sci. Rev.* **64**, v–92.

Pavlov, A. A., Toon, O. B., Pavlov, A. K., Bally, J. and Pollard, D. (2005). Passing through a giant molecular cloud: 'snowball' glaciations produced by interstellar dust. *Geophys. Res. Lett.* **32**, L03705(4).

Peacock, J. A. (1999). *Cosmological Physics*. Cambridge: Cambridge University Press.

Peebles, P. J. E. (1993). *Principles of Physical Cosmology*. Princeton, NJ: Princeton University Press.

Pegg, D. T. (1971). Night sky darkness in the Eddington–Lemaître universe. *Mon. Not. R. Astron. Soc.* **154**, 321–327.

Penrose, R. (1989). *The Emperor's New Mind*. Oxford: Oxford University Press.

Perlmutter, S. *et al.* (1999). Measurements of omega and lambda from 42 high-redshift supernovae. *Astrophys. J.* **517**, 565–586.

Pierson, T. (2006). SETI Institute as a model for managing interdisciplinary science. *Acta Astronaut.* **58**, 478–484.

Pigliucci, M. (2010). *Nonsense on Stilts: How to Tell Science from Bunk*. Chicago: University of Chicago Press.

Piran, T. (2000). Gamma-ray bursts: a puzzle being resolved. *Phys. Rep.* **333/334**, 529–553.

Poe, E. A. [1847] (1912). *Tales of Mystery and Imagination*. London: J. M. Dent and Sons.

Popper, K. (1972). *Objective Knowledge: An Evolutionary Approach.* Oxford: Oxford University Press.

Popper, K. (1974). Darwinism as a metaphysical research programme. In *The Philosophy of Karl Popper, Vol. I*, ed. by P. A. Schilpp. La Salle, IL: Open Court, pp. 133–143.

Price, H. (1996). *Time's Arrow and Archimedes' Point.* Oxford: Oxford University Press.

Quine, W. V. (1976). Worlds away. *J. Philos.* **73**, 859–863.

Rampino, M. R. (1997). The Galactic theory of mass extinctions: an update. *Celest. Mech. & Dyn. Astron.* **69**, 49–58.

Rampino, M. R. (2002). Supereruptions as a threat to civilizations on Earth-like planets. *Icarus* **156**, 562–569.

Raulin-Cerceau, F. (2010). What possible life forms could exist on other planets: a historical overview. *Orig. Life Evol. Biosph.* **40**, 195–202.

Raup, D. M. (1991). *Extinction: Bad Genes or Bad Luck?* New York: W. W. Norton.

Raup, D. M. (1992). Nonconscious intelligence in the universe. *Acta Astronaut.* **26**, 257–261.

Raup, D. M. (1999). *The Nemesis Affair: A Story of the Death of Dinosaurs and the Ways of Science.* 2nd edition. New York: W. W. Norton.

Raup, D. M. and Valentine, J. W. (1983). Multiple origins of life. *Proc. Natl. Acad. Sci. USA* **80**, 2981–2984.

Ray, T. S. (2003). An evolutionary approach to synthetic biology: Zen in the art of creating life. In *Advances in Evolutionary Computing*, ed. by A. Ghosh and S. Tsutsui. New York: Springer, pp. 479–517.

Rees, M. J. (1997) *Before the Beginning: Our Universe and Others.* Cambridge, MA: Helix.

Rees, M. J. (2001). *Our Cosmic Habitat.* Princeton, NJ: Princeton University Press.

Rescher, N. (1985). Extraterrestrial science. In *Extraterrestrials, Science and Alien Intelligence*, ed. by E. Regis Jr. Cambridge: Cambridge University Press, pp. 83–116.

Roberge, W. and Dalgarno, A. (1982). The formation and destruction of HeH^+ in astrophysical plasmas. *Astrophys. J.* **255**, 489–496.

Robles, J. A. *et al.* (2008). A comprehensive comparison of the Sun to other stars: searching for self-selection effects. *Astrophys. J.* **684**, 691–706.

Robson, J. (2005). *Natural History.* New York: Bantam Books.

Rose, C. and Wright, G. (2004). Inscribed matter as an energy-efficient means of communications with an extraterrestrial civilization. *Nature* **431**, 47–49.

Rospars, J.-P. (2011). Terrestrial biological evolution and its implication for SETI. *Acta Astronaut.* **67**, 1361–1365.

Roy, K. I., Kennedy III, R. G. and Fields, D. E. (2004). Shell worlds: an approach to making large moons and small planets habitable. *AIP Conference Proceedings* **699**, 1075–1084.

Rubtsov, V. V. (1991). Criteria of artificiality in SETI. In *Bioastronomy The Search for Extraterrestial Life*, Lecture Notes in Physics, Vol. 390, ed. by J. Heidmann and M. J. Klein. Heidelberg: Springer-Verlag, pp. 306–310.

Ruse, M. (1977). Karl Popper's philosophy of biology. *Philos. Sci.* **44**, 638–661.

Russell, D. A. (1983). Exponential evolution: implications for extraterrestrial intelligent life. *Adv. Space Res.* **3**, 95–103.

Russell, D. A. (1995). Biodiversity and time scales for the evolution of extraterrestrial intelligence. In *Progress in the Search for Extraterrestrial Life*, ed. by G. Seth Shostak. San Francisco: ASP Conference Series, Vol. 74, pp. 143–151.

Sagan, C. (1973). On the detectivity of advanced Galactic civilizations. *Icarus* **19**, 350–352.

Sagan, C. and Walker, R. G. (1966). The infrared detectability of Dyson civilizations. *Astrophys. J.* **144**, 1216–1218.

Saitoh, T. R. *et al.* (2008). Toward first-principle simulations of galaxy formation: I. How should we choose star-formation criteria in high-resolution simulations of disk galaxies? *Publ. Astron. Soc. Jpn.* **60**, 667–681.

Sandberg, A. (2000). The physics of information processing superobjects: daily life among the Jupiter brains. *J. Transhumanism* **5** (now *J. Evol.Technol.* at http://transhumanist.com/volume5/Brains2.pdf).

Scalo, J. and Wheeler, J. C. (2002). Astrophysical and astrobiological implications of gamma-ray burst properties. *Astrophys. J.* **566**, 723–737.

Schmitt-Kopplin, P. *et al.* (2010). High molecular diversity of extraterrestrial organic matter in Murchison meteorite revealed 40 years after its fall. *PNAS* **107**, 2763–2768.

Schulze-Makuch, D. and Darling, D. (2010). *We Are Not Alone: Why We Have Already Found Extraterrestrial Life*. Oxford: Oneworld.

Schrödinger, E. (1944). *What is Life?* Cambridge: Cambridge University Press.

Schwarz, D. J. and Terrero-Escalante, C. A. (2004). Primordial fluctuations and cosmological inflation after WMAP 1.0. *J. Cosmol. Astropart. Phys.* **JCAP08**, 003.

Sciabà, A. (ed.) (2011). Worldwide LHC computing grid. CERN-LCG-GDEIS-722398 document (https://edms.cern.ch/file/722398/1.4/gLite-3-UserGuide.html).

Sciama, D. W. (1955). On the formation of galaxies in a steady state universe. *Mon. Not. R. Astron. Soc.* **115**, 3–14.

Segerstråle, U. (2000). *Defenders of the Truth: The Sociobiology Debate*. Oxford: Oxford University Press.

Sergi, V., Galletta, G., Celotti, L. and Turatto, M. (2001). Nearby supernova explosions and radiation effects on the Earth's environment. In *Proceedings of the First European Workshop on Exo- and Astro-Biology*, ed. by P. Ehrenfreund, O. Angerer and B. Battrick, , ESA SP-496, pp. 409–412.

Shanahan, T. (2004). *The Evolution of Darwinism: Selection, Adaptation, and Progress in Evolutionary Biology*. Cambridge: Cambridge University Press.

Shapiro, R. and Feinberg, G. (1982). Possible forms of life in environments very different from the Earth. In *Extraterrestrials, Where are They?* ed. by M. H. Hart and B. Zuckerman. New York: Pergamon Press, pp. 113–120.

Shaviv, N. J. (2002a). Cosmic ray diffusion from the Galactic spiral arms, iron meteorites, and a possible climatic connection. *Phys. Rev. Lett.* **89**, 051102(4).

Shaviv, N. J. (2002b). The spiral structure of the Milky Way, cosmic rays, and ice age epochs on Earth. *New Astron.* **8**, 39–77.

Shermer, M. (2002). *In Darwin's Shadow: The Life and Science of Alfred Russel Wallace*. Oxford: Oxford University Press.

Shklovskii, I. S. (1962). *Universe – Life – Intelligence*. Moscow: AN SSSR (in Russian).

Shklovskii, I. S. and Sagan, C. (1966). *Intelligent Life in the Universe*. San Francisco: Holden-Day.

Signor III, P. W. and Lipps, J. H. (1982). Sampling bias, gradual extinction patterns, and catastrophes in the fossil record. In *Geological Implications of Impacts of Large Asteroids and Comets on the Earth*, ed. by L. T. Silver and P. H. Schultz. Boulder, CO: Geological Society of America Special Publication, Vol. 190, pp. 291–296.

Silk, J. (2000). Galaxy formation: what the future may hold. *Publ. Astron. Soc. Pac.* **112**, 1003.

Simpson, G. G. (1964). The nonprevalence of humanoids. *Science* **143**, 769–775.

Slysh, V. I. (1985). A search in the infrared for astroengineering activity. In *The Search for Extraterrestrial Life: Recent Developments*, ed. by M. D. Papagiannis. Dordrecht: IAU, Reidel Publishing Co., pp. 315–319.

Smith, D. S., Scalo, J. and Wheeler, J. C. (2004). Importance of biologically active aurora-like ultraviolet emission: stochastic irradiation of Earth and Mars by flares and explosions. *Orig. Life Evol. Biosph.* **34**, 513–532.

Smolin, L. (1997). *The Life of the Cosmos*. Oxford: Oxford University Press.

Smolin, L. (2007). Scientific alternatives to the anthropic principle. In *Universe or Multiverse?* ed. by B. Carr. Cambridge: Cambridge University Press, pp. 323–366.

Sober, E. (2010). Natural selection, causation, and laws: what Fodor and Piattelli-Palmarini got wrong. *Philos. Sci.* **77**, 594–607.

Spengler, O. (1918). *Decline of the West*. 1996 edition. New York: Alfred A. Knopf Publisher.

Stapledon, O. (1937). *Star Maker*. London: Methuen.

Starkman G. and Trotta, R. (2006). Why anthropic reasoning cannot predict Λ. *Phys. Rev. Lett.* **97**, 201301(4).

Steckline, V. S. (1983). Zermelo, Boltzmann, and the recurrence paradox. *Am. J. Phys.* **51**, 894–897.

Stein, H. (1970). On the paradoxical time-structures of Kurt Gödel. *Philos. Sci.* **37**, 589–601.

Sterelny, K. (2005). Another view of life. *Stud. Hist. Philos. Biol. & Biomed. Sci.* **36**, 585–593.

Stozhkov, Y. I. (2003). The role of cosmic rays in the atmospheric processes. *J. Phys. G: Nucl. Part. Phys.* **29**, 913–923.

Stozhkov, Y. I., Okhlopkov, V. and Pokrevsky, P. (2001). Cosmic rays tell us on supernova explosion in the nearby interstellar space. *Nucl. Phys. B (Proc. Suppl.)* **97**, 207–210.

Summers, M. E., Lieb, B. J., Chapman, E. and Yung, Y. L. (2002). Atmospheric biomarkers of subsurface life on Mars. *Geophys. Res. Lett.* **29**, 24–1.

Susskind, L. (2003). The anthropic landscape of string theory. Preprint at http://arXiv: hep-th/0302219.

Susskind, L. (2006). *The Cosmic Landscape: String Theory and the Illusion of Intelligent Design*. New York: Back Bay Books.

Tarter, J. (2001). The Search for Extraterrestrial Intelligence (SETI). *Annu. Rev. Astron. Astrophys.* **39**, 511–548.

Teerikorpi, P. (1997). Observational selection bias affecting the determination of the extragalactic distance scale. *Annu. Rev. Astron. Astrophys.* **35**, 101–136.

Tegmark, M. (1998). Is 'the theory of everything' merely the ultimate ensemble theory? *Ann. Phys.* **270**, 1–51.

Tegmark, M. (2003). Parallel universes. *Sci. Am.* (May), 41–51.

Tegmark, M. (2008). The mathematical universe. *Found. Phys.* **38**, 101–150.

Tegmark M. and Rees, M. J. (1998). Why is the cosmic microwave background fluctuation level 10^{-5}? *Astrophys. J.* **499**, 526–32.

Thomas, B. C. *et al.* (2005). Terrestrial ozone depletion due to a Milky Way gamma-ray burst. *Astrophys. J.* **622**, L153–L156.

Thomas, B. C., Melott, A. L., Fields, B. D. and Anthony-Twarog, B. J. (2008). Superluminous supernovae: no threat from η Carinae. *Astrobiology* **8**, 9–16.

Thompson, D'Arcy, W. (1917). *On Growth and Form*. Cambridge: Cambridge University Press.

Thorsett, S. E. (1995). Terrestrial implications of cosmological gamma-ray burst models. *Astrophys. J.* **444**, L53–L55.

Thucydides (1972). *The History of the Peloponnesian War*. Revised edition. Translated by R. Warner. New York: Penguin.

Tilgner, C. N. and Heinrichsen, I. (1998). A program to search for Dyson spheres with the Infrared Space Observatory. *Acta Astronaut.* **42**, 607–612.

Timofeev, M. Yu., Kardashev, N. S. and Promyslov, V. G. (2000). Search of the IRAS database for evidence of Dyson spheres. *Acta Astronaut.* **46**, 655–659.

Tipler, F. J. (1980). Extraterrestrial intelligent beings do not exist. *Q. J. R. Astron. Soc.* **21**, 267–281.

Tipler, F. J. (1981a). Additional remarks on extraterrestrial intelligence. *Q. J. R. Astron. Soc.* **22**, 279–292.

Tipler, F. J. (1981b). A brief history of the extraterrestrial intelligence concept. *Q. J. R. Astron. Soc.* **22**, 133–145.

Tipler, F. J. (1982). Anthropic-principle arguments against steady-state cosmological theories. *Obs.* **102**, 36–39.

Tipler, F. J. (1988). The anthropic principle: a primer for philosophers. In *Proceedings of the Biennial Meeting of the Philosophy of Science Association*, Vol. 2, pp. 27–48.

Tipler, F. J. (1994). *The Physics of Immortality*. New York: Doubleday.

Tipler, F. J. (2003). Intelligent life in cosmology. *Int. J. Astrobiol.* **2**, 141–148.

Tirard, S., Morange, M. and Lazcano, A. (2010). The definition of life: a brief history of an elusive scientific endeavor. *Astrobiology* **10**, 1003–1009.

Tofler, A. (1990). *Powershift: Knowledge, Wealth and Violence at the Edge of the 21st Century*. New York: Bantam Books.

Torbett, M. V. (1986). Injection of Oort cloud comets to the inner Solar System by Galactic tidal fields. *Mon. Not. R. Astron. Soc.* **223**, 885–895.

Tough, A. (1990). A critical examination of the factors that might encourage secrecy. *Acta Astronaut.* **21**, 97–101.

Toulmin, S. and Goodfield, J. [1966] (1982). *The Discovery of Time*. Chicago: University of Chicago Press.

Trotta, R. (2007). Dark matter: probing the Arche-fossil (interview). *Collapse* **II**, 83–169.

Tsokolov, S. A. (2009). Why is the definition of life so elusive? Epistemological considerations. *Astrobiology* **9**, 401–412.

Turnbull, M. C. *et al.* (2006). Spectrum of a habitable world: earthshine in the near-infrared. *Astrophys. J.* **644**, 551–559.

Turner, D. D. (2011). Gould's replay revisited. *Biol. Philos.* **26**, 65–79.

Udry, S. *et al.* (2007). The HARPS search for southern extra-solar planets: XI. Super-Earths (5 and 8 $M_{\oplus}$) in a 3-planet system. *Astron. & Astrophys.* **469**, L43–L47.

Ulvestad, E. (2002). Biosemiotic knowledge: a prerequisite for valid explorations of extraterrestrial intelligent life. *Sign Syst. Stud.* **30**, 283–292.

Valdes, F. and Freitas, R. A., Jr. (1983). A search for objects near the Earth–Moon Lagrangian points. *Icarus* **53**, 453–457.

Valdes, F. and Freitas, R. A., Jr. (1986). A search for the tritium hyperfine line from nearby stars. *Icarus* **65**, 152–157.

Valencia, D., O'Connell, R. J. and Sasselov, D. D. (2007). Inevitability of plate tectonics on super-Earths. *Astrophys. J.* **670**, L45–L48.

Van de Kamp, P. (1982). The planetary system of Barnard's star. *Vistas Astron.* **26**, 141–157.

Vermeij, G. J. (2006). Historical contingency and the purported uniqueness of evolutionary innovations. *PNAS* **103**, 1804–1809.

Vilenkin, A. (1992). Did the Universe have a beginning? *Phys. Rev. D* **46**, 2355–2361.

Vilenkin, A. (1995). Predictions from quantum cosmology. *Phys. Rev. Lett.* **74**, 846–849.

Vilenkin, A. (2006). *Many Worlds in One*. New York: Hill and Wang.

Vilenkin, A. (2007). Freak observers and the measure of the multiverse. *J. High Energy Phys.* **01**, 092.

Vukotić, B. (2010). The set of habitable planets and astrobiological regulation mechanisms. *Int. J. Astrobiol.* **9**, 81–87.

Vukotić, B. and Ćirković, M. M. (2007). On the timescale forcing in astrobiology. *Serb. Astron. J.* **175**, 45–50.

Walker, M. A. and Ćirković, M. M. (2006). Anthropic reasoning, naturalism and the contemporary design argument. *Int. Stud. Philos. Sci.* **20**, 285–307.

Wallace, A. R. (1903). *Man's Place in the Universe: A Study of the Results of Scientific Research in Relation to the Unity or Plurality of Worlds.* London: Chapman & Hall.

Wallace, A. R. (1907). *Is Mars Habitable? A Critical Examination of Professor Percival Lowell's Book 'Mars and Its Canals'. With an Alternative Explanation.* London: Macmillan & Co.

Wallis, M. K. and Wickramasinghe, N. C. (2004). Interstellar transfer of planetary microbiota. *Mon. Not. R. Astron. Soc.* **348**, 52–61.

Walters, C., Hoover, R. A. and Kotra, R. K. (1980). Interstellar colonization: a new parameter for the Drake equation? *Icarus* **41**, 193–197.

Waltham, D. (2004). Anthropic selection for the Moon's mass. *Astrobiology* **4**, 460–468.

Ward, P. (2005). *Life as We Do Not Know It.* New York: Viking.

Ward, P. D. and Brownlee, D. (2000). *Rare Earth: Why Complex Life Is Uncommon in the Universe.* New York: Springer.

Ward, P. D. and Brownlee, D. (2002). *The Life and Death of Planet Earth.* New York: Times Books.

Watson, A. J. (2008). Implications of an anthropic model of evolution for emergence of complex life and intelligence. *Astrobiology* **8**, 175–185.

Wayne, J. (1978). *Musical Version of the War of the Worlds.* Columbia Records (audio CD).

Webb, S. (2002). *Where is Everybody? Fifty Solutions to the Fermi's Paradox.* New York: Copernicus.

Weinberg, S. (1972). *Gravitation and Cosmology.* New York: Wiley-VCH.

Weinberg, S. (1987). Anthropic bound on the cosmological constant. *Phys. Rev. Lett.* **59**, 2607–2610.

Weinberg, S. [1977] (1993). *The First Three Minutes: A Modern View of the Origin of the Universe.* New York: Basic Books.

Weinberg, S. (1992). *Dreams of a Final Theory.* New York: Pantheon.

Weinberg, S. (2008). *Cosmology.* Oxford: Oxford University Press.

Wells, H. G. [1898] (2006). *The War of the Worlds.* Clayton, DE: Prestwick House.

Wesson, P. S. (1990). Cosmology, extraterrestrial intelligence, and a resolution of the Fermi–Hart paradox. *Q. J. R. Astron. Soc.* **31**, 161–170.

Wheeler, J. A. (1975). The universe as home for man. In *The Nature of Scientific Discovery*, ed. by O. Gingerich. Washington DC: Smithsonian Institution Press, pp. 261–296.

Wickramasinghe, C. (2005). *A Journey with Fred Hoyle: The Search for Cosmic Life.* Singapore: World Scientific.

Wilczek, F. (1993). Weinberg: a philosopher in spite of himself. *Phys. Today* (April), 59–60.

Wilson, E. O. (1975). *Sociobiology: The New Synthesis.* Cambridge, MA: Harvard University Press.

Wilson, E. O. (1998). *Consilience: The Unity of Knowledge.* New York: Alfred Knopf.

Wilson, P. A. (1994). Carter on anthropic principle predictions. *Br. J. Philos. Sci.* **45**, 241–253.

Wilson, R. C. (2003). *Blind Lake.* New York: Tor Books.

Wilson, T. L. (1984). Bayes' theorem and the real SETI equation. *Q. J. R. Astron. Soc.* **25**, 435–448.

Wilson, T. L. (2001). The search for extraterrestrial intelligence. *Nature* **409**, 1110–1114.

Wittgenstein, L. (1953). *Philosophical Investigations.* Oxford: Blackwell.

Wolfe, A. J. (2002). Joshua Lederberg, exobiology, and the public imagination, 1958–1964. *Isis* **93**, 183–205.

Zank, G. P. and Frisch, P. C. (1999). Consequences of a change in the Galactic environment of the Sun. *Astrophys. J.* **518**, 965–973.

Zdravković, S. (2008). *Perception*. Zrenjanin, Serbia: Gradska narodna biblioteka (in Serbian).

Index

Page numbers in *italic* denote figures.